NATURAL RESOURCE MANAGEMENT AND POLICY

Series Editors:

David Zilberman
Department of Agricultural and Resource Economics
University of California, Berkeley
Berkeley, CA, USA

Renan Goetz
Department of Economics
University of Girona, Spain

Alberto Garrido
Department of Agricultural Economics and Social Sciences
Technical University of Madrid, Spain

For further volumes:
http://www.springer.com/series/6360

EDITORIAL STATEMENT

There is a growing awareness to the role that natural resources, such as water, land, forests and environmental amenities, play in our lives. There are many competing uses for natural resources, and society is challenged to manage them for improving social well-being. Furthermore, there may be dire consequences to natural resources mismanagement. Renewable resources, such as water, land and the environment are linked, and decisions made with regard to one may affect the others. Policy and management of natural resources now require interdisciplinary approaches including natural and social sciences to correctly address our society preferences.

This series provides a collection of works containing most recent findings on economics, management and policy of renewable biological resources, such as water, land, crop protection, sustainable agriculture, technology, and environmental health. It incorporates modern thinking and techniques of economics and management. Books in this series will incorporate knowledge and models of natural phenomena with economics and managerial decision frameworks to assess alternative options for managing natural resources and environment.

The Series Editors

Karinne Ludlow • Stuart J. Smyth • José Falck-Zepeda
Editors

Socio-Economic Considerations in Biotechnology Regulation

Springer

Editors
Karinne Ludlow
Faculty of Law
Monash University
VIC
Australia

José Falck-Zepeda
International Food Policy Research Inst.
Washington, DC
District of Columbia
USA

Stuart J. Smyth
University of Saskatchewan
Saskatoon
Canada

ISBN 978-1-4614-9439-3 ISBN 978-1-4614-9440-9 (eBook)
DOI 10.1007/978-1-4614-9440-9
Springer New York Dordrecht Heidelberg London

Library of Congress Control Number: 2013954721

Printed on acid-free paper

Springer is part of Springer Science+Business Media (www.springer.com)

Abbreviations

CBD	Convention on Biological Diversity
CPB	Cartagena Protocol on Biosafety
FAO	Food and Agriculture Organization
GM	Genetically modified
GMO	Genetically modified organism
HT	Herbicide tolerant
IP	Intellectual property
IPR	Intellectual property right
LMO	Living modified organism
SEC	Socio-economic consideration
SSA	Sub-Saharan Africa
TK	Traditional knowledge

Preface

The pace of change seems boundless. New technologies, new products and new ideas emerge on such a rapid basis that simply staying abreast of the changes not only seems, but proves, daunting. Innovations in the tripartite fields of information, technology and communications (ITC) continue to advance based on Moore's Law. Some have boldly predicted, if not the end of this rapid rate of change, a slowdown in the pace of innovation, however, these predictions are all for naught. It is not out of reason that the twenty-first century will witness a greater degree of innovation than the innovation that occurred in all the centuries preceding it.

Accompanying innovation is change and change brings a multitude of responses. Change can be welcomed or rejected by some or can bring uncertainty for others, leading to trepidation, nervousness and in some instances fear. Although modern society is based on innovation, it is important to keep in mind that adoption occurs at varying rates for all sorts of reasons, with some countries and cultures leading the way, with others closely following, and yet others languish. Innovation displaces current practices and processes, most of these changes being beneficial, but not all. One thing common to all innovations is that they create considerable discourse. Agricultural biotechnology and the resulting genetically modified products have triggered rigorous international discussions, debates and dilemmas. Response to the innovation of genetically modified organisms (GMOs) has been diverse; with some jurisdictions preferring to rely on the market to determine the success of GMOs, while others have rationalized a more centralist approach and heavily regulated GM crops.

The global debate regarding the acceptance of GM crops has grown particularly acrimonious of late. Most opponents and advocates of the technology continue to talk past each other, with progress on resolving the regulation and trade of GM crops and food products moving at a glacial pace at best. The discussions are sporadically populated with evidence and facts, however, they are frequently rift with myths, rumours and innuendos. One way to distill the issue to its core is to observe that for the large part, many of the countries that have adopted and produced GM crops, have largely done so using science-based regulations, while non-adopters have often included socio-economic considerations (SECs) in domestic biosafety regulations, as is allowed under Article 26 of the Cartagena Protocol on Biosafety (CPB).

Given the global context of the debate on the regulation of GM crops and food products, there is a strong need for a book that compiles expert assessments of the issues relevant to SEC assessment of GMOs which are, ultimately, also fundamental for decisions regarding whether to undertake such assessments at all. To this end, we have produced an authored and edited book that provides an overview of the inclusion of SEC assessment in the regulation of GMOs, that:

1. Looks at the rationale for the inclusion of SECs, in the context of the existing science-based risk assessment systems;
2. Through the use of a chapter template, reviews the various factors that can and have been suggested for inclusion in SEC assessment; and
3. Provides a meaningful dialogue about the contrasts, benefits and tradeoffs that are, and will be, created by the potential move to the inclusion of SECs in the regulation of GMOs.

It is not intended to assess whether the inclusion of SECs should or should not occur. That is a policy decision to be made by policy-makers in each of the federal jurisdictions contemplating the inclusion of SECs into their domestic regulatory frameworks. Instead, this book is intended to assist in the development of best practice, methods and policy guidelines for SEC evaluation implementation and inclusion in decision-making. The compilation of materials found in this book will mean that countries and policy-makers will have a clearer, more consistent understanding of the issues raised by each SEC and what is required for the evaluation of them so that better informed decisions can be made regarding the inclusion of SECs in biotechnology regulation and decision-making.

This book is intended to be a resource for a broad audience, not just an audience that supports or opposes GM agriculture. The book focuses on agricultural GM plants and animals. Nevertheless, it can be expected that much of the learning will be relevant to other GMOs and sectors. The invited authors represent a spectrum of disciplines and jurisdictions, providing readers with a thorough discussion of relevant issues. It is our hope that the compilation of a volume containing this information will be a resource for all delegates participating in the Intergovernmental Committee meetings for the CPB. In addition, we hope that this information will be of value to policy-makers in the countries that will be formulating national and regional policy, as well as to industry and non-governmental organizations on these issues.

Of course, the production of this large undertaking required many hands and the authors are deeply appreciative of the contributions from the invited experts. Without their contributions, the content of the book would be a pale imitation of this highly insightful and knowledgeable compendium.

Contents

Contributors

Leslie J. Butler Department of Agricultural & Resource Economics University of California, Davis, CA, USA

Alexandra Coe Goucher College, Department of Cultural Sustainability, MD, USA

Siglinde Fischer BioAlliance AG, 60313, Frankfurt, Germany

Marnus Gouse Department Agricultural Economics, Extension and Rural Development, University of Pretoria, Pretoria, South Africa

Guillaume P. Gruère International Food Policy Research Institute, Washington, DC, USA

Charles Lawson Australian Centre for Intellectual Property in Agriculture, Griffith Law School, Griffith University, Gold Coast, QLD, Australia

Karinne Ludlow Faculty of Law, Monash University, Clayton, Victoria, Australia

Martina Newell-McGloughlin University of California, Davis, USA

Vivian Moses Division of Nutritional Sciences, King's College, London, UK

Latha Nagarajan Department of Agriculture, Food and Resource Economics, Rutgers University, New Jersey, USA

International Fertilizer Development Center (IFDC), Washington, DC, USA

Ari Novy United States Botanic Garden, Washington, DC, USA

Smithsonian Institution, National Museum of Natural History, Washington, DC, USA

Peter WB Phillips University of Saskatchewan, Saskatoon, Canada

Melinda Smale Michigan State University, East Lansing, MI, USA

Richard Smart Research Department of Agriculture Economics, Center of Life and Food Sciences Weihenstephan, Technische Universität München, Freising-Weihenstephan, Germany

Stuart J. Smyth University of Saskatchewan, Saskatoon, Canada

Debra M. Strauss Fairfield University, Fairfield, CT, USA

Paul B. Thompson Department of Philosophy, Department of Community Sustainability and Department of Agricultural, Food and Resource Economics, Michigan State University, East Lansing, MI USA

Crina Viju Institute of European, Russian and Eurasian Studies, Carleton University, Ottawa, Canada

Justus Wesseler Research Department of Agriculture Economics, Center of Life and Food Sciences Weihenstephan, Technische Universität München, Freising-Weihenstephan, Germany

Patricia Zambrano International Food Policy Research Institute (IFPRI), Washington, DC, USA

José Falck-Zepeda International Food Policy Research Institute (IFPRI), Washington, DC, USA

Part I
Governance of Agricultural Biotechnology

Those countries that have adopted agricultural biotechnology are familiar with using science-based risk assessment processes to regulate GMOs. This process involves gathering scientific data and applying standard methodologies to determine whether a risk is present, and if one is, whether that particular jurisdiction is willing to take it in allowing the deliberate release of GMOs into the environment. However, this process is generally used with respect to safety concerns (such as risk to human, plant and/or animal health and safety) or environmental damage. It is not that these concerns are 'science-based', but rather that society is comfortable with using science to measure risk regarding them. Inclusion of SECs into the GMO regulation process means there is now a need for a broader range of factors to be the subject of consistent, clear and objective assessment processes and decision-making rules. This is necessary if very significant problems are to be avoided. Such problems include the avoidance of claims that SECs are being used (or abused) as a blanket justification for rejecting the technology without the need to support claims or debating the strengths of evidence presented to support such claim. International trade is also facilitated if there is consistency in approach or at least use of important terms, between jurisdictions and data and research can also be more easily shared and developed. Such clarity and transparency is also important to maintain the trust of both stakeholders and other interested parties.

This section frames the present global environment for agricultural biotechnology and SECs. The Cartagena Protocol on Biosafety came into effect 10 years ago when the 50th nation ratified the Protocol in 2003, thus allowing it to enter into force. Over the past decade though, there has been an increasing global disconnect in approaches to agbiotech. For example, Argentina, Brazil, Canada and the USA are all strong adopters of GM varieties of canola, corn, cotton and soybean, while the EU and its historical trading partners in Africa have predominantly avoided the technology. This trans-Atlantic gap in attitudes to agbiotech shows no signs of abating in the near future and the intent of this section is to document the current status of varying approaches to the regulation of GMOs.

Chapter 1
Introduction to Socio-Economic Considerations in the Regulation of Genetically Modified Organisms

Karinne Ludlow, Stuart J. Smyth and José Falck-Zepeda

1.1 Introduction

Just over 20 years ago, an innovation occurred in agriculture—one with global implications. In 1992, the first commercial planting of a genetically modified (GM) crop was undertaken. This involved about 100 acres of GM tobacco production in China (James and Krattiger 1996). From this modest beginning, agricultural biotechnology and GM crops have become nothing short of the most rapidly, and widely, adopted innovation in the history of commodity agriculture. However, the innovative wave of biotechnology has not been without controversy. A diverse array of opponents to agricultural biotechnology and GM crops quickly began championing an equally diverse array of criticisms regarding the impacts of the technology. As with any innovation, there have been challenges and setbacks. Ultimately, though, farmers, be they in developed or developing countries, have rapidly adopted GM crops and are benefiting from their production.

James (2010) estimated that by 2010 GM crops were grown on 148 million ha around the world, in 29 different countries, made up of 10 industrial and 19 developing countries. Within those countries, 15.4 million farmers grew the crops, more than 90% or 14.4 million of whom were farmers in developing countries. The dominant GM crop is soya bean, followed by maize, cotton and canola. However, other GM crops are being introduced to agriculture all the time and now include fruits and

K. Ludlow (✉)
Monash University, Clayton, Australia
e-mail: karinne.ludlow@monash.edu

S. J. Smyth
University of Saskatchewan, Saskatoon, Canada
e-mail: stuart.smyth@usask.ca

J. Falck-Zepeda
International Food Policy Research Institute, Washington, DC, USA
e-mail: j.falck-zepeda@cgiar.org

K. Ludlow et al. (eds.), *Socio-Economic Considerations in Biotechnology Regulation*, Natural Resource Management and Policy 37, DOI 10.1007/978-1-4614-9440-9_1,

vegetables, flowers and spices. GM livestock are also being approved by regulatory authorities, including pigs, cows, sheep and goats.

Herbicide tolerance has consistently been the dominant trait introduced into crops by the modification. However, as second- and third-generation GM crops (and now animals) are released, a wider spectrum of characteristics is being introduced. These characteristics include pest and disease resistance, improvements to nutritional qualities, changes to the plant or animal's growing patterns such as time of maturity or colour and changes to reduce the plant or animal's environmental impact or production costs. The ever-growing range of modified organisms and the traits modified create an immeasurable array of both national and individual attitudes to genetically modified organisms (GMOs).

Farmers, when they enjoy the option to produce GM crops in a market without burdensome conditions, are choosing to do so. Many other farmers have expressed the desire to produce GM crops, as is evidenced by the regular, illegal production of GM crops in unapproved markets. Presently, one of the leading challenges for the innovation of agricultural biotechnology and GM crops is that posed by market access. Numerous governments are using regulatory frameworks that govern the assessment and approval for new plant varieties as a means to increase the barriers for commercialization. Early adoption countries simply modified their existing science-based regulatory frameworks to accommodate the innovation of agricultural biotechnology (see Chap. 2 for more details), while other countries are currently engaged in the process of increasing the regulatory requirements for products of biotechnology, many of which are a move away from science-based risk assessments. Many of the new regulatory requirements involve precaution and include socio-economic aspects and as such are less concerned about risks measurable using science, but more about aspects of the innovation that cannot be assessed in scientific terms.

There is a growing need for clarification of socio-economic considerations (SECs) associated with biotechnology, particularly GMOs. At their sixth meeting in October 2012, more than 100 countries that attended the Conference of the Parties to the *Convention on Biological Diversity* (CBD) serving as the meeting of the Parties to the Cartagena Protocol on Biosafety (COP/MOP 6) agreed on the need for more research and discussion on the issue of including SECs in regulatory frameworks.

The Cartagena Protocol on Biosafety (CPB), an implementing agreement of the Convention, establishes the rights of recipient countries to be notified of and to approve or reject the domestic import and/or use of LMOs. Decisions regarding import and use are to be on the basis of a biosafety assessment. Article 26.1 of the CPB allows for the optional inclusion of SECs into that biosafety assessment process.

Some countries have already included SECs assessments in their national biosafety frameworks, policies and/or regulations (Falck-Zepeda 2009). However, the implications resulting from the inclusion of SECs in biosafety decision making are poorly understood. This is not surprising given that there is very little information available on important aspects of SECs in biotechnology decision making. While there is much discussion of the issue and there has been some work by certain jurisdictions, there is a lack of information on crucial matters. These include the content (or meaning) of particular SECs, methodologies to measure such SECs and their advantages, disadvantages and costs and the legal repercussions of including them in biotechnology regulation, such as compatibility with existing international agreements and institutions.

This lack of understanding and detailed information complicates the policy setting environment. There is therefore a strong need for a book that compiles expert assessments of the issues relevant to SEC assessment of GMOs and fundamentals for decisions regarding whether to undertake such assessments at all. This edited book aims to meet that need by providing an overview of SEC assessment in biotechnology regulation. Through the use of a consistent chapter template, it reviews the various factors that can and have been suggested for inclusion in SEC assessment in the context of agricultural biotechnology regulation. It then evaluates and discusses the contrasts, benefits and tradeoffs that are, and will be, created by the move to the inclusion of SECs in biotechnology regulation. This book uses both the terms GMOs and living modified organisms (LMOs), although the CPB itself uses LMO, because both terms are commonly used in the literature.

It is not intended to assess whether the inclusion of SECs should or should not occur. That is a policy decision to be made by policy makers in each of the federal jurisdictions contemplating the inclusion of SECs into their domestic regulatory frameworks. Instead, it is intended to assist in the development of best practice, methods and policy guidelines for SEC evaluation implementation and inclusion in decision making. The development of such material in this book will provide countries and policy makers with a clearer and more consistent understanding of the issues raised by each SEC and what is required for the evaluation of them so that better informed decisions can be made. It should also assist in answering perhaps the most central question for policy makers and one that may be answered differently by different policy makers—what is the purpose or goal of including SECs in the regulatory framework.

This book is intended to be a resource for a broad audience, not just an audience that supports or opposes GM crops and agricultural biotechnology. The book focuses on agricultural GM plants and animals. Nevertheless, it can be expected that much of the learning will be relevant to other LMOs and sectors. The editors and authors are from a spectrum of disciplines and jurisdictions, providing readers with a thorough discussion of relevant issues. The compilation of this information will be a resource for policy makers in countries formulating national and regional policy on these issues, as well as to industry and non-governmental organizations.

1.2 The Conflicted Field of Agricultural Biotechnology Regulation

International trade in products of agricultural biotechnology is a two-sided coin. In parallel with growing international trade in such products, there is a growing threat (and reality) of the spread of plant and animal pests and diseases, particularly in cases of intercontinental trade. Therefore, as international trade law aimed at encouraging trade was introduced, international measures to protect plant, animal and human health also emerged. The first international plant health control measure was signed in 1878 and was aimed at preventing the spread of an American vine louse to European vineyards. Amongst the measures it introduced was the need for a written assurance to be given of the pest-free provenance of vine material traded internationally (Ebbels 2003). This began a pattern of competing regulatory priorities in agricultural trade.

International trade law is now codified under the rules of the World Trade Organization (WTO), the main function of which is ensuring non-discriminatory trade. Amongst the principal WTO measures aimed at protecting plant and animal health and food safety is the Agreement on the Application of Sanitary and Phytosanitary Measures (SPS Agreement), the main aim of which is to prevent the abuse of health protection measures for trade protectionist purposes. The multilateral approach to trade taken by the WTO (and plant and animal health in particular under the SPS Agreement) is primarily aimed at enhancing the global interest by requiring all countries to be treated equally.

However, the regulatory field for agricultural products is crowded with concerns beyond trade and human, plant and animal health. Agricultural products are unique when compared with other traded goods because as living things (or goods derived from living things), they are an important part of the world's biological diversity. The provisions of the CBD, one of the main purposes of which is to address conservation of biological diversity, are therefore also relevant. The lens of the CBD is also attracted by the growing competition between agriculture and the natural environment for limited resources fundamental to life, such as land and water.

Tightening the focus of our concern to agricultural biotechnology does not exclude these competing regulatory regimes. Whilst the CBD has no provisions specifically regulating biotechnology, it does provide for a protocol on biosafety, in the context of the aims of the CBD. That Protocol, the CPB referred to above, creates a regime for the regulation of transboundary movements of LMOs. In the WTO context, LMOs can be the subject of particular regulation provided it is justified through scientific assessment of the risks to food safety or to plant or animal health protection. The use of regulations under the Biosafety Protocol to advance aims other than protection of plant or animal health would not fall within the terms of the SPS Agreement but could well be in pursuance of the aims of the CBD.

A final participant in this crowded field of agricultural biotechnology regulation is domestic regulations aimed at social purposes. Schefer defines 'social' trade regulations as laws or policies enacted 'for purposes that are related to the furtherance of political or ideological goals—praiseworthy goals such as the protection of the environment, the raising of health standards, or the increased protection of human rights (whether political, civil, economic, cultural, or social), as well as less commendable goals…of imposing the sender's "way of doing things" on the target' (Schefer 2010, p. 2). Agriculture and the food it produces are intimately important to individuals and societies—foods' characteristics and other attributes are sometimes central in a community's culture or religion. Matters such as food security and the importance of a country's agricultural industry to broader national political and economic debates are central to a nation's sovereignty. But the pursuit of such goals inevitably creates conflict between WTO objectives and domestic social trade regulations intended to advance goals relevant to the jurisdiction imposing them. The lack of success of countries seeking to justify social trade regulations in the WTO dispute resolution process illustrates the preference of that forum to prevent interference with the global interest in international trade by the disparate values held by different societies even where they are linked to plant and animal health concerns.

It is not surprising then that the Biosafety Protocol, with its provision for SECs to be included in a country's agricultural biotechnology regulatory matrix, is a flashpoint for debate.

1.3 Bringing SECs into LMO Regulation

Amongst other things, Article 19 of the CBD provides for Parties to the Convention to consider the need for and modalities of a protocol on biosafety. The meaning of 'biosafety' itself is controversial but in this context one possible meaning is that given by the Convention's own website—the safe use of biotechnology and its products, in particular those precautionary practices intended to ensure the safe transfer, handling, use and disposal of LMOs and their products.

While Article 19 is important, what is creating considerable international debate and discussion is the inclusion of SECs within domestic regulatory frameworks as is provided for in Article 26 of the Biosafety Protocol. Article 26 reads as follows:

1. The Parties, in reaching a decision on import under this Protocol or under its domestic measures implementing the Protocol, may take into account, consistent with their international obligations, socio-economic considerations arising from the impact of living modified organisms on the conservation and sustainable use of biological diversity, especially with regard to the value of biological diversity to indigenous and local communities.
2. The Parties are encouraged to cooperate on research and information exchange on any socio-economic impacts of living modified organisms, especially on indigenous and local communities.

As noted above, more than 100 countries attended the sixth meeting of the Conference of the Parties to the CBD serving as the meeting of the Parties to the Protocol (COP/MOP 6) in October 2012. It was agreed at that meeting to develop conceptual clarity on what constitutes SECs under Article 26, in order to assist Protocol Parties that may wish to consider SEC factors in reaching decisions on LMO imports. To assist with that clarification, it was agreed that an Ad Hoc Technical Expert Group (AHTEG) be established, with the ultimate aim being the future development of guidelines.

Leading up to the COP/MOP 6, second national reports were submitted by the vast majority of Parties. These were reviewed by the Executive Secretary together with the results of regional online conferences and a workshop held by the Executive Secretary.[1] Of those countries that responded to the question in the second national reports regarding whether, in making a decision on LMO import, the country had ever taken into account SECs arising from the impact of the LMO on the conservation and sustainable use of biological diversity, the majority (56 % of 72 responses) responded they had not. Of those that had done so, 15 % responded this was only done sometimes, and for the remainder it was not clear whether this was done always or not. Looked at by region, members of the Africa region were most likely to have taken into account SECs (33 %), members of the Western Europe and Others region were second most likely (29 %), 14 % in both the Asia-Pacific region and the Group of Latin American and the Caribbean countries did so and the least likely to take SECs into account were those countries in the Central and Eastern Europe region (10 %).[2]

The International Institute for Sustainable Development (IISD) analysis of other information provided in the national reports concludes that there is a low rate of implementation of biosafety frameworks by Parties and that only a very small number of countries oppose LMOs entirely (IISD 2012). The IISD put this down to the reality that the biotechnology landscape is a fast changing one and that as more countries move from simply importing LMOs to also producing and exporting them, the more their biosafety frameworks must balance the aim of protection from risks associated with LMO imports with that of the least disruption to international trade to advance their exporter interests (IISD 2012).

1.4 What Are the Relevant SECs?

Whilst Article 26 allows parties to take SECs into account in their LMO decision making, it does not define either the term 'SEC' or provide any guidance on what issues would be included with that term. Its reference to SECs in Article 26.1 to '[SECs] arising from the impact of living modified organisms on the conservation and sustainable use of biological diversity', is the basis for debate about the proper interpretation of the article and whether a narrow focus of SECs should be used. At the 2012 Conference of the Parties to the Protocol, it was noted that any guidelines developed following the work of the AHTEG will need to respond to the debate between 'those who felt that broader [SECs] should not be dealt with under the Protocol, because it is limited to transboundary movements; and participants who wanted to address potential negative socio-economic impacts of LMOs, such as the loss of agricultural varieties that have cultural value' (IISD 2012). A similar schism exists around whether the impact must be to indigenous or local communities to be relevant (Jaffe 2005).

Nevertheless, others, such as the Third World Network, argue that because the Protocol sets only minimum standards countries can take into account much broader

[1] UNEP/CBD/BS/WS-SEC/1/3 (18 October 2011)

[2] UNEP/CBD/BS/COP-MOP/6/15 Socio-Economic Considerations (Article 26) 28 July 2012.

SECs (Third World Network, post 22 March 2011 on online conference, theme 1, #1996) and clearly some countries, including developed and developing countries, provide for a broad range of SECs in their biotechnology regulation. But whichever view is accepted, SECs in the context of agricultural biotechnology are a 'broad spectrum of concerns about the actual and potential consequences of biotechnology' (La Vina and Fransen 2004).

Various bodies are beginning to compile possible lists. The European Directive 2001/18/EC[3] on the deliberate release of GMOs into the environment requires the European Commission to report every 3 years on Member States' experiences with implementing the Directive, including an assessment of the socio-economic implications (defined to mean the advantages and disadvantages) of deliberate releases and placing on the markets of GMOs. The 2004 report raised the lack of experience with the issue as a hurdle in making an assessment of the SECs. Therefore, further work on such assessments was undertaken by the European Commission and Member States, resulting in a 2011 report and accompanying working paper which provides a useful snapshot of what EU Member States are doing in respect of the implications of GMOs. But it does not provide any detail on how SECs are or could be included in regulatory frameworks. EU Member State reports have a similar deficiency. For example, a 2009 report by the Netherlands Commission on Genetic Modification (COGEM 2009) on the potential role of socio-economic assessments as part of the approval process for LMOs suggests nine criteria as 'building blocks' for an assessment framework on the socio-economic and sustainability aspects of LMOs. The COGEM report concludes that the criteria most relevant to Europe but which have not yet been included in an assessment framework could be summarized down to three themes: (1) benefit to society, (2) economics and prosperity and (3) cultural heritage.

A more detailed list of SECs has been compiled as part of the CPB process (UNEP/CBD/BS/COP-MOP/5/INF/10 17 September 2010). The 2008 Protocol's Conference of the Parties/Meeting of the Parties (COP/MOP 4) decided further technical guidance on SECs in biosafety assessments was needed. In 2009, countries with decision-making systems that allowed for SECs were asked to identify the SECs that could be taken into account in their countries' decisions on LMOs and to rank SECs after they were provided with a list of 20 possible issues. From that information the list of SECs referred to above was drawn up. Alternative lists come from the Third World Network (2008), Chatham House, UK (Jarvis 2009) and a briefing paper by Brookes (2009). A very influential report in developing countries is that published by the World Resources Institute (Fransen et al. 2005).

After considering these lists, this book uses the following list of SECs to focus discussion:

- Benefits to producers and society
- Consumer choice
- Environmental impacts
- Ethics/equity

[3] OJ L 106, 17 April 2001.

- Food security and safety
- Health impacts
- Impacts on biodiversity
- Traditional knowledge
- Intellectual property rights
- Labour impacts
- Market access and trade
- National trade interests
- Producer choice
- Culture and religion
- Animal welfare

Even when an acceptable list of SECs is compiled, the content of any particular SEC must be defined. This definition is difficult because the SECs are tightly interwoven. As the recent 2011 European Commission Report to the European Parliament and the Council on socio-economic implications of GMO cultivation noted, '[t]he understanding of the meaning and scope of the socio-economic dimension of GMO cultivation varies widely among the Member States and the stakeholders. … several participants regretted that the terms, indicators and baseline for comparison (conventional and/or organic sectors) were not sufficiently defined' (European Commission, COM (2011), p. 3, para 1.2). So, for example, 'ethics' is commonly cited as a SEC that should be included in the decision-making process regarding the release of LMOs and has been included in this book's list. However, there is no consistent meaning of what is included in that concern. For example, some commentators include equity with respect to developing countries in the term and others include morality of behaviour only in the jurisdiction under consideration as part of this SEC. Further still, others include cultural concerns in 'ethics'. For example, the New Zealand regulator has noted that '[e]thics' is defined for its purposes as the values held by individuals and collectives, and is deemed to incorporate the consideration of ethical and cultural issues, since both are concerned with values and beliefs. This is a broader term than that used by academia where 'ethics' is more narrowly defined as a sub-branch of philosophy (New Zealand ERMA Technical Guide 1999). This book has split ethics and cultural issues into two separate chapters, grouping ethics and equity together and culture and religion in another chapter. But admittedly that classification is arbitrary.

Content of any particular SEC is also very dependent on context. Different countries and groups within countries (such as consumers, producers and industry) interpret and are impacted by any particular SEC in different ways. Impact and what methodologies are therefore appropriate to measure that impact will vary because of different cultural and religious values, political and economic capacity and forms of agriculture and biotechnology research undertaken in particular countries. How a particular SEC is relevant to a specific country and what methodologies are appropriate in an assessment of that will also vary depending on, for example, whether the country is a developed or developing nation because that determines the infrastructure and human resources available and the usual governance patterns used in the nation. This array of interpretations and impacts poses an additional problem—defining a SEC broadly enough to be inclusive of at least a representation of this array so that the definition is justifiable as being shared by most in the community, but avoiding using a definition that is so broad that it is likely to become ambiguous (Hope 2001, p. 444). But responding the other way when defining a SEC by diluting homogeneity so that individual attitudes and values are more accurately reflected gives rise to the real risk that many of those other attitudes and values will be excluded from the decision-making process (Hope 2001, p. 444).

1.5 SEC Measurement, Assessment and Inclusion in Decision Making

When the relevant SECs have been agreed upon and defined, important decisions must then be made regarding what a functional SEC assessment regime requires and how the results of that assessment should be included in the LMO regulatory decision-making process. So, for example, in regards to the construction of a functional SEC assessment, regime decisions must be made as to the institution most suited to undertaking the assessment, who should have input into the process or parts of it, at what stage of the LMO's introduction should the assessment be undertaken (ex-ante (before) or ex-post (after) release into the environment), how will the implications of a SEC be measured, will only negative or also positive implications of a particular SEC be taken into account and will the assessment be at a local, national or international scale?

This policy milieu is complicated because the examination is, in fact, of complex interactions between democratic societies' right to decide how to proceed, producers' freedom to operate and consumers' right and freedom to know. It also creates an interface that must be negotiated between an approach focusing on the importance of individual autonomy in choosing to adopt or reject a new technology with one that puts the sum total of individual decisions and what these mean for society as a whole as the priority. As Moses has explained in a more general context '[a] new technology carries with it new possibilities, and these can potentially conflict with existing social, environmental and cultural value' (Moses 2007, p. 248). These impacts are the result of 'innumerable individual decisions to develop individual technologies for individual purposes without explicit attention to what all these decisions add up to for society as a whole and for people as human beings' (Moses 2007, p. 248).

Given the background of the WTO and its primary aim of non-discriminatory trade and consistency, many nations, but particularly export-driven nations, are accustomed to using science-based risk assessment processes using data gathered and measured using standard methodologies. This approach has then been adapted for the regulation of LMOs. Scientific data are gathered and standard methodologies applied to determine whether the risk(s) of environmental release of a LMO is one that the particular jurisdiction is willing to take. The 'risks' assessed in this process are generally limited to safety concerns, such as risk to human and/or animal and plant health and safety or environmental damage. Risks or benefits to other concerns such as SECs or market implications are excluded, as is the use of social factors in the evaluation of the human health or safety or environmental risks.

Nevertheless, methodologies and literature with respect to assessment of economic concerns regarding LMOs is well developed, at least in some areas. For example, the economic impacts of LMOs at the farmer level is well developed, at least for some nations, such as the USA and Canada. However, for social concerns this is not the case. As the New Zealand regulator has observed (New Zealand ERMA, Technical Guide 2009, p. 34):

> Other forms of knowledge in addition to science are necessary to understand the values embedded in and transmitted through the social, economic and cultural dimensions of the environment, and how these things may be affected by or in turn affect a particular use of a technology. The key issue is how to relate these other forms of knowledge to scientific knowledge in the decision making process.

Literature about assessment of SECs already exists. Indeed as discussed later in this book, some countries or regions already include (or could include) SEC assessment in their LMO regulation. However, as between different jurisdictions and even within the one jurisdiction, there is considerable uncertainty and inconsistency in the current approach to assessment of SECs. It may be that there is not sufficient consensus on what approach to take but there is a fundamental need for this broader range of concerns to be the subject of evidence-based, predictable and objective measurement and assessment processes. This is necessary if very significant problems are to be avoided. Such problems include the avoidance of claims that SECs are being 'used (or abused) as a blanket justification for rejecting the technology without the need to support claims or debating the strengths of evidence presented to support such claim' (Falck-Zepeda et al. 2010, p. 8). International trade is also facilitated if there is consistency in approach or at least use of important terms, between jurisdictions and data and research can also be more easily shared and developed. Such clarity and transparency is also important to maintain the trust of both stakeholders and other interested parties (New Zealand ERMA, Technical Guide 1999). It is hoped that this book will assist in the move towards certainty and consistency.

Even when the individual SECs are appropriately assessed though, a functioning regulatory framework is needed to balance and prioritize competing SECs and other concerns, including the health, safety and environmental risks already assessed in biotechnology regulation. Inevitable conflicts will arise between the different SECs and their potential, probable and future positive and negative effects. Balancing these in a consistent and transparent way will be crucial to creating regulatory frameworks that are both democratically inclusive but cost-effective, tread the line between risk and progress and satisfy the tension between national sovereignty and international trade obligations. This work is particularly important because in the context of agricultural biotechnology regulation, where there are as yet no international standards, guidelines or recommendations on SEC assessment, any best practice developed under the Cartagena Protocol may become the de facto international protocol (Smyth and Falck-Zepeda 2012, p. 18).

1.6 Structure of This Book

This book is made up of three sections. The first section of the book, 'Governance of Agricultural Biotechnology', provides a broad overview of the context and setting for the discussion of SECs in biotechnology regulation. This is done through three chapters, the first of which is this, the Introduction. The next two chapters consider the two ends of the spectrum of GMO regulatory frameworks. Chapter 2 addresses the evolution and current state of 'science-based' GM crop regulatory frameworks, using examples from existing regimes. The role of international institutions such as the WTO and IPPC which have been important in the development of the international governance will also be discussed. Chapter 3 examines GM crop regulatory frameworks that already include SECs. Again, the evolution of such an approach and examples of existing regimes will be considered.

The second section of the book, 'Analysis of Socio-Economic Considerations', comprises 15 chapters. Each chapter concerns a different SEC taken from the list given above and is written by an expert in the particular field. The chapters in this second section of the book use a template to analyse the common factors routinely included in the discussion about SECs. The purpose of the template is to enhance consistency across the chapters to better enable comparison by readers across the different SECs. Generally, each chapter therefore includes an introduction that defines the particular SEC and explains how that SEC is relevant to the production and import of LMOs. An explanation of existing methodologies that are used or could be adapted to measure and account for the SEC and a critical assessment of the methodology or methodologies is then provided. Each chapter then undertakes an analysis of how such methodology fits with existing international agreements and protocol obligations and the administrative consequences for both international and national institutions that administer or coordinate decisions regarding the SEC. Each chapter then ends with a brief policy summary and synthesis of the chapter in dot-point form so readers can quickly gain an appreciation of the most significant issues raised by the SEC concerned.

The third and final section of the book, 'Navigating the Challenges', contains a practical synthesis of the practicalities and options relevant to the assessment of SECs and their introduction and implementation into the agricultural biotechnology decision-making process. The section highlights the methodological issues that arise and provides a sense of the matters that need to be developed for the introduction of a regime where SECs are included as part of the regulation of the import and production of GMOs and why it matters that inclusion be done right.

References

UNEP/CBD/BS/WS-SEC/1/3 (18 October 2011) Synthesis of information on national experiences with socio-economic considerations in decision-making on living modified organisms

UNEP/CBD/BS/COP-MOP/5/INF/10 (17 September 2010) Summary Report on the Survey on the Application of and Experience in the Use of Socio-Economic Considerations in Decision-Making on Living Modified Organisms
Brookes G (2009) Socio-economic impacts of GM crop technology: 'second round' impacts. Briefing note, UK
COGEM (2009) Socio-economic aspects of GMOs. Building Blocks for an EU Sustainability Assessment of Genetically Modified Crops. COGEM Report CGM/090929-01. http://www.cogem.net/main-adviesdetail-signaleringEN.aspx?pageid=54&loc=2&version=&mode=&id=517. Accessed 7 June 2013
Ebbels DL (2003) Principles of plant health and quarantine. CABI Publishing http://www.corpoica.org.co/SitioWeb/WebBac/Documentos/Principlesofplanthealthandquarantine.pdf. Accessed 7 June 2013
European Commission, COM (2011) Final Report from the Commission to the European Parliament and the Council on socio-economic implications of GMO cultivation on the basis of Member States contributions, as requested by the Conclusions of the Environmental Council of Dec 2008
Falck Zepeda JB (2009) Socio-economic considerations, Article 26.1 of the Cartagena Protocol on Biosafety: what are the issues and what is at stake? AgBioForum 12(1):90–107. http://agbioforum.org/v12n1/v12n1a09-falckzepeda.htm. Accessed 7 June 2013
Falck Zepeda J, Wesseler J, Smyth S (2010) The current status of the debate on socio-economic assessments and biosafety highlighting different positions and policies in Canada and the US, the EU and Developing Countries. Paper presented at World Congress of Environmental and Resource Economics (WCERE), Montreal, Canada, 29 June
Fransen L, La Vina A, Dayrit F et al (2005) Integrating socio-economic considerations into biosafety decisions: the role of public participation. World Resources Institute, Washington, DC, USA
Hope J (2001) New Zealand Royal Commission on genetic modification. Environmental and Planning Law Journal 18(5):441–444
IISD (International Institute for Sustainable Development) (2012) Sixth meeting of the Conference of the Parties to the Convention on Biological Diversity (CBD) serving as the meeting of the Parties to the Cartagena Protocol on Biosafety (COP/MOP 6). Earth Negotiations Bulletin 1–5 October, http://www/iisd.ca/download/pdf/enb09585e.pdf. Accessed 30 Oct 2012
Jaffe G (2005) Implementing the Cartagena Biosafety Protocol through national biosafety regulatory systems: an analysis of key unresolved issues. J Publ Aff 5:299–311
James C (2010) Global status of commercialized biotech/GM crops: 2010. ISAAA Brief No 42. ISAAA, Ithaca, NY, USA
James C, Krattiger AF (1996) Global review of the field testing and commercialization of transgenic plants, 1986–1995: the first decade of crop biotechnology. ISAAA Brief 1. International Service for the Acquisition of Agri-biotech Applications, Ithica, NY, USA
Jarvis A (2009) A way forward for GMO cultivation in Europe? Chatham House Programme Paper, Nov
La Vina A, Fransen L (2004) Integrating socio-economic considerations into biosafety decisions: the challenge for Asia. Paper for IUCN-IDRC Meeting on Biosafety, Sri Lanka
Moses LB (2007) Recurring dilemmas: the law's race to keep up with technological change. U Ill JL Tech & Policy 7:239–285
New Zealand ERMA (1999) Technical Guide Identifying Risks for applications under the Hazardous Substances and New Organisms Act 1996 (ER-TG-01-1 9/99)
New Zealand ERMA (2009) Decision making. A technical guide to identifying, assessing and evaluating risks, costs and benefits. ER-TG-05-02 03/09
Schefer KN (2010) Social regulation in the WTO. Trade policy and international legal development. Edward Elgar, UK
Smyth S, Falck-Zepeda J (2012) Socio-economic considerations and international trade agreements. International Political Science Association Conference, Madrid, Spain, 8–12 July
Third World Network (2008) Assessing the socio-economic, cultural and ethical impacts of GMOs. Third World Network Briefings for MOP 4#3

Chapter 2
The State of Science-Based Regulation and Genetically Modified Crops

Stuart J. Smyth, José Falck-Zepeda and Karinne Ludlow

2.1 Introduction

Mankind's relationship with risk has changed in a multitude of ways and degrees over the millennia. While the likelihood of being attacked, killed and eaten by a wild animal has dramatically decreased, the probability of being killed in an automobile accident has increased. Life expectancies at the dawn of the twentieth century in Europe ranged from the mid-30s to the high 40s but by the close of the century, life expectancies had risen to the mid-70s for men and low 80s for women (Kinsella 1992). Clearly, the nature of the risks that humans face has changed over time and so have the incidences of life-threatening risks. At the beginning to the twenty-first century, mankind has mitigated many risks that have previously been life threatening, especially when it comes to food and food security.

This mitigation has been more successful in industrialized nations than developing ones, but even in developing nations risks are being successfully addressed. For example, in the centuries past, millions and millions of people in the world died from a disease simply known as "consumption" or, more colorfully, galloping consumption. Today this disease is known as tuberculosis and, in the industrial world, it has been virtually eradicated. In Canada, for example, the risk of dying from tuberculosis reached a high of 7% of all deaths in 1926 (when death from other infectious diseases was 5% of all deaths) and, by 1990, deaths from any infectious disease contributed less than 1% of all deaths (Canadian Lung Association 2004). The developing world is also making significant progress in reducing this risk of

S. J. Smyth (✉)
University of Saskatchewan, Saskatoon, Canada
e-mail: stuart.smyth@usask.ca

J. Falck-Zepeda
International Food Policy Research Institute, Washington, DC, USA
e-mail: j.falck-zepeda@cgiar.org

K. Ludlow
Monash University, Melbourne, Australia
e-mail: karinne.ludlow@monash.edu

K. Ludlow et al. (eds.), *Socio-Economic Considerations in Biotechnology Regulation*,
Natural Resource Management and Policy 37, DOI 10.1007/978-1-4614-9440-9_2,

death, with the World Health Organization reporting that tuberculosis incidences have fallen globally in several years and that the mortality rate has fallen by 41 % since 1990 (World Health Organization 2012).

While the actual risks and the magnitude of them change over time, total human exposure to risks is relatively constant. Exposure to a specific risk or even a class of risks may appear and then increase or wane over time, but the overall exposure to risk stays relatively constant through the ages as other risks appear and are addressed. So, for example, while exposure to risk in areas of food safety, nuclear contamination and climate change may be increasing, the risks of exposure to global warfare, starvation and enslavement have greatly decreased. Returning to the tuberculosis example, as the risk of tuberculosis has waned, global concerns about overall infectious disease risks are once again growing because of the effect that climate change may be having on the habitat of disease pathogens. One example of this is the recent rapid spread of the West Nile virus in North America.

Nevertheless, although the actual level of risk has stayed about the same, the perceived magnitude of risks in modern societies has risen to previously unimaginable levels. More often than not, it is our perception of exposure to risks that has grown rather than the actual risks. The risk of living in modern, industrial, and largely urban societies directly parallels the risks experienced in the societies of our grandparents and even that of their grandparents. Whilst the variety of risks that members of modern societies are exposed to are certainly larger, the absolute incidence of risk may not be substantially changed. Technological advances throughout the past century have, for example, introduced the risk of death due to airplane travel. While this risk is not trivial, it is actually lower than risks posed by many historical or traditional forms of transportation.

The pace of change of the variety of risks we face, however, has increased rapidly over the past century and is expected to continue into the future in response to technological and scientific advances. For example, within the past decade, the fields of computers, robotics and genetics have combined to dramatically affect human reproduction, agricultural biotechnology and nanotechnology. Technological innovations in these areas have allowed genetic modification of crop varieties; arguably the most important and successful innovation in the modern history of agriculture.

One fundamental unchanged factor when responding to risk is the need for a governance strategy adapted to the ever-changing nature of risk. Risk governance strategies are inevitably diverse, both in their objectives and their implementation. This chapter offers insights into the governance challenge by examining the development of risk assessment frameworks that have manifested themselves within and around the technology of agricultural biotechnology and the role that science-based regulation plays in biotechnology regulation. As will be seen, the original risk assessment frameworks were developed to respond to risks to human health posed by factors such as chemicals and human, plant and animal pathogens, matters which are particularly amenable to scientific assessment. That approach continues to be used by many nations today, particularly North America.

2.2 Origins of the Risk Analysis Framework

Regulations have been used to address innovation for practically as long as there have been innovations. Building codes, labor laws and unions, all in part, owe their origins to methods of dealing with innovation. Aspects of all of these social features can be related to improvements in economies, environments, and/or health. Given our ancestral history of coping with the innovation of the day and the resulting impacts on societies, it is no wonder that modern society is increasingly fixated on the mitigation of and, to a large extent, the eradication of risk. It is this latter concept of the eradication of risk that creates multiple challenges in modern societies as opponents of a particular innovation will argue that the innovation should have zero risk if it is to be commercialized. The problem they ignore is the technology that the innovation is replacing, or at least competing with, frequently, has a higher degree of risk associated with it. Further, in today's marketplace, no scientist or manufacturer will declare that any product is 100% safe, because there is always a degree of risk associated with every product that is purchased, be it for entertainment (a television), communication (a cell phone) or to eat (a tomato).

Over time, the framework for understanding and responding to risk in relation to innovation became increasingly codified. This was, especially, the case following the technological innovations of the post-war years (Phillips et al. 2006). It was during this period that regulators developed a structured format for risk analysis as a regulatory response to public policy problems.

This process officially became part of government regulation with the American National Research Council's 1983 report to the United States federal government. This report has become known as the Redbook on the Risk Analysis Framework (RAF). The report identifies the Council's mission as "a search for the institutional mechanisms that best foster a constructive partnership between science and government, mechanisms to ensure that government regulation rests on the best available scientific data and judgments in the unavoidable collision of the contending interests that accompany most important regulatory decisions" (National Research Council 1983, p. 1). While this is a broadly interpreted objective, the mandate was more focused in that it sought to "examine whether alterations in *institutional arrangements or procedures*, particularly the organizational separation of risk assessment from regulatory decision-making and the use of uniform guidelines for inferring risk from available scientific information, can improve federal risk assessment activities [original emphasis]" (National Research Council 1983, p. 17). This process was done within the scope of examining the possible risks of cancer from exposure to the increased use of chemicals in the environment.

The RAF Redbook definitions have become the cornerstone for modern RAFs. The Redbook defines risk assessment as "the use of the factual base to define the health effects of exposure of individuals or populations to hazardous materials and situations" (National Research Council 1983, p. 3). Risk management is defined as "the process of weighing policy alternatives and selecting the most appropriate regulatory action, integrating the results of risk assessment with engineering data and with social, economic, and political concerns to reach a decision" (National Research Council 1983).

Most risk assessments comprise some or all of the following aspects: hazard identification, dose–response assessment, exposure assessment, and risk characterization. Hazard identification determines if the element in question is causally linked, or not linked, to known health effects. Dose–response assessment determines the relationship between the magnitude of the exposure and the probability of detrimental health effects. Exposure assessment determines the extent of human exposure prior to the onset of detrimental health effects. Finally, risk characterization describes the nature and magnitude of the risk, such as low, high, or uncertain.

The development of the RAF based on these factors was derived from a variety of the US federal risk assessment guidelines developed by numerous federal regulatory agencies or institutions dating back to the early 1960s. Some of these guidelines met with greater success than others. In the late 1970s, efforts were undertaken by the Interagency Regulatory Liaison Group (IRLG) to reduce inconsistencies and duplication among the various federal agencies while working to improve coordination (National Research Council 1983). In 1979, the IRLG developed guidelines for carcinogens that were adopted by the President's Regulatory Council. These guidelines were used as a starting point for the development of the RAF under the premise that if structured properly, uniform guidelines should be used for risk assessment by all federal agencies. The report of the IRLG served to develop the uniform risk analysis framework, which has gone on to be widely adapted and adopted. The next section examines how the RAF has evolved over the past 30 years.

2.3 Advancement of the Risk Analysis Framework

The risk evaluation systems operating in most developed Organisation for Economic Co-operation and Development (OECD) member countries generally are scientifically based processes that combine the identification and characterization of hazards with assessments of exposure to evaluate risk. In essence, they purport to objectively assess the probabilistic outcomes of discrete adverse events, abstracting from issues related to risk management or risk communication. In practice, governments establish a risk threshold for products or classes of products, allowing those with acceptable impacts to enter the market and prohibiting new products with unacceptable risks. Over time, the ability to accurately and reliably test for the presence of adventitious materials in food products has dropped from the detection of parts-per-million to parts-per-billion.

The Food and Agriculture Organization (FAO) of the United Nations defines risk assessment as a scientifically based process consisting of the following steps: hazard identification, hazard characterization, exposure assessment, and risk characterization (Food and Agriculture Organization 2012). Powell (2000) offers an elaboration of the system by combining the United States NRC model of risk assessment with the observations of Lammerding and Paoli (1998). In this model, hazard identification is the determination of whether a particular element in the food system is, or is not, causally linked to particular health effects. This includes determining the link between disease and the presence of the food pathogen, as well as the conditions where the pathogen survives, grows, causes infection, and dies. As such, this stage often involves epidemiological and surveillance data, challenge testing, and scientific studies. These macro results need to be scaled to sub-populations in society. Exposure assessment, sometimes called dose–response assessment, involves determining the relation between the magnitude of exposure and the probability of occurrence of the health effects in question. Therefore, by necessity, a range of responses in the population to a pathogen must be examined. This often involves examining sub-groups of consumers that might be most at risk (e.g., old, young, and immunosuppressed). The combination of hazard identification and characterization provides a theoretically supported rationale for a causal relationship between exposure and response.

Exposure assessment, in contrast, is the determination of the extent of human exposure both in the absence of and, with the application of, post-release regulatory controls. This includes a description of the pathways through which a hazard is introduced, distributed, and challenged in the production, distribution, and consumption of food. In short, it is assigning a probability to the event based on extensive situational analysis of how the food system operates and how it would relate to a hazard. Finally, risk characterization entails describing the nature and often the magnitude of human risk, including aspects of uncertainty. As such, this is the final stage of the analysis where the hazard, exposure, and variability of the results are combined to estimate the potential risk of a new or transformed product.

Traditional risk assessment theory (Isaac 2002) suggests that risk is a combined measurement of the degree of exposure multiplied by the hazard, which is the level of adverse effects of the agent on other organisms. This can be expressed by the following formula:

$$\text{RISK scientific} = \text{HAZARD} \times \text{EXPOSURE}.$$

Science has used this formula to evaluate whether initial research findings should proceed or be halted. If the assessment revealed a level of risk higher than was scientifically safe, then government agencies would not approve the technology or product for commercial release. Although the hazard would appear to be quite objectively derived through risk assessment by the global scientific community, the acceptable levels and the estimated relative level of risk for a product could vary widely between intended uses (e.g., countries or markets). It is not unreasonable to expect to see different levels of risk accepted in different circumstances.

There has been significant effort put into understanding the divergence between the old model of objectively assessed risks and what many are calling socially constructed risks. Sandman (1994) believes the old formula underestimated the actual level of risk because it ignored the public response to a risk, which he termed outrage. He argues that regulators should instead use the following formula for understanding consumer perceptions of risk:

$$\text{RISK socially constructed} = \text{HAZARD} \times \text{OUTRAGE}.$$

Sandman argues that public concern is focused on whether the risk is acceptable rather than on the scientifically perceived incidence of that risk. Although that model accommodates areas where outrage dominates, it does not fully account for the interaction between expert opinion on exposure and public concerns.

Perhaps a better configuration of the risk analysis framework for new technologies is one that incorporates all elements of the perspectives, that is, hazard identification and characterization, exposure assessment and consumer/citizen response, or outrage. Thus:

$$\text{RISK modern} = \text{HAZARD} \times \text{EXPOSURE} \times \text{OUTRAGE}.$$

Hazard and exposure would be as elucidated in the scientifically derived measure of risk, but the outrage factor would be a socially derived measure (Phillips et al. 2006). When outrage factors are very high (or very low), industry, non-government organizations (NGOs) and international agencies and organizations may either take suitable actions to position themselves to exploit those divergences or they may seek to ameliorate the divergences through lobbying or engagement in public processes.

Ultimately, the risk assessment system ought to be designed to make the right decisions—that is accepting safe products and rejecting unsafe products. As with any human system there is potential for error, especially when a new class of products is being considered where there is no accepted body of empirical evidence. Although the system is, and should be, designed to avoid making Type I errors—that is accepting something that is not safe—it has to be mindful of the trap of making Type II errors—that is rejecting safe products and activities. While we can tally up the cost of Type I errors in lost lives or damaged ecosystems, we cannot convincingly estimate the cost of foregone opportunities and all of the attendant benefits that could flow from them. The difficulty is that social amplification of risk (reflected in high outrage factors) significantly raises the potential of making a Type II error, thereby diminishing the flow of new and innovative products and progress in a knowledge-based economy. In one sense, risk amplification increases the probability that a Pareto potential improvement—net welfare enhancement—in our production system might not be realized because of the uncertainties of how the market might respond. Fear, in and of itself, can raise the potential and cost of Type II errors.

This underlying set of overlapping and, at times, conflicting processes and objectives provides the baseline against which new products or technologies are assessed. Deviations from expectations, especially those reflecting an outrage factor, provide the basis for legal, political and socio-economic discussions about liabilities and compensation. Moreover, if the degree of outrage is not included explicitly in the risk analysis, when there is a Type I error, the liability may be even higher due to punitive damages.

2.4 International Governance of Risk

Safety, especially as it relates to the safety of food and agricultural commodities, has a history of being subject to manipulations (Smyth et al. 2009, 2011). The improper use of sanitary and phytosanitary measures as the rationale for restricting trade in agricultural products is difficult for those countries, or firms, adversely affected to document and substantiate. Therefore, such measures are attractive to governments seeking to protect national interests. The ability of individual governments to manipulate sanitary and phytosanitary standards to serve the interests of particular segments of domestic agricultural markets created an international agricultural trade market that was fraught with frustration and uncertainty. Throughout the course of the twentieth century, numerous international institutions sought to harmonize the standards pertaining to international trade in agricultural products to ensure a more "level" playing field. At present, there are five different international institutions that stake a claim to coordinating and regulating the food and environmental/health safety of biotechnologically developed bioproducts, foods and crops. Table 2.1 provides a summary of these institutions. While many of these institutions and agreements have a broader range of mandates, for relevant purposes, only those that relate to agricultural biotechnology will be discussed.

The WTO does not, and has not, established regulations governing trade in products of biotechnology, but it does adjudicate international trade disputes, based upon the standards established by three standards setting organizations: *Codex Alimentarius* Commission (Codex); the Office International des Epizooties (OIE), which is now known as The World Organization for Animal Health, and the International Plant Protection Convention (IPPC). A nation that enacts a regulation or standard that contravenes the standards of any of Codex, the OIE or the IPPC, can be subject to any other WTO member nation filing a claim that argues that the regulation or standard is an unfair barrier to trade and, therefore, seeks compensation for lost trade opportunities. The Agreement on Sanitary and Phytosanitary Measures (SPS) of the WTO establishes the use of science as the decision-making criteria for justifying barriers to trade for the protection of the environment or human, animal and plant health. The SPS Agreement allows for the adoption or enforcement of measures necessary to protect the environment or human, animal or plant life or health. However, criteria are specified as to the application of any such measures. The SPS specifies that: (1) any measure(s) should not discriminate between countries; (2) standards which conform to international standards developed by international organizations (i.e. Codex, OIE, IPPC) are presumed to be consistent with the obligations outlined in the SPS Agreement; (3) standards that are in excess of established international standards or where no international agreement exists must be based on scientific principles and the completion of a risk assessment; and (4) measures shall not constitute a disguised restriction on international trade.

Codex, the OIE and the IPPC provide the technical standards framework for the SPS. If an International Standard for Phytosanitary Measures (ISPM) established by the IPPC allows for a trade barrier, then every member country of the WTO is

Table 2.1 International institutions regulating biotechnology. (Source: Adapted from Buckingham and Phillips 2001; updated by authors)

Institution	Date	Coverage	Member states	DSM	Orientation
World Trade Organization	1947	Trade in all goods and most services	159	Binding	Establish rules for transparency and dispute settlement through TBT and SPS agreements
International Plant Protection Convention	1952	Pests and pathogens of plants and plant products	178	Non-binding, sets WTO standards via SPS S.3.4	International standards for plant measures involving quarantines
Organisation for Economic Co-operation and Development	1961	Harmonization of regulatory requirements, standards and policies	34	None	Consensus documents; special commissions and events to seek common ground
Codex	1962	Food labeling and safety standards	185	Non-binding, sets WTO standards via SPS S.3.4	International standards to provide guidance for the food industry and protection for consumer health
Cartagena Protocol on Biosafety	2003	Transboundary movements of GMOs	166	None	Treaty creates rules for the transboundary movement of GMOs

TBT technical barriers to trade, *SPS* sanitary and phytosanitary, *DSM* dispute settlement mechanism, *GMOs* genetically modified organisms

allowed to implement this standard into their domestic regulatory framework without fear of challenge. If a WTO member country implements a regulatory standard that contravenes the IPPC standards, then that country may be accused of using a trade barrier in a case brought to the WTO by any other member country. Countries may have higher standards than the IPPC but only if there is a scientific justification and a risk assessment that satisfies SPS commitments.

The IPPC is a multilateral treaty that seeks to protect natural flora, cultivated plants and plant products from the spread of pathogens through international trade. Through collaboration between regional and national plant protection organizations, it provides a forum for international cooperation, dialogue, harmonization and technical exchange of plant information. The IPPC has addressed the regulation of biotechnology and genetically modified (GM) crops through several of the ISPMs.

To determine the relationship between socio-economic considerations (SECs) and the IPPC, one must look to the International Standards for Phytosanitary Measures (IPSMs). There are 24 different ISPMs, none of which provide an allowance for SECs. However, ISPM No. 5, Supplement No. 2 provides guidelines relevant to understanding the potential economic importance and the related terms of reference for environmental considerations. Economically unacceptable impacts and/or environmental damage relating to the unintended introduction of a plant pest are compensable. Three criteria are required to be documented before economic compensation from plant pests can be sought: first, the potential for the plant pest to be introduced; second, the potential for the pest to spread; and third, the potential for harmful impacts on crops (lower yield or quality), the environment (damage to ecosystems, habitats or species) and other values (tourism or recreational activities). Based on the definition of economic damage provided by ISPM No. 5 of the IPPC and therefore as part of the SPS Agreement of the WTO, any country that incorporates SECs into their domestic regulatory system that do not address risk reduction of the environment or human, plant or animal health, should know that this will be deemed a barrier to trade and said country should expect to have a disputes case brought to the WTO to have this barrier removed.

Codex develops international food standards that identify a processed food product and its essential composition and quality factors, identifies additives and potential contaminants, sets hygiene requirements, provides labeling requirements and establishes the scientific procedures used to sample and analyze the product. Jackson and Jansen (2010) provide a detailed discussion of the science-based risk assessment process for food safety and its relationship to WTO dispute cases. It commonly takes in excess of six years to develop a Codex standard. Upon a Codex standards being adopted, each member country is encouraged to incorporate it into any relevant domestic rules and legislation but they may unilaterally impose more stringent food safety regulations for consumer protection, provided the different standards are scientifically justifiable. Codex plays an important role in the agri-food trade because its standards, guidelines and recommendations are acknowledged in the SPS and TBT Agreements of the WTO. There are currently no Codex standards in place for products of biotechnology; however, there has been significant effort on behalf of Codex to develop a standard for the labeling of food products derived from biotechnology. The Codex Committee on Food Labeling was tasked in 1993 to initiate work on the development of a standard on the labeling of GM-derived foods and for nearly 20 years the Committee's efforts were gridlocked. However, in 2011 the USA relented on its opposition to the labeling of GM food products and, in 2012, Codex adopted the principles for a risk analysis of foods derived from biotechnology, which establishes that if a risk is identified, labeling is an appropriate management strategy. Codex stresses that any risk analysis of biotechnology derived foods has to be science-based and that these principles do not address "environmental, ethical, moral and socio-economic aspects…" (Codex Alimentarius Commission 2012, p. 1). It is important to note that this is a Codex principle on risk analysis of foods derived from biotechnology and not the standard on the labeling of GM foods that the Committee was tasked with 20 years ago.

The OECD has actively assisted in the harmonization of international regulatory requirements, standards and policies related to biotechnology, beginning in 1995. The OECD has undertaken a number of projects aimed at making regulatory processes more transparent and efficient, to facilitate trade in the products derived from biotechnology and to provide an information exchange and dialogue with non-OECD countries. The OECD leads efforts to develop Consensus Documents that set out the biology of the crop plant, introduced trait or gene product and to provide a common base to be used in the regulatory assessment of an agricultural or food product derived from biotechnology. These Consensus Documents focus on the biology of the organism, containing the technical knowledge that is utilized in risk assessment of products derived from biotechnology, that are becoming embedded in the regulatory frameworks for Member States and are to be mutually recognized by other Member States.

The Cartagena Protocol on Biosafety (CPB) is representative of recent efforts to provide a comprehensive international structure to ensure the protection of biodiversity and to facilitate considerations of non-scientific concerns, so-called SECs. The CPB is a new international institution negotiated specifically to deal with trade in the products of biotechnology. The CPB, which was concluded in Montreal in 2000 and came into effect in 2003, provides rules for the trans-boundary movements of LMOs intended for environmental release and those destined for the food chain. The CPB only applies to the nations that have ratified the agreement, and the challenge of using the CPB to govern production and trade of GM crops is that many of the leading producers of GM crops are not signatories to the CPB, let alone having plans to ratify the agreement. None of Argentina, Canada or the USA is signatory to the CPB and these three countries represent three of the top five producers of GM crops (James 2011). This creates considerable challenges for the CPB.

As illustrated by the above discussion, most of the international institutions that hold a stake in the governance of products of biotechnology have existed for 50 years or greater and use regulations and principles to respond to risks that science can measure and assess. The recent CPB takes a different approach. That approach has been led by the efforts of numerous European nations and can be seen as a means of reducing global reliance on the WTO as the dominant institution that governs biotechnology products and trade. To a large extent, it has been the CPB that has created the impetus for the development and incorporation of concerns that science has not (and perhaps cannot) measured and assessed (the so-called SECS) into domestic regulatory frameworks for GM crops. This international divergence of regulatory strategies, with Europe following a socio-economic/precautionary principle-based approach which is discussed further in the next chapter, versus the science-based approach utilized in nations such as those in North America, has created a trans-Atlantic gap in the international governance of products from biotechnology. Many African nations have opted to follow the regulatory path of the European Union and its open aversion to agricultural biotechnology, while many South and Latin American countries are choosing to base their GM crop regulations on science. There have been many costs to this trans-Atlantic gap as illustrated by the January 2012 announcement by BASF that it was moving its research division from Europe to the USA due to the delays in regulatory decisions in Europe (BASF 2012). The increased polarization of attitudes towards the regulation of biotechnology across the Atlantic is not likely to be resolved in the near future as is witnessed by the 20 years of Codex dialogue regarding the labeling of GM food products.

2.5 Structure of Science-Based Biotechnology Regulatory Frameworks

The advent of biotechnology products has triggered a spirited debate about what risks we should assess, how we should assess risks, who should undertake the task of defining what is acceptable, what rules we should draw upon and where we should vest the authority to decide. For the most part, the Americas have increasingly moved towards the "scientific evidence-informed" or what is referred to as the "science-based" assessment of risks, where the risks being governed are those posed to environmental and human health and safety capable of measurement and assessment by science. This was initiated by Canada and the USA in the early years of the 1990s, when their domestic biotechnology regulatory frameworks were created by adapting existing regulations and institutions. This section provides a concise overview of the science-based approach to the regulation of biotechnology in both Canada and the USA. For greater details on the Canadian and American regulatory systems, see Smyth and McHughen (2008) and McHughen and Smyth (2008).

2.5.1 Canada's Institutional Regulatory Framework

Two departments of the Canadian federal government are responsible for the regulatory framework for GM crops, the Canadian Food Inspection Agency, and Health Canada.

2.5.1.1 The Canadian Food Inspection Agency (CFIA)

In Canada, all commercialized GM plants to date have been considered to contain novel traits and, therefore, have been assessed for additional safety factors. Because of this, government regulators carefully assess potential impacts before these modified plants can be released into the environment on a case-by-case basis. The regulation of plants with novel traits (PNTs) is the responsibility of the Plant Biosafety Office within the CFIA. Environmental safety assessments examine five broad categories of possible impacts of a PNT. These are:

- The potential of the plant to become a weed or to be invasive of natural habitats;
- The potential for gene flow to wild relatives;
- The potential for a plant to become a plant pest;
- The potential impact of a plant or its gene products on non-target species; and

- The potential impact on biodiversity (Canadian Food Inspection Agency 2011a).

Every GM crop variety application that the CFIA receives is treated as a PNT and is assessed for safety under the following CFIA directives:

- Directive 94-08: Assessment Criteria for Determining Environmental Safety of Plants with Novel Traits;
- Directive 95-03: Guidelines for the Assessment of Novel Feeds: Plant Sources;
- Directive 96-13: Import Permit Requirements for Plants with Novel Traits, and their Products; and
- Directive 2000-07: Guidelines for the Environmental Release of Plants with Novel Traits within Confined Field Trials in Canada.

Using these directives, the CFIA assesses all PNT applications for environmental release and use as animal feed. There are three stages in the assessment process for a PNT variety. In Stage 1 of the development of a new PNT variety that is intended for unconfined environmental release and/or use as a livestock feed, the plants are required to be grown in a contained facility (i.e., greenhouse or laboratory growth chamber). Growing conditions in these types of facilities follow biosafety guidelines that have been established by Health Canada and the Medical Research Council. Research institutions may develop and require that codes of practice be followed in addition to the above.

In Stage 2, the PNT variety developer must submit an application to the CFIA and receive authorization to conduct confined field trials in Canada. Directive 2000–2007 is used to establish how many trials are allowed in Canada, the size of the plot and isolation distances that are required. The CFIA notifies each province where field trials applications have been received and provincial authorities are given a 30-day comment period. Field trials are conducted over several years in various locations that represent potential adoption regions and the data produced by these trials is used to provide information to the CFIA for the safety assessments in Stage 3.

Stage 3 is designed to address the five priority categories listed above. To provide the necessary information to satisfy these questions, the product developer is required to submit scientific data that has been gathered from the field trials. The CFIA has undertaken scientific studies and uses these to assist in review and may commission additional studies if required. Peer-reviewed journal articles are also utilized as sources of relevant information. The scientific data that are required for the CFIA to undertake the safety assessment includes: identification and classification of the PNT; modification methods; description of the novel trait(s); environmental data; and livestock feed data that is comprised of nutritional, toxicity, and allergenicity data (Canadian Food Inspection Agency 2011b).

Following the review of the scientific data, a decision document is drafted and sent to the product developer as well as posted on the CFIA's website. This document explains how the review took place and provides a basis for rendering the final decision. If, at any point following this, additional scientific information becomes available regarding the crop variety, the product developer is required to report this information to the CFIA who will undertake a re-evaluation based on the information.

At this point, the CFIA regulation process is complete and the successful product developer receives notification from the CFIA granting unconfined release status.

When unconfined release is granted, the developer then submits the variety to the specified variety recommending committee so that it can be evaluated based on agronomic attributes, the same as conventional varieties. Each crop commodity has its own variety recommendation committee that evaluates a variety of agronomic attributes for all varieties put forward for consideration. If the agronomic attributes are sufficient, compared to a "check" variety, the recommending committee sends a letter recommending variety approval to the Variety Registration Office (VRO) of the CFIA. The VRO accepts the recommendations of the variety recommending committee and grants final variety approval to all new crop varieties, GM or conventional.

2.5.2 Health Canada (HC)

Unlike the CFIA, which uses a product trigger, Health Canada defines novel foods as foods resulting from a process not previously used for food, products that do not have a history of safe use as a food or foods that have been modified by genetic manipulation, genetically engineered foods or biotechnology-derived foods (Health Canada 2011a). Health Canada assesses the safety of all GM and other novel foods proposed for sale in Canada. Companies are required to submit detailed scientific data for review and approval by Health Canada, before such foods can be sold.

Health Canada is also responsible for the environmental assessment of products regulated under the Food and Drugs Act, including novel foods, but only those that are not subject to environmental safety assessment by the CFIA. This activity is required by the New Substance Notification Regulations of the Canadian Environmental Protection Act (1999). Two branches within Health Canada, the Health Products and Food Branch (HPFB), and the Healthy Environment and Consumer Safety Branch (HECSB), started working in 2001 to develop new regulations to assess the impact on the environment and on human health of new substances used in these products. This process was known as Health Canada's Environmental Impact Initiative (Health Canada 2011b) and was finalized in 2010.

Health Canada does not review all foods new to the Canadian market but only those that are deemed novel. Therefore, the concept of prior safe use as a food was introduced to exclude foods new to the Canadian market which have a history of safe food use in other countries, from being the target of a novel food notification. Secondly, the concept of "major change" was introduced into the novel food definition in order to avoid the potential of a minor processing change triggering a novel food notification. This approach is intended to restrict novel food notifications due to the introduction of new processes to only those that are truly new and cause substantial changes in the composition of the food.

A major change with respect to food is defined as a change peripheral to the manufacturer's experience or generally accepted nutritional or food science theory. This would place the modified food outside the accepted limits of natural variations for that food with regard to the following:

- Composition, structure or nutritional quality of the food or its generally recognized physiological effects;
- Manner in which the food is metabolized in the body; or
- Microbiological safety, the chemical safety or the safe use of the food. (Health Canada 2011c).

Regulators at Health Canada receive safety assessment data from the product developer relating to novel foods. This is when the nutritional, toxicity, and allergenicity data are reviewed and assessed. Data are also needed to satisfy safety assessments regarding dietary exposure, metabolization, and microbiological safety. One salient feature of the Health Canada regulatory process is the use of experience from other jurisdictions. If a novel food product has a history of safe production and consumption in another country, then this history is admissible as data for regulatory approval in Canada. Health Canada is more accepting of this data than the CFIA, although the CFIA will allow data from other jurisdictions to be included in a submission package, provided it has a valid scientific rationale.

Health Canada has established criteria for the assessment of novel foods that provides information to establish the safety of the novel food. Written notification is required at least 45 days prior to the sale or advertising for sale of any novel food. Health Canada is required to respond within 45 days of receipt of the notification regarding its acceptability for sale. If additional information is required to properly establish the safety of the product, such information will be requested in writing and "the clock" is stopped, thus extending the period. The applicant is not permitted to sell or advertise the product until the additional information requirement is fulfilled and the Department has agreed to the acceptability of the product.

Once the Novel Foods Section of Health Canada receives the application for a new food product from the product developer, there are four assessments involved at this stage. The product developer has to address environmental safety, chemical safety, nutritional changes/stability and microbial hazards.

Once the scientific review of data is complete, Health Canada can request additional information, which then requires another scientific review of the new data. If there are no requests at this point, a draft ruling is developed by the Novel Foods Section that then goes up the bureaucratic ladder for review. Senior management within Health Canada has the right to request additional information from the product developer at this stage, and this process would trigger another scientific review. If the drafted proposal is acceptable, then a letter is sent to the product developer informing them of this, and the Decision Document is posted on Health Canada's website. At this point, the product developer may market the novel food.

2.5.3 The United States Institutional Regulatory Framework

In 1986, the White House established a committee at the Office of Science and Technology Policy (OSTP) to recommend how best to regulate the quickly advancing technology known as biotechnology. The resulting publication outlined several important factors for the regulation of biotechnology in the USA (Office of Science and Technology Policy 1986). One of the most vital findings was that the OSTP concluded that rDNA was not inherently risky and that regulations should therefore focus on the risks of products, not on the processes used to develop them. Existing legislation and regulations designed for current products could be adapted to deal with products of biotechnology. The coordinated framework also recognized the concept that GMOs were not inherently riskier than other, non-modified organisms. Finally, the OSTP document assigned regulatory priority amongst the relevant federal agencies: the US Department of Agriculture (USDA), the Food and Drug Agency (FDA), and the Environmental Protection Agency (EPA). Under the coordinated framework, the USDA was to be the lead agency in the evaluation of plants as potential pests of agriculture, the FDA was to review GMOs as potential threats to the food and feed supply and the EPA was to take priority in evaluating new GMOs with pesticidal properties.

The National Environmental Policy Act (NEPA) of 1970 requires most federal regulatory agencies to investigate the potential for environmental impacts prior to making certain decisions or taking certain actions that could pose environmental risks. The relevant agency starts by asking "Is the action we are being asked to regulate likely to have significant environmental effects?" and then sets out to answer this question. The simplest finding is a categorical exclusion, which includes items or actions that do not pose significant effects on the environment. After ascertaining that no extraordinary circumstances exist (due to, e.g., possible interactions with unique regional features or endangered species) the agency approves the application, which is known in the USA as deregulation. If the proposal presents significant environmental effects, the agency conducts and publishes an environmental assessment (EA).

The EA is a critical analysis of the environmental consequences of conducting the proposed activity or releasing the product. After reviewing the various relevant factors, the agency can conclude that either the proposed activity/product demands additional analyses and (1) issues a Notice of Intent (NOI) to prepare a more elaborate Environmental Impact Statement (EIS) or (2) that the proposed activity/product poses insignificant risk and prepares another document, the Finding of No Significant Impact (FONSI). The FONSI summarizes the EA and justifies and provides rationale, using the data presented in the EA, why the agency came to the conclusion that the activity/product was deemed environmentally benign. Both the EA and the FONSI are public documents and the NEPA Act provides a parallel opportunity for citizen review of agency decision-making. When industry petitions to a regulatory agency for deregulation of a biotech crop, the resulting agency's EA process includes solicitation of public comment regarding draft assessment and the deregulation petition. It is during this public comment process that private individuals may raise or voice concerns.

If the EA suggests that the proposed activity or product might present a significant environmental impact, the agency can publish the NOI in the Federal Register. The NOI includes information on the proposed activity/product, outlines how the agency plans to proceed with an EIS and how the public can contribute along with contact information at the agency. The plan identifies specific relevant issues for in-depth investigation and a timeline for completion.

The EIS is a major analysis document, requiring careful deliberation and active wide consultation. When the agency completes a draft EIS, a Notice of Availability (NOA) is published in Federal Register, which opens the draft to public comment. For at least 45 days, anyone can read the draft and provide input to the agency, which may additionally provide other fora (such as public meetings) to solicit broad public input. The agency is required to take public comments seriously and respond to all reasonable input in preparing the final EIS. When the final EIS is completed, the agency publishes another NOA in the Federal Register, which signals another 30-day (or more) waiting period before a final decision is made. Eventually, the agency publishes a Record of Decision (RoD), the final step in the whole process. The RoD summarizes and discusses the issues investigated in the proposed activity/item prior to making the final decision and is made publicly available.

2.5.3.1 United States Department of Agriculture (USDA)

Within the USDA, oversight for GM crops is the responsibility of the office of Animal and Plant Health Inspection Service (APHIS). The primary concern is with protecting agriculture and the environment (broadly interpreted) from potential pests (also broadly interpreted). The USDA regulates all GM plants prior to environmental release, including during import, interstate movement, small and large field trials and commercial production. Initially, legislative authority was distributed among several statutes, including the Plant Quarantine Act, the Federal Plant Pest Act and the Federal Noxious Weed Act, but this regulatory authority was consolidated in 2000 in the federal Plant Protection Act (PPA).

Most field trials are approved under a notification procedure, which is a process designed for the simplest or most familiar GM plants. Usually, notification involves submitting a letter to APHIS, documenting how the proposed GM plant meets six specified criteria and designated performance/characteristic standards. The criteria include such considerations as the GM plant not being of a noxious weed species and not transformed with human or animal pathogenic sequences. The notification can be used for field trial approval as well as importation and transport of specified GM plants within the USA.

APHIS issues permits for the production of field trials and is primarily concerned with biosafety, that is, the unintended release and spread of a potential plant pest. The permit procedure is much more elaborate than the notification process and requires greater amount of information and data.

In March 2003, in response to concerns surrounding non-food substances in transgenic plants and a series of highly publicized permit violations, APHIS announced that they would strengthen mandatory permit conditions for field-testing transgenic crops. The number of site inspections would increase to five during the trial and two, the following season. The permits for pharmaceutical trials with transgenic corn (a common host plant species) imposed several conditions, including that no corn can be grown within one mile of the trial site, that no food or feed crop can be grown on the site the following season and the size of the buffer zone was doubled.

2.5.3.2 Food and Drug Administration (FDA)

The FDA has responsibility for ensuring the safety and security of human food and the supply of animal feed. In 1992, the FDA issued a policy statement establishing its authority to regulate new foods and feeds under the Federal Food, Drug and Cosmetic Act (FFDCA), irrespective of the method of breeding (Food and Drug Administration 1997). Under this Act, the FDA considers the food or feed composition relative to currently available counterparts, looking especially at the presence of allergens and toxins and any changes in levels of nutritional and anti-nutritional substances. Foods containing unexpected or novel substances, or usual substances falling outside normal ranges for that kind of food, are considered adulterated and subject to FDA regulatory action. Foods and feeds identical, or nearly identical, in composition to regular versions are not considered adulterated and do not trigger FDA review, even if they were produced using rDNA technology. The FFDCA establishes that the FDA is concerned for food and feed safety, and that safety is a function of substances present (or of nutrients absent) from the food in question. If foods or feeds produced from or with GMOs are composed of the same substances and in the same amounts and relative proportions, there is no basis for a safety concern (above and beyond whatever safety concerns may ordinarily reside in that food or feed), and no need to invoke the "adulteration" action trigger. This is why some people consider the FDA review to be "voluntary". Because most foods and feeds from GM plants are compositionally identical (or nearly so) to regular versions, the FDA does not require mandatory regulatory assessment. The FDA regulates food and feed based on the objective changes in product composition.

Although called "voluntary," all GM foods and feeds currently on the US market have undergone what is called a FDA "consultation", in which the developer submits a dossier of compositional data relating to the putative "identical" food or feed, and FDA scientists evaluate the composition in comparison with the composition of the regular food and feed. The data submitted include such information as genetic stability of the plant, compositional analyses, nutritional assessment, as well as allergenicity and toxicology of any substances ordinarily present in the food or feed, along with such assessments of the introduced gene products. The FDA published guidelines to assist developers in compiling the dossier in 1997 (National Academy of Sciences 2000). This procedure is beneficial to all parties, as it provides some assurance to consumers that a government agency is evaluating a new food or feed product prior to commercial release. It gives the developer an opportunity to have an independent third party (FDA) cast expert eyes over the data to ensure no potential problems were overlooked, and it keeps the FDA up to speed on new foods and feeds coming through the development pipeline. Even without a compulsion, all developers of GM foods and feeds on the US market have completed the FDA consultation, largely because it is relatively simple, straightforward and prudent to do so. Nevertheless, some people demand that the FDA adjust their policy to make the procedure mandatory.

2.5.3.3 Environmental Protection Agency (EPA)

The EPA enjoys broad regulatory authority over substances with pesticidal characteristics, with particular concern for threats to human health and the environment. In addition to regulating the pesticides themselves, the EPA regulates according to changes in pesticidal properties or pesticide usage. Importantly, the EPA claims not to regulate GM plants per se, but rather regulates the pesticidal properties associated with GM plants. This trigger captures plants such as GM virus resistant plants, even though there is no pesticidal substance necessarily sprayed (or synthesized internally), as well as the more obvious herbicide tolerant GM plants, where the crop is designed to be sprayed with a new pesticidal substance. The EPA also captures GM plants that produce their own substances with pesticidal properties, the plant incorporated protectant (PiP), which means GM plants expressing, for example, Bt or other insecticidal substance.

The EPA has regulatory authority to regulate the pesticidal properties in GM plants under the Federal Insecticide, Fungicide and Rodenticide Act (FIFRA). Under the coordinated framework, the EPA published their proposed regulations in 1994 and began acting on those in 1995. The EPA's working definition of a PiP was "… a pesticidal substance produced in a living plant and the genetic material necessary for the production of that pesticidal substance, where the substance is intended for use in the living plant" (National Academy of Sciences 2000, p. 127).

In 1994, the EPA proposed exempting several low-risk categories. One would be those plant pesticides in which the genetic material originates in a sexually compatible species. That is, if the pesticidal trait could be crossed through ordinary breeding, the resulting novel pest-protected plant would be exempted under FIFRA. A second exemption category included those using physical barriers (and similar mechanisms such as inactivating toxic substances) to preclude the pest from attaching to or invading the plant. The third category included plants expressing viral coat proteins as means to provide virus resistance. The proposals also included language to circumvent, as required under FFDCA, the establishment of a tolerance limit for such substances (National Academy of Sciences 2000). By 2001, the EPA issued final rules exempting the previously captured sexually compatible PiPs, as well as exemptions for residues of the pesticidal substances and genetic material (DNA, RNA).

In accordance with the coordinated framework, the EPA evaluates each submission on a case-by-case basis, so the focus of the concerns with novel herbicide uses will differ from those with novel insect protection. To date, all GM PiP plants evaluated by the EPA produce proteins, mainly the Bt endotoxin and viral resistance proteins, such as coat proteins. In addition to data requirements related to product characterization, the EPA also requires data on mammalian toxicity, non-target organisms' effects and environmental metabolism. For Bt products, the EPA also demands an insect resistance management program. For herbicide-resistant GM plants, the EPA coordinates with the USDA and the FDA. The EPA emphasizes that it does not regulate the GM plant per se, but the herbicide used on or with the GM plant.

Like APHIS, the EPA is also concerned with gene flow issues. However, unlike APHIS, where gene flow interest is driven by concern for potential increase in weediness or plant pest characteristics, the EPA's interest in gene flow is due to the possibility of expanding exposure to novel pesticidal substances. The EPA is required by FIFRA to consider adverse environmental impacts attributable to possible gene flow, and by FFDCA to exempt or issue tolerances for the pesticidal substances that might enter the food and feed supply. So far, the EPA has analyzed several plant species with Bt constructs and all have received exemptions. The EPA has prohibited the unregulated sale and cultivation of Bt cotton, however, in some areas (Hawaii, Florida, Puerto Rico and the Virgin Islands), due to the local presence of interfertile relatives or feral cotton populations, as they present a recipient sink and opportunity for greater uncontrolled Bt exposure.

By the same reasoning, the EPA seeks to preclude gene flow between GM plants and wild or feral relatives as that is a primary means of gene escape, invasion and possible establishment of undesirable plants. To date, this policy has not posed great hardship (except possibly to growers in Hawaii, Florida, Puerto Rico, or the Virgin Islands wishing to grow Bt cotton) but may take on greater significance with the increasing interest in biofuels made from GM versions of energy crops such as switchgrass. At present, in spite of considerable research and development of technologies to limit gene flow (via, e.g., pollen disabling genes), no such gene flow mitigation technology is 100 % effective (National Academy of Sciences 2004).

The EPA is also concerned with effects of PiPs on non-target organisms in the environment. The requirements involve an initial assessment of potential toxicity and exposure to non-target species, followed, where warranted, by up to four tiers of testing on the relevant species. Finally, the EPA considers the environmental fate of PiP substances, for example of Bt endotoxin in the soil, and how soil biota respond to the Bt deposited by transgenic plant roots, decaying leaf matter, pollen settling, etc.

The EPA is also concerned about organisms—particularly insects—developing resistance to pesticides, and so the EPA considers management strategies to minimize and delay the onset of resistance in pest populations. Pests are known to develop such resistance to pesticides, antibiotics and other such substances based on exposure and intensity. Because Bt is an important insect control chemical to many farmers—even organic farmers—the onset of resistant insect pest populations is a concern for all. The EPA takes the lead in requiring appropriate insect resistance management (IRM) strategies, and farmers are required to follow the IRM practice regulations. For Bt, these practices include areas of on farm refugia to allow Bt sensitive and resistant insects to mate in the absence of Bt selection pressure. The exact size and locations of the refugia will vary depending on the crop, the particular pest and the nature of the pesticide being used. Other factors, such as nearby alternate refugia or PiP crop species, may also influence the optimum presentation of the refugia.

2.6 Conclusions

Regulatory oversight can never bring perfect foresight of risks. Essentially, we see through a dark glass and can only hope to pick out every potential risk requiring preventive action—through new designs, new warnings or, in some cases, a ban on the emerging technology pending thorough assessment.

In Canada, the use of science-based regulations results in a regulatory framework that provides consistent and repeatable regulatory decisions on GM crops. In turn, this provides firms in the industry with confidence about the regulatory system and these firms continue investing in the development of new crop varieties. This is not to say that the biotechnology industry is without complaints about the regulatory system, but what complaints do exist can be addressed via consultative processes. The triggering of PNT status on some mutagenic varieties does frustrate variety developers, especially given that there is no way of knowing in advance of a submission as to whether or not the CFIA will deem the submitted variety as a PNT. The ability of science to test, and to test with confidence, at smaller and smaller percentages, means that the system responds to new discoveries. The adaption of the previous regulatory framework for plant agriculture to regulate GM crops has resulted in a continual and efficient process that has approved new GM crops for nearly 20 years.

Like the Canadian experience, the US regulatory framework is based on the adaptation of existing regulations. Given the scientific basis of the regulations, consistent and repeatable regulatory decisions result from the American regulatory system. The USDA has suffered somewhat of a setback at the end of the past decade with courts regarding GM alfalfa, bent grass and sugar beets. In each case, the USDA was found to have failed to undertake a thorough EIS, which is in violation of NEPA. The USDA will have to become more rigorous in its approach to the conducting of environmental impacts as it deals with future variety assessments.

However, despite the success of the North American regulatory frameworks, there is no definitive set of "science" facts that can underpin decision-making for technologies like GM crops. Even confining the assessed risks to those relevant to environmental or human health and safety does not provide a closed set of facts to be measured. With harm to biodiversity in particular, the harm depends largely on complex questions of ecological science. Detecting and documenting all the lives that comprise biodiversity and monitoring and measuring negative impacts to such biodiversity present daunting technical challenges given limits of knowledge and understanding. If the assessed risks are expanded beyond these traditional concerns, the difficulties greatly increase. However, this does not mean that science-based systems should be replaced—rather, it highlights the need to more clearly and concretely define the appropriate human institutions that can develop and use the "scientific consensus" that is needed to make decisions that further the public interest. In essence, the efforts to develop an effective regulatory and trade system, one that facilitates optimal market adoption, is not complete. The integration of agriculture into the WTO, the adoption of a binding dispute settlement mechanism and the incorporation of the international standards setting organizations (Codex, IPPC, OIE) into the regulatory and trade systems were all vital steps. But, they should not be viewed as the end of the process. Rather, they provide the launching pad for a process to develop a more highly integrated and effective science-based regulatory system for new agri-food technologies.

References

BASF (2012) News release: BASF to concentrate plant biotechnology activities on main markets in North and South America. http://www.basf.com/group/pressrelease/P-12-109. Accessed 10 June 2013

Buckingham D, Phillips PWB (2001) Hot potato, hot potato: regulating products of biotechnology by the international community. J World Trade 35(1):1–31

Canadian Food Inspection Agency (2011a) Directive 94-08: assessment criteria for determining environmental safety of plants with novel traits. http://www.inspection.gc.ca/english/plaveg/bio/dir/dir9408e.shtml. Accessed 10 Dec 2011

Canadian Food Inspection Agency (2011b) Regulation of agricultural biotechnology in Canada: an educator's resource. http://www.inspection.gc.ca/english/sci/biotech/gen/educ_postsece.shtml. Accessed 10 Dec 2011

Canadian Lung Association (2004) Resurgence: is TB a problem in Canada today? http://www.lung.ca/tb/tbtoday/resurgence. Accessed 10 June 2013

Codex Alimentarius Commission (2012) Principles for the risk analysis of foods derived from modern biotechnology

Food and Agriculture Organization (2012) FAO/WHO Codex training package: section four—scientific basis for Codex work. 4.1 Risk analysis in the framework of Codex. ftp://ftp.fao.org/es/esn/food/capacity_building/4_1.pdf. Accessed 10 June 2013

Food and Drug Administration (1997) Guidance on consultation procedures; foods derived from new plant varieties. http://vm.cfsan.fda.gov/~lrd/consulpr.html. Accessed 10 Dec 2011

Isaac GE (2002) Agricultural biotechnology and transatlantic trade: regulatory barriers to GM crops. CABI Publishing, Wallingford

Health Canada (2011a) Genetically modified (GM) foods and other novel foods. http://www.hc-sc.gc.ca/fn-an/gmf-agm/index_e.html. Accessed 10 Dec 2011

Health Canada (2011b) Environmental impact initiative. http://www.hc-sc.gc.ca/ewh-semt/contaminants/person/impact/about-impact/index_e.html. Accessed 10 Dec 2011

Health Canada (2011c) Guidelines for the safety assessment of novel foods. http://www.hc-sc.gc.ca/fn-an/legislation/guide-ld/nf-an/guidelines-lignesdirectrices_e.html. Accessed 10 Dec 2011

Jackson LA, Jansen M (2010) Risk assessment in the international food safety policy arena. Can the multilateral institutions encourage unbiased outcomes? Food Policy 35:538–547

James C (2011) Global status of commercialized biotech/GM crops: 2011. ISAAA Brief 43. http://www.isaaa.org/resources/publications/briefs/43/executivesummary/default.asp. Accessed 10 June 2013

Kinsella KG (1992) Changes in life expectancy 1900–1990. Am J Clin Nutr 55:1196–1202

Lammerding AM, Paoli GM (1998) Quantitative risk assessment: an emerging tool for emerging food borne pathogens. Emerg Infect Dis 3(4):483–487

McHughen A, Smyth S (2008) US regulatory system for genetically modified (GMO, rDNA or transgenic) crop cultivars. Plant Biotechnol J 6(1):2–12

National Academy of Sciences (2000) Genetically modified pest protected plants: science and regulation. National Academies Press, Washington, DC

National Academy of Sciences (2004) Biological confinement of genetically engineered organisms. National Academies Press, Washington, DC
National Research Council (1983) Risk assessment in the Federal Government: managing the process. National Academy Press, Washington, DC
Office of Science and Technology Policy (1986) Coordinated framework for regulation of biotechnology: announcement of policy and notice for public comment. Fed Regist 51(123):23302–23350
Phillips PWB, Smyth SJ, Kerr WA (eds) (2006) Governing risk in the 21st century: lessons from the world of biotechnology. Nova Science Publishers, Hauppauge
Powell D (2000) Risk based regulatory responses in global food trade: a case study of Guatemalan raspberry imports into the United States and Canada, 1996–1998. In: Doern GB, Reed T (eds) Risky business: Canada's changing science-based policy and regulatory regime. University of Toronto Press, Toronto
Sandman PM (1994) Mass media and environmental risk: seven principles. Risk Health Saf Environ 5(3):1–7
Smyth S, McHughen A (2008) Regulating innovative crop technologies in Canada: the case of regulating genetically modified crops. Plant Biotechnol J 6(3):213–225
Smyth S, Phillips PWB, Kerr WA (2009) Global governance quandaries regarding transformative technologies for bioproducts, crops, and foods. J World Trade 43(6):1299–1323
Smyth SJ, Kerr WA, Phillips PWB (2011) Recent trends in the scientific basis of sanitary and phytosanitary trade rules and their potential impact on investment. J World Invest Trade 12(1):5–26
World Health Organization (2012) Global tuberculosis report 2012. www.who.int/tb/publications/global_report/gtbr12_main.pdf. Accessed 10 June 2013

Chapter 3
International Context of Socioeconomic Considerations and the Regulation of Genetically Modified Organisms

José Falck-Zepeda, Karinne Ludlow and Stuart J. Smyth

3.1 Introduction

Many countries have made significant investment in developing broad agricultural biotechnology capacity focused on developing plant and animal breeding, advanced genomics, tissue culture, and genetic transformations (Falck-Zepeda et al. 2009). Developing countries have been particularly interested in supporting their resource poor smallholder farmers while addressing multiple productivity challenges and cultural diversity and operating in agricultural and natural ecosystems which may be mega biodiverse. These challenges are made more difficult to overcome because developing countries also face significant institutional and policy challenges.

However, investments in innovation have yielded a growing portfolio of GM technologies made by developing countries for developing countries. These technologies are likely to address agricultural productivity issues relevant to their context. Some have reached a relatively advanced state of development and may have moved along the biosafety regulatory pipeline but most are simply not reaching farmers (Atanassov et al 2004; Stein and Rodriguez-Cerezo 2009). This state of affairs will continue unless countries address regulatory hurdles that unnecessarily slow or stop the deployment of potentially valuable technology (Nature Biotechnology 2012). Investigating the regulatory hurdles affecting biosafety may help develop a robust innovation system responsive to the needs of farmers in developing and developed countries.

J. Falck-Zepeda (✉)
International Food Policy Research Institute, Washington, DC, USA
e-mail: j.falck-zepeda@cgiar.org

K. Ludlow
Monash University, Melbourne, Australia
e-mail: karinne.ludlow@monash.edu

S. J. Smyth
University of Saskatchewan, Saskatoon, Canada
e-mail: stuart.smyth@usask.ca

K. Ludlow et al. (eds.), *Socio-Economic Considerations in Biotechnology Regulation*,
Natural Resource Management and Policy 37, DOI 10.1007/978-1-4614-9440-9_3,

Over time, diverse actors and stakeholders have developed conceptual frameworks, models, and positions with regard to agriculture, food production, and the role of science and technology in addressing productivity. Actors and stakeholders have interacted in different policy arenas with contrasting outcomes which have influenced policies and decisions made in the international context. Policy outcomes range from countries making proactive investments in agricultural biotechnology serving as a major driver of agricultural productivity to countries that have passed informal or formal bans or moratoriums.

In this chapter, we propose that over time conceptual and theoretical models and frameworks have shaped the international policy milieu. The policy milieu in turn has molded many developed and developing countries' political and decision-making landscape with regard to GM crops. In this chapter, we attempt to describe some of the conceptual drivers that help explain and describe the international context related to GMOs.

3.2 Models

We have identified four knowledge strands in the literature that are useful in defining the policy pathways taken by countries toward GM technologies and thus, in turn, help shape the international context. The four strands are (1) productivity and endogenous growth theory and models, (2) technological and sector dualism, dependence theory and the structure of the agricultural sector, (3) the political economy of agricultural policy and innovation models, and (4) post-normal science frameworks.

3.2.1 Productivity and Endogenous Growth

Productivity and endogenous growth theories and subsequent models focus on technology's role in promoting growth and externalities as a source of endogenous growth (Solow 1957; Lucas 1988). The induced innovation models (Hayami and Ruttan 1985; Binswanger and Ruttan 1978) proposed that technical innovation is directly correlated by the incentive environment farmers face while making decisions. Schultz (1964) proposed the "poor but efficient hypothesis" which suggested that most developing country farmers lack technological and financial resources necessary to improve their welfare even when being technologically efficient with the resources they have available. Technology adoption and its impact garnered a prominent role in economic development. Seminal papers by Griliches (1957), Feder et al. (1985), de Janvry (1981), and Feder (1985) introduced the concept of technology adoption dynamics and its critical impact on development. One outcome of these productivity and endogenous growth models is that producers in developing countries will need significant external investments in technology and complementary inputs in order to improve production and productivity. This will include investments in developing appropriate improved technologies to benefit those that adopt the technology.

The previous models were tempered in their embrace of technology by models proposed in the literature. One such model is the "technology treadmill" (Cochrane 1958; Gardner 2002). Application of these models showed that in many small open economies most gains from technology adoption would accrue to consumers. In the long run, producers tend to not gain from technical change, and are forced to perpetually pursue newer technologies to capture short-term gains in productivity which are then bid away by the market.

Productivity and endogenous growth hypotheses and models provide the conceptual framework that supported, and continues to support, significant investments by multilateral development agencies, developed and developing countries' governments and the private sector. For example, the outcome from investment in organizations such as the Consultative Group of International Agricultural Research (CGIAR) and its research centers around the world gave rise to the Green Revolution in the 1950–1970s and its impact on agriculture.

3.2.2 Technological and Sector Dualism, Dependency, and the Structure of the Agricultural Sector

The development economics literature introduced models based on the concept of dualism and dependence. Lewis (1954) proposed that growth in developing economies could be described and might be limited by the existence of dual sectors in the economy: a modern and capitalistic sector coexisting with a traditional one. Producers in the modern sector frequently have access to credit, technical services, technology, and complementary inputs while participating in market operations. Rural producers in the traditional sector tend to be subsistence farmers, frequently poor, and often excluded from formal markets. The basic Lewis model was expanded to consider technology dualism and technology adoption and use, such as those by Eckaus (1955) and Higgins (1956).

Dualism has also been expanded to explain dissimilarities between developed and developing countries' pathways and the deteriorating terms of trade described by Singer (1970), Myrdal (1968), Prebisch (1950), and others. These models, collectively known as dependency theory, indicated that poor countries are not poor because they lack technologies or resources, or whether they are integrated to the global and modern economic system, but rather how they are integrated in such system.

The persistent poverty, the existence of dualistic sectors, and the economic inequality in many developing countries is perhaps one of the main drivers for the existing policy and decision-making positions, pressure groups, and other formal and informal organizations that have opposed the deployment of GM technologies. Many of these agents may oppose most "modern" technologies because they are viewed as a causal agent reinforcing duality and poverty. They also assert that modern technologies may be irrelevant or not economically justifiable under the set of decision-making parameters available to traditional producers. As poverty tends to be associated with ethnicity and race, duality can also be a driver for the indigenous and ethnically based organizations and movements which may oppose GM and other modern technologies.

3.2.3 Political Economy and Institutional Analysis Considerations

The Institutional Analysis and Decision (IAD) Framework proposed by Ostrom (2005) and other political economy approaches are quite similar in their focus and objectives. Duncan and Williams (2010) describe the centrality of politics while focusing on institutions to describe the incentives frameworks that guide behavior. Both approaches focus on understanding the intricacies of the systems or country situations and thus base analysis and potential strategies for improvement on a strong contextual base. Furthermore, both approaches describe the factors that shape the political and institutional processes and the arenas for interaction. Whether a study emphasizes politics or institutions may determine whether it is labeled one or the other.

3.2.4 Post-normal Science and the Social Interpretation of Science

Post-normal science is a conceptual framework developed by Funtowicz and Ravetz (1991, 1992, 1993). Post-normal science attempts the characterization of an extension and modification of the scientific approach as a methodology of inquiry appropriate in those cases where according to the authors "facts are uncertain, values in dispute, stakes high and decisions urgent." The primary use of this approach is regarding long-term issues such as climate change where less information is available creating uncertainty requiring short-term decisions with uncertain outcomes. Proponents of this approach propose that this conceptual framework is appropriate in such circumstances especially when potential outcomes are quite costly to society. Some authors have proposed that biosafety and environmental evaluation are a post-normal science (Aslaksen et al. 2013; Myhr 2010).

The main implication of the implementation of the post-normal framework is the absolute critical need to implement the "extended peer community" that includes all stakeholders that may be affected by an issue, prepared to enter into a dialogue about the issue. The extended peer community would be supported by an expanded definition of what constitutes evidence and facts about the issue. The later usually implies departing from a reliance on experimental and field research, scientific peer review, and not applying such concepts as burden of proof, causality, and statistical significance in favor of inclusion of social review, local knowledge, and other materials.

Table 3.1 Risk analysis framework and the regulatory trajectories for biosafety: general regulatory issues. (Source: Adapted from Isaac (2002, 2004))

	Risk analysis framework	
Attribute	Scientific rationality	Social rationality
Belief	Technological progress	Technological precaution
Type of risk	Recognized and hypothetical	Recognized, hypothetical, *and* speculative
Substantial equivalence	Accepts substantial equivalence	Rejects substantial equivalence on the basis of the precautionary approach
Science or other in risk assessment	Safety Health	Safety Health Quality Other legitimate factors
Burden of proof	Innocent until proven guilty	Guilty until proven innocent
Risk tolerance	Minimum risk	Zero risk
Science or other in risk management	Safety or hazard basis: Risk management is for risk reduction and prevention only	Broader SECs: Risk management is for social responsiveness
	Science makes the regulatory decision	Science only informs the regulatory decision

SEC socioeconomic consideration

Discussion of the different knowledge strands above helps understand why countries have developed different approaches to science, technology, biosafety, and other regulations and agricultural and food policy issues. In the next section, we explain in more detail the general approaches that countries have taken with regard to biosafety and thus inclusion of socioeconomic considerations (SECs) in decision-making.

3.3 Current Approaches to Biosafety

Isaac (2004) discusses two significantly different international approaches to biosafety in the context of the US and the EU regulatory frameworks. These approaches can be roughly grouped into two distinct categories, the scientific and the social rationality, which can be used to examine regulatory paradigms and risk analysis. A summary of both interpretations is presented in Tables 3.1 and 3.2, and shows significant differences between both trajectories (Isaac's terminology) in their approach to risk analysis and other regulatory paradigms. Isaac (2004) proposes that the substantial equivalence approach is closely associated with the scientific rationale for viewing reality. In turn the precautionary approach has been linked to the social rationale for examining reality.

According to Isaac, the main difference between both rationalities is their fundamental belief about the role of science and technology in society. Scientific rationality posits that innovation and technology are vital to enhance productivity and maximize efficiency. The outcome of an innovative process is the maximization of society's

Table 3.2 Risk analysis framework and the regulatory trajectories for biosafety: specific regulatory issues. (Source: Adapted from Isaac (2002, 2004))

	Risk analysis framework	
Attribute	Scientific rationality	Social rationality
Precautionary principle	Scientific interpretation	Social interpretation
Focus	Product-based, novel applications	Process- or technology-based
Structure	Vertical, existing structures	Horizontal, new structures
Participation	Narrow, technical experts	Wide, "social dimensions"
	Judicial decision-making	Consensual decision-making
Mandatory labeling strategy	Safety or hazard based	Consumers' right to know based

welfare and thus an increased interest in food production and other related outcomes. Countries that follow the scientific rationality of regulation therefore tend to implement regulations that support or encourage innovation. This strong support of science, technology, and innovation tends to yield regulations based in the scientific method and in risk analysis processes. This approach requires compilation of existing data to estimate objective risk or, if data is not available, the use of subjective risk estimates, usually by a community of experts. Substantial equivalence is used as a regulatory decision-making rule, where if a new technology is deemed to have the same level of risk as existing technologies; the technology is approved for use by consumers. If a technology is deemed as not "substantially equivalent" then it is considered novel and extensive testing is required to demonstrate safety. The risk assessment approach that uses substantial equivalence follows an approach that estimates the likelihood and magnitude of the impact based on quantitative approaches. The scientific rationale will tend to only consider SECs when they affect a technology's safety profile.

The social rationality approach views science, technology, and innovation as one more component of society. Regulatory decisions are not only taken based on the biophysical impacts on public health and the environment, but also on the multiple relationships and impacts with other aspects relevant to society. There is then greater inclusion of societal concerns consistent with proponents view that science cannot explain all issues related to humanity. Quantitative risk assessments are viewed as incomplete and not sufficiently broad to consider multiple and often non-linear impacts. Thus countries which have adopted a social rationality approach to regulation will tend to support inclusion of SECs in their decision-making, including ethical, philosophical, religious, and other broad considerations.

3.4 Institutions and Organizations

3.4.1 Historical Context

An understanding of the history behind the CPB discussions and country positions is necessary for a better understanding of the relationship between SECs, biosafety, and GM technologies. During the negotiations leading to the CPB, inclusion of SECs in biosafety decision-making was a divisive issue between countries. In general, most developing countries supported their inclusion, whereas developed countries took the position that inclusion of SECs should fall under the purview of domestic measures and were not appropriate for consideration in an international protocol. However, these positions were not monolithic and changed over time. Thus, the next subsection provides a more detailed review of such positions.

SECs were first considered by the CPB Parties after a decision of the first Conference of the Parties serving as a Meeting of the Parties (COP-MOP decision BS-I/12) and in other Protocol documents, such as medium-term plans. The second meeting of the Parties (COP-MOP2) requested feedback from the Parties in terms of position reviews and case studies regarding socioeconomic impacts while providing incentives for information sharing among parties (COP-MOP decision BS-II-12). The fourth meeting of the Parties (COP-MOP4) in 2008 considered the feedback from parties, non-parties, and other relevant international non-governmental organizations. COP-MOP decision BS-IV/16 invited the intersessional coordination meetings on capacity building to further discuss the issues.

The formal agenda at the Protocol's Conference of the Parties/Meeting of the Parties in 2008 and 2010 (COP-MOP4, COP-MOP5) included some discussion on socioeconomics usually under the agenda item on capacity building. In both meetings, discussion on socioeconomics was postponed to allow more inputs from expert groups and to define more conceptual clarity within the Parties. In turn, in COP-MOP6 in Hyderabad, the formal agenda included discussions on SECs.

At the conclusion of the COP9-MOP 6 meeting, however, CPB Parties decided to postpone any decision with regard to SECs to pursue further deliberations, online consultations, and expert opinions (including the creation of so-called Ad-Hoc Technical Expert Group or AHTEG) that would help parties obtain conceptual clarity. Inclusion of SECs into GMO biosafety decision-making is controversial and has been extensively debated by Parties and non-parties to the CPB.

3.4.2 *Interests and Institutions*

Table 3.3 introduces a list of potential agents relevant to the GM debate.[1] In Table 3.3 the Miami Group, composed mostly of those countries which are innovators and/or large users of GM crops, tend to oppose the formal and/or mandatory inclusion of SECs in risk assessment and decision-making. According to Gupta (1999) the Miami Group and Industry viewed the CPB as a "vehicle by which to institutionalize predictability in decision-making with regard to genetically modified organisms (predictability premised upon decisions being based on sound science)."

[1] Kikulwe et al. (2010, 2011) reminds us there may be significant heterogeneity among different groups and agents in society in their positions toward GM crops. These differences may shape perceptions and influence policy and thus understanding each group of countries and agents is warranted.

Table 3.3 Agents and groups involved in the biosafety and socioeconomic policy debate. (Source: Author's compilation from CBD Secretariat Conference of the Parties serving as a Meeting of the Parties (COP-MOP) meeting documents)

Actors	Examples
The Miami Group	Australia, Argentina, Canada, Chile, Uruguay, USA
European Union bloc	27 member states
Like-minded countries	G77 and China
African Bloc	47 member states
Compromise Group	Japan, Mexico, Norway, Singapore, South Korea, Switzerland and, occasionally, New Zealand
Latin American and Caribbean Countries Group (GRULAC)	30 member states
Central and Eastern European bloc of countries (CEE)	23 member states
Green, environmental and socially responsible groups (nongovernmental organizations)	Third World Network, Greenpeace, European Network of Scientists for Social and Environmental Responsibility (ENSSER), GenØk—Centre for Biosafety Norway, ECOROPA, Worldwide Fund for Nature (WWF), Friends of the Earth, Council for Responsible Genetics, Edmonds Institute at Washington State University, Washington Biotech Action Council
Industry	Global Industry Coalition (GIC), CropLife International, Biotechnology Industry Organization (BIO), Individual company representatives
Scientists (public and private sector)	Public Research and Regulation Initiative (PRRI)
Intergovernmental organizations (IGOs)	United Nations Environmental Program-Global Environmental Facility (UNEP-GEF)
Supranational treaties	World Trade Organization, CODEX Alimentarius
Government agencies	Ministries of Environment, Agriculture, Science and Technology, Foreign Affairs, International Trade and Commerce, Food and Drug, National Biosafety Committees/Agencies

The implication of this statement is that biosafety would focus on science-based issues only.

Gupta (1999) indicated that in turn the European Union (EU) bloc, like-minded countries, some Green and environmental groups viewed the CPB as a "vehicle by which to institutionalize flexibility in decision-making with regard to genetically modified organisms (a flexibility premised on the ability to implement the precautionary approach)." Inclusion of socioeconomics within a social view of regulation that considers biosafety as a post-normal science implies that socioeconomics would be a critical aspect in the regulation of GM crops. The Compromise, GRULAC, and CEE groups seem to have taken a somewhat neutral position or in some cases a wait-and-see approach to socioeconomics.

It is important to note that the EU has not been a heavy supporter of such inclusion—compared to the like-minded countries which includes the African group—apparently attempting to address the issue through domestic measures (Tewolde

Berhan Gebre Egziabher 2000, 2009). Gupta (1999), however, indicates that although negotiations leading to the deployment of the CPB may have been portrayed as protecting the interests of developing countries, in the end the Protocol is more like a conflict between divergent factions on biosafety and technology governance within the OECD. We tend to concur but we expand this narrow interpretation to include other actors which may have used both positions to advance their own agendas and positions as described in Table 3.3.

We do not want to convey the message that the EU positions have been monolithic nor constant over time. Rhinard and Kaeding (2006) have indicated that EU efforts in promoting and negotiating a biosafety protocol have evolved over time toward convergence but not unanimity of positions. Positions with regard to biosafety and SECs have changed over time, although there seems to be a tendency toward convergence.

In turn other Green and environmental groups viewed the Protocol as a way to stop the deployment of agricultural biotechnologies and other technologies. Many NGOs had a critical role in supporting the passage of the CPB and of several issues purported to be of importance to developing countries including SECs and liability and redress issues. The supporters of NGOs such as Third World Network, Friends of the Earth, Greenpeace, and the Edmonds Institute at Washington State University, among others, were critical in putting forward issues such as the precautionary principle and other issues viewed as protective to countries in the South (Tewolde Berhan Gebre Egziabher 2000, 2009). According to Tewolde Berhan Gebre Egziabher, the lead negotiator of the African Bloc for the negotiation leading to the Protocol, this support was made possible due to a large coalition of like-minded NGOs and research units in educational institutions connected through the internet, which tended to oppose GMOs.[2]

Government agencies may have different positions or even in some cases conflicting positions with regard to GM technologies and the issue of SECs. This state of affairs may be due to the mission, objectives, and time horizons that each agent has in a country. Agriculture or Science and Technology Ministries may have a different position than Environmental and Natural Resources Ministries, which may be different to that of Foreign Affairs or Trade Ministries. It is also important to note that the CPB, being part of the CBD, has until recently been under the purview of the Ministries of Environment or Natural Resources. Although limited, agents such as Ministries of Agriculture and public sector scientists have only recently begun to express their positions in the matter.

---

[2] A paper by Takeshima and Gruere (2010) suggests that anti-GMO lobbying efforts may be more successful in those countries where conditions may not be favorable for the introduction of a GMO in the first place. This hypothesis seems to be confirmed by results obtained in the analysis conducted in the paper. We posit that this hypothesis may also apply to the particular case of policy debate spaces, yet this remains an unexplored research area.

3.5 Examples of Existing Regulatory Frameworks That Include SECs

The following examples of countries are from those who have included socioeconomics through regulations, policies, through informal or ad hoc procedures, and/or where there has been pressure from actors to their inclusion. This is not meant to be an exhaustive list of countries who have taken socioeconomics into consideration, rather to illustrate the fact that there are different approaches by which to do so in practice.

3.5.1 Australia and New Zealand

The Australian national regulatory framework for gene technology, created by the *Gene Technology Act 2000* (Cth), does not address SECs, focusing only on human health and environmental impacts. However, some individual states have called for the inclusion of SECs in the risk assessment process and have introduced state legislation dealing with SECs. Moratoria on certain GMO releases pursuant to that legislation have now ended in four states but continue in two others. While the purpose of the state legislation was to protect markets and provide time for consideration of SECs, no explicit details were given regarding which, or how, SECs were to be assessed or included within a broader assessment regime.

3.5.2 Norway

Norway is an example of a country that explicitly includes SECs in its national biosafety framework. The Gene Technology Act of 1993, Section 10 explicitly mentions:

> The deliberate release of genetically modified organisms may only be approved when there is no risk of adverse effects on health or the environment. In deciding whether or not to grant an application, considerable weight shall also be given to whether the deliberate release will be of benefit to society and is likely to promote sustainable development.

The Norwegian Biotechnology Advisory Board has prepared a document "describing the implementation of the concepts in the Gene Technology Act" (Norwegian Biotechnology Advisory Board 2009, Preface), which includes information for the assessment of benefits to the community. These elements are seen "as a source of inspiration for how the Biotechnology Advisory Board will implement the requirements of 'benefit to the community'" (Norwegian Biotechnology Advisory Board 2009, p. 15). Interestingly, the SECs are not necessarily only relevant to imported GMOs—Norway currently does not produce any GMOs—but also to the conditions of producers in the exporting country.

3.5.3 Argentina

Argentina, although not a party to the CPB, has formally included SECs in its biosafety approval process. Argentinian regulations limit the scope though to impact on Argentinean productivity with a focus on exports (SAGPyA Argentina 2003, 2006, 2007). This situation may change as more products enter the regulatory pipeline for internal use, where producer productivity may become an additional aspect in the SEC assessment for regulatory purposes.

3.5.4 Brazil

Brazil approved its Biosafety Law No. 11.105 in 2005. That law establishes three distinct bodies for GMO regulation (GoB 2005): two technical bodies, the national biosafety committee (CTNBio) and the Institutional Biosafety Committee, and the National Biosafety Council (CNBS). CTNBio is considered the competent regulatory authority and performs the GMO risk assessment in conjunction with the proponent, considering human and animal health and environmental impacts. The CNBS is composed of government ministers. CNBS has commercialization approval power. If any social or economic issue is raised during the risk assessment process, CNBS has the power to commission a socioeconomic study by a third party. An important note is that the Brazilian law explicitly sets the different risk analysis, communication, and management roles that all actors may take which are involved in the biosafety regulatory process.

In Brazil's case, several advantages derive from separating the functions of the technical body from those of the body that examines non-biosafety related issues. The main advantage is that risk assessment is in essence separated from political issues that may obscure the technical assessment, thus allowing the former to be completed as needed. An important observation is that the Brazilian system is similar to the one used by the EU, the main difference being that the political process deals only with one country rather than many as in the EU.

3.5.5 Mexico

In Mexico, the Biosafety Law and other related law instruments explicitly refer to the inclusion of SECs. For example, Article 64 of the National Biosafety Law of 2005 (GEUM 2005) and Chap. III Article 16 Section V(d) of the implementing regulations (GEUM 2008) strongly encourage inclusion of SECs.

Chapter II Article 5.XIV of the 2006 decree (CONACYT 2007) defines making a decision on the socioeconomic studies needed to analyze the impact of GMOs as one role of the competent authority (Comite Intersecretarial de Bioseguridad de los Organismos Geneticamente Mejorados—CIBIOGEM). The 2006 decree indicates that the proponent must perform the SEC assessment, while the CIBIOGEM conducts a detailed review and can ask for additional information as needed. The law or policy instruments do not, however, define any methodologies or decision-making standards on which to base a decision.

3.5.6 India and China

In India, current rules from 1989 (GoI 1989) based on the Environment Protection Act (GoI 1986) guide the manufacture, use, import, export, and storage of hazardous microorganisms and/or genetically engineered organisms. The 1989 rules do not explicitly require inclusion of SECs, although in past biosafety evaluations the competent regulatory authority, the Genetic Engineering Advisory Committee (GEAC), has commissioned economic studies assessing the potential impact from the introduction of GM crops in India. There is very little clarity on the extent of these evaluations, how they have been considered for decision-making or their impact.

In China, the 2002 Decrees 8, 9, and 10 of the Ministry of Agriculture (MoA PRC 2002a,b,c) and the 2002 Decree 304 of the State Council of the People's Republic of China (SC-PRC 2002), and other regulations govern the application of biosafety and use of GMOs in China, although these decrees seem to be in a state of review (USDA-FAS 2012). The text of available regulations only considers the technical aspects of biosafety assessment procedures made by the competent regulatory authority. The final decision for commercial approval lies in the Chinese central government where other considerations including SECs, such as impacts on foreign trade, may play a role in the decision-making process. The process of assessing SECs is not explicitly defined in any of these regulations.

3.6 Critical Assessment

There are two defined and contrasting country positions in the GM debate although many countries have not taken a formal position at this time. One position taken by some countries focuses on the scientific risk assessment of GMOs. In this view, SECs have a limited role in risk assessment, except perhaps on those issues which may affect risk management plans and strategies. This position is based on the right reaffirmed in CPB Article 26 to not include a mandatory requirement for SEC assessment and that biosafety assessments focus on environmental and public health issues only. Countries which do not support SEC inclusion as a formal regulatory step tend to be innovators, developers, and/or broad-scale users of GMOs.

The second position taken by some countries is that SECs are critical for protecting biodiversity, particularly that relevant to indigenous and local communities in developing countries especially those classified as mega-diverse. Countries proposing this view tend to focus on potential negative impacts of GMOs including, in some instances, hypothetical and even speculative risks that may be associated with their use. This approach is firmly based on the post-normal science framework and the social rationality approach to science and regulation discussed above. Most countries that pursue this approach tend to be net importers, have limited investments in GM innovation, and have significant limitations in terms of science, technology, and innovation capacity that includes the capacity to implement biosafety assessments.

These two seemingly conflicting positions can lead to delays in the implementation of functional biosafety systems and to the deployment of GM technologies that may be of value, especially in developing countries. The tensions between these opposing positions on SECs can disrupt further discussion defining technical issues including the implementation of the risk assessment itself which are mandatory for CPB parties.

The fact that CPB Article 26 dealing with SECs is of a voluntary nature and that some countries have already acted to exclude SECs from their processes, points to the need for those countries that have or wish to include SECs in decision-making to undertake internal discussions. This conversation should focus on achieving conceptual clarity, evaluate existing national financial and human resources, and seek support from the international community and supranational organizations when there is very little or no capacity for implementation, rather than attempting to develop an internationally agreed response that would be mutually agreeable to all parties, particularly when, as is the standard procedure in the CPB, this would require consensus.

3.7 Concluding Comments

National policy spaces often define the international context and vice versa. In both spheres, agents and organizations can have a critical role in influencing policy pathways. We propose that this has been the case for biosafety debates and for the potential inclusion of SECs in decision-making. Critical drivers shaping the international context of biosafety and SEC inclusion have been development models and frameworks, economic development history, agents and organizations, and the precautionary principle as a unifying theme. Most signatories to the CPB incorporate a version of the precautionary principle in their biosafety regulations, although version of such principle vary from weak to strong.

We encourage those countries that have explicitly indicated that they will move forward with a policy and regulations requiring inclusion of SECs to focus on implementation issues as a way to move forward. Focusing on identification of methods and research approaches, timing, scope, definitions, rules, and decision-making standards, and on innovative ways to comply with such regulatory requirements with a proactive view of improving time and cost efficiencies will contribute to the development of a functional biosafety system nationally and internationally. The next section in this book will help define these implementation issues in SEC assessments.

References

Aslaksen J, Glomsrød S, Myhr AI (2013) Post–normal science and ecological economics: strategies for precautionary approaches and sustainable development. Int J Sustain Dev 16(1–2):107–126. doi:10.1504/IJSD.2013.053793

Atanassov A, Bahieldin A, Brink J et al (2004) To reach the poor. Results from the ISNAR-IFPRI Next Harvest Study on Genetically Modified Crops, Public Research, and Policy Implications, EPTD Discussion Paper 116. International Food Policy Research Institute, Washington DC

Binswanger H, Ruttan VW (1978) Induced innovation: technology, institutions and development. Johns Hopkins University Press, Baltimore

Cochrane WW (1958) Farm prices: myth and reality. University of Minnesota Press, Minneapolis

CONACYT (2007) 'Acuerdo por el que se expiden las Reglas de Operación de la Comisión Intersecretarial de Bioseguridad de los Organismos Genéticamente Modificados'. Consejo Nacional de Ciencia y Tecnología, Estados Unidos Mexicanos, December 5 2007. http://www.cibiogem.gob.mx/Docum_interes/Reglas_Op_05dic07.pdf. Accessed June 1, 2013.

de Janvry A (1981) The agrarian question and reformism in Latin America. Johns Hopkins University Press, Baltimore

Duncan A, Williams G (2010) Making development assistance more effective by using political economy analysis: what has been done and what have we learned? The Policy Practice, Brighton. http://www.thepolicypractice.com/papers/16.pdf. Accessed 6 June 2013

Eckaus RS (1955) The factor proportions problem in underdeveloped areas. Am Econ Rev 45(4):539–565

Falck-Zepeda JB, Falconi C, Sampaio-Amstalden MJ et al (2009) La biotecnología agropecuaria en América Latina: Una visión cuantitativa. IFPRI Discussion Paper 860SP, International Food Policy Research Institute (IFPRI), Washington, DC, USA. http://www.ifpri.org/sites/default/files/publications/ifpridp00860sp.pdf. Accessed 6 June 2013

Feder G (1985) The relation between farm size and farm productivity: the role of family labor, supervision and credit constraints. J Dev Econ 18(2–3):85–101

Feder G, Just RE, Zilberman D (1985) Adoption of agricultural innovations in developing countries: a survey. Econ Dev Cult Change 33(2):255–298

Funtowicz SO, Ravetz JR (1991) A new scientific methodology for global environmental issues. In: Costanza R (ed) Ecological economics: the science and management of sustainability. Columbia University Press, New York

Funtowicz SO, Ravetz JR (1992) Three types of risk assessment and the emergence of post-normal science. In: Krimsky S, Golding D (eds) Social theories of risk. Praeger, Westport

Funtowicz SO, Ravetz JR (1993) Science for the post-normal age. Futures 25(7):739–755

Gardner BL (2002) American agriculture in the twentieth century: how it flourished and what it cost. Harvard University Press, Cambridge

GEUM (2005) 'Ley de Bioseguridad de los Organismos Genéticamente Mejorados'. Gobierno de los Estados Unidos Mexicanos, March 18 2005. http://www.diputados.gob.mx/LeyesBiblio/pdf/Ley_BOGM.pdf

GEUM (2008) 'Reglamento de la Ley de Bioseguridad de los Organismos Genéticamente Mejorados.' Gobierno de los Estados Unidos Mexicanos, March 19 2008. http://www.cibiogem.gob.mx/Norm_Leyes/Reglamento_LBOGM.pdf

GoB (2005) 'Decree 5.591/2005, which regulates Law 11.105/2005.' Government of Brazil. November 2005. http://bch.biodiv.org/database/attachedfile.aspx?id=600

GoI (1986) Environment (Protection) Act, 1986. http://dbtbiosafety.nic.in/. Accessed 10 May 2013

GoI (1989) Rules for the manufacture, use, import, export an storage of hazardous microorganisms, genetically engineered organisms or cells. Notification No. 621 Official Gazette of Government of India, December 5, 1989. http://dbtbiosafety.nic.in/. Accessed 10 May 2013

Griliches Z (1957) Hybrid corn: an exploration in the economics of technological change. Econometrica 25(4):501–522

Gupta A (1999) Framing biosafety in an international context. ENRP Discussion Paper E-99-10, Kenneday School of Government, Harvard University, Cambridge
Hayami Y, Ruttan V (1985) Agricultural development: an international perspective. Johns Hopkins University Press, Baltimore
Higgins B (1956) The dualistic theory of underdeveloped areas. Econ Dev Cult Change 4(2):99–115
Isaac GE (2002) Agricultural biotechnology and transatlantic trade: regulatory barriers to GM crops. CABI Publishing Inc, Oxon
Isaac GE (2004) The interaction between levels of rule making in international trade and investment: the case of sanitary and phytosanitary measures. Discussion Paper Prepared for the Workshop on the Interaction Between Levels of Rule Making in International Trade and Investment UNU CRIS/LSE ITPU Project, Brussels, Belgium, December
Kikulwe EM, Birol E, Wesseler J et al (2010) A latent class approach to investigating demand for genetically modified banana in uganda. Agr Econ 42(5):547–560. doi:10.1111/j.1574-0862.2010.00529.x. Accessed 6 June 2013
Kikulwe EM, Wesseler J, Falck-Zepeda J (2011) Attitudes, perceptions, and trust: insights from a consumer survey regarding genetically modified banana in uganda. Appetite 57(2):401–413
Lewis WA (1954) Economic development with unlimited supplies of labor. Manchester School of Economic and Social Studies, Manchester
Lucas RE (1988) On the mechanics of economic development. J Monet Econ 22:2–42
MoA PRC (2002a) 'Implementation Regulations on Safety Assessment of Agricultural Genetically Modified Organisms Decree 8'. Ministry of Agriculture, PR China January 5 2002. http://bch.biodiv.org/database/attachedfile.aspx?id=559
MoA PRC (2002b) 'Implementation Regulations on Safety Assessment of Agricultural Genetically Modified Organisms Decree 9'. Ministry of Agriculture, PR China, January 5 2002. http://bch.biodiv.org/database/attachedfile.aspx?id=561
MoA PRC (2002c) 'Implementation Regulations on Safety Assessment of Agricultural Genetically Modified Organisms Decree 10'. Ministry of Agriculture, PR China, January 5 2002. http://bch.biodiv.org/database/attachedfile.aspx?id=563
Myhr AI (2010) The challenge of scientific uncertainty and disunity in risk assessment and management of GM crops. Environ Values 19(1):7–31
Myrdal KG (1968) Asian drama: an inquiry into the poverty of nations. Pantheon, New York
Nature Biotechnology (2012) Agnostic about agriculture: Editorial. Nat Biotechnol 30(3):197
Norwegian Biotechnology Advisory Board (2009) Sustainability, benefit to the community and ethics in the assessment of genetically modified organisms: implementation of the concepts set out in sections 1 and 10 of the Norwegian gene technology act, 2nd revised edn. The Norwegian Biotechnology Advisory Group, Oslo
Ostrom E (2005) Understanding institutional diversity. Princeton University Press, Princeton
Prebisch R (1950) The economic development of Latin America and its principal problems. United Nations, New York
Rhinard M, Kaeding M (2006) The international bargaining power of the European Union in mixed competence negotiations: the case of the 2000 Cartagena Protocol on Biosafety. J Common Mark Stud 44(5):1023–1050. http://ssrn.com/abstract=950067 or http://dx.doi.org/10.1111/j.1468-5965.2006.00672.x. Accessed 6 June 2013
SAGPyA Argentina (2003) 'Resolución 39 Liberación al Medio de Organismos Vegetales Genéticamente Modificados (OVGM)'. Secretaria Agricultura, Ganadería, Pesca y Alimentos, Republica de Argentina, July 11, 2003. http://www.sagpya.mecon.gov.ar/new/0-0/programas/conabia/resolucion39.PDF
SAGPyA Argentina (2006) Resolución 212: Producción de Semilla de Maíz Genéticamente Modificada en Etapa de Evaluación en Republica Argentina Secretaria Agricultura, Ganadería, Pesca y Alimentos, Republica de Argentina, May 15 2006. http://www.sagpya.mecon.gov.ar/new/0-0/programas/conabia/res212.pdf
SAGPyA Argentina (2007) 'Resolución 60: Régimen para la Comercialización de Organismos Genéticamente Mejorados que Contengan Eventos Acumulados'. Secretaria Agricultura,

Ganadería, Pesca y Alimentos, Republica de Argentina, February 5, 2007. http://www.sagpya.mecon.gov.ar/new/0-0/programas/conabia/Resolucion_60_2007.pdf
Schultz TW (1964) Transforming traditional agriculture. Yale University Press, New Haven
SC-PRC (2002). 'Regulations on Safety of Agricultural Genetically Modified Organisms'. State Council of the People's Republic of China. May 9, 2001.
Singer HW (1970) Dualism revisited: a new approach to the problems of the dual society in development countries. J Dev Stud 7(1):60–75
Solow RM (1957) Technical change and the aggregate production function. Rev Econ Stat 39(3):312–320
Stein AJ, Rodríguez-Cerezo E (2009) The global pipeline of new GM crops: implications of asynchronous approval for international trade. JRC Technical Report EUR 23486 EN. Office for Official Publications of the European Communities, Luxembourg. http://ipts.jrc.ec.europa.eu/publications/pub.cfm?id=2420. Accessed 6 June 2013
Takeshima H, Gruère G (2011) Pressure group competition and GMO regulations in Sub-Saharan Africa: insights from the Becker model. J Agric Food Ind Organ 9(1). http://dx.doi.org/10.2202/1542-0485.1325. Accessed 6 June 2013
Tewolde Berhan Gebre Egziabher (2000) Civil society and the Cartagena Protocol on Biosafety. Paper presented at the Montreal International Forum FORUM 2000, Montreal, Canada. http://fimforum.org/en/library/TEgziabher2000.pdf. Accessed 1 June 2013
Tewolde Berhan Gebre Egziabher (2009) The Cartagena Protocol on Biosafety: history, content and implementation from a developing country perspective. In: Traavik T, Ching LL (eds) Biosafety first. Holistic approaches to risk and uncertainty in genetic engineering and genetically modified organisms. Tapir Academic Press, Trondheim
USDA-FAS (2012) China—People's Republic of—Agricultural Biotechnology Annual. GAIN Report Number CH132046, USDA FAS, Washington, DC

Part II
Analysis of Socio-Economic Considerations

Many countries have already included SECs assessments in their national biosafety frameworks, policies or regulations. However, there is very little understanding of all the implications resulting from the inclusion of SECs in biosafety and/or biotechnology decision-making. This lack of information on important aspects of SECs in biotechnology decision-making complicates the policy decision environment. While there is much discussion of the issue and has been some work by certain jurisdictions, there is a lack of information on the content (or meaning) of particular SECs, methodologies to measure SECs, including the advantages, disadvantages and costs of such methodologies, and the legal repercussions of including particular SECs in biotechnology regulation, such as compatibility with existing international agreements and institutions.

Perhaps the assessment of the broader group of concerns, the so-called SECs, could go so far as using a science-based assessment but at the least the assessment needs to be objectively measurable, evidence based and identifiable in advance. Literature about assessment of SECs already exists. However, as between different jurisdictions and even within the one jurisdiction, there is considerable uncertainty and inconsistency in the current approach to assessment of SECs. Content is also very dependent on context. Different countries and groups within countries (such as consumers, producers and industry) interpret and are impacted by any particular SEC in different ways. Impact differs because of factors such as different cultural and religious values and forms of agriculture practiced in particular countries. How a particular SEC is relevant to a specific country and what methodologies are appropriate in an assessment of that may also vary depending on, for example, whether the country is a developed, developing or least developed nation. This section is intended to assist in the move towards certainty and consistency.

Using a standard template, the chapters in this section provide an informative, factual and concise synthesis of 15 different SECs. Authors of the chapters in this section were tasked with identifying possible models that could be utilized to assess the various SECs and to then provide a summary of how the models are applied, and what data requirements are necessary. Some SECs methodologies are extremely challenging to identify, while others have several options.

Chapter 4
Benefits to Producers and Society

José Falck-Zepeda and Melinda Smale

4.1 Introduction

Most studies conducted to date about the adoption and impacts of genetically modified organisms (GMOs) have examined the direct, on-farm benefits to producers (Qaim 2009; Smale et al 2009; Pontifical Academy of Sciences 2010; Potrykus and Ammann 2010; Areal et al. 2012). To estimate on-farm benefits, applied researchers have most often relied on farm data collected through survey interviews to test hypotheses about changes in yield, use of labour and other inputs, costs and returns. The same data can be aggregated to represent benefits to a sector and to society.

Researchers have also used estimates of net profits per hectare from survey data, trial data or data obtained from companies to estimate overall returns to producers (for a complete discussion, see Smyth et al., forthcoming). Benefits to society have been generally estimated via the well-known economic surplus approach or the Global Trade Analysis Project (GTAP) and similar approaches that consider market supply and demand for a specific sector or for the economy as a whole. Use of trade analysis reflects the interest of decision-makers in understanding whether a specific country stands to gain from the potential adoption of GMOs. Researchers have also modified the economic surplus approach to address issues such as temporary monopoly conferred to innovators through intellectual property, heterogeneity among producers, uncertainty and the postponement of benefit flows due to regulatory processes. Benefits to society are often expressed as the size and distribution of economic benefits accruing to producers, consumers and innovators.

This chapter discusses methods and approaches used to evaluate benefits to producers and society, and their limitations and implications for national and

J. Falck-Zepeda (✉)
International Food Policy Research Institute (IFPRI), 2033 K Street NW,
Washington, DC 20006-1002, USA
e-mail: j.falck-zepeda@cgiar.org

M. Smale
207 Agriculture Hall, Michigan State University, East Lansing, MI 48824, USA
e-mail: msmale@msu.edu

K. Ludlow et al. (eds.), *Socio-Economic Considerations in Biotechnology Regulation*,
Natural Resource Management and Policy 37, DOI 10.1007/978-1-4614-9440-9_4,

international regulatory environments. Studies examining producer and societal benefits are more relevant to formal technology evaluation processes than the environmental and/or the food/seed risk assessment procedures used in biosafety regulatory processes. This is largely because very few aspects related to the measurement of benefits to producers and to society are directly related to risk considerations. Some exceptions to this state of regulatory and decision-making affairs may be the financial or production risks connected to market and post-release environments.

4.2 Methodologies

A fundamental distinction concerns ex ante as compared to ex post analyses. Methods described here can be used in either. Ex ante impact assessment refers to analysis that estimates the potential impact of the adoption and diffusion of a GM innovation. Practitioners customarily perform ex ante analysis to evaluate competing project options and allocate resources based on a priority setting exercise. In an ex ante analysis, the researcher proposes plausible values for key parameters in the model chosen.

In contrast, ex post impact assessment refers to analysis that evaluates past performance and achievements. This is an after-the-fact analysis that examines use of inputs and seeks to provide information to policy-makers. In an ex post analysis, the researcher collects data on key parameters from primary or secondary sources. The probability of success, adoption rates and information about production performance are known, elicited from experts, or can be estimated from different sources.

One critical issue of ex post (and to a certain degree ex ante) analysis is making the appropriate comparison between the scenarios that depict conditions 'with' and 'without' the innovation (the counterfactual). Agronomic and other life sciences experiments customarily deal with the counterfactual by comparing a group of individuals (plants, persons) that received the proposed treatment (a treatment group) with a group that did not (a control group). In the case of quasi-experimental, economic impact studies conducted among farmers, it is not feasible to identify the control group. We observe those farmers who received a treatment (adopted a technology) after they adopted, and those who did not, but we cannot observe adopters had they not adopted. Adopters and non-adopters are likely to differ in important ways. Similarly, the researcher cannot observe the prices and quantities that would have prevailed had technical change not occurred. The next best solution is to use information available for the conventional (older) technology. The researcher estimates 'without-innovation' prices and quantities using formulas derived from the system of supply and demand equations.

We group analytical approaches into three distinct sub-categories, according to their objectives: (1) cost/benefit analysis, (2) partial and general equilibrium analysis, and (3) regulatory impact assessment.

4.2.1 Cost/Benefit and Net Present Value Methods

The purpose of Cost/Benefit Analysis (CBA) is to evaluate the consequences of decisions using consistent procedures. Often the literature distinguishes two types of CBA. The financial approach considers only cash costs and benefits, and is typically used by private firms and individuals. The decision rule is simply that a project is undertaken as long as benefits exceed costs. This decision rule is equivalent to positing that net benefits are greater than or equal to zero. The second, or economic approach, adds the cost of alternatives (opportunity cost) and external influences on society. Alternative costs are customarily valued using 'shadow prices', which include all the cost incurred by society in order to supply a good in a specific market.

To reflect the time value of money, researchers discount the flow of future net benefits using the appropriate discount factor or interest rate. Discounting is a way to estimate the present value of benefits and costs realized during the course of adoption and use of a technology. The process of discounting assumes that money spent today is more valuable than money spent in the future because today's money can be invested, generating income until it is spent in the future. In addition, discounting assumes that most people prefer to consume now rather than later.

The net present value (NPV) is the sum of the discounted stream of annual net benefits. All costs necessary to bring the project into existence are considered. An alternative measurement is the internal rate of return (IRR), which is simply the rate of interest which, when applied to discount the stream of net benefits, equates the NPV to zero. The analyst compares the IRR to an existing benchmark rate of interest, usually the prevailing bank-lending rate. If the IRR is greater than the benchmark rate of return, the project is accepted.

4.2.2 Partial and General Equilibrium Analysis

Partial and general equilibrium analysis—including economic surplus approaches—seeks to estimate net additional benefits to the economy due to the adoption of an innovation. The economic surplus methodology is based on the principle that supply and demand for a particular good reaches an equilibrium point. Equilibrium represents the combination of prices and quantities at which the quantity demanded by individuals exactly equals the amount supplied by producers. Changes in the equilibrium quantity and price occur because of external shocks to the system of supply and demand functions (e.g. introduction of a biotechnology innovation). In particular, a technology innovation may cause a per-unit cost reduction (increase) or equivalently, more (less) output produced with the same amount of inputs.

Economic surplus is composed of consumer and producer surplus, both measured as changes with respect to a counterfactual. Change in consumer surplus arises when adoption of the technology causes a shift in supply and a decrease in product price, and is calculated by multiplying the price change by the quantity of the good consumed. Change in producer surplus results from the increase in benefits

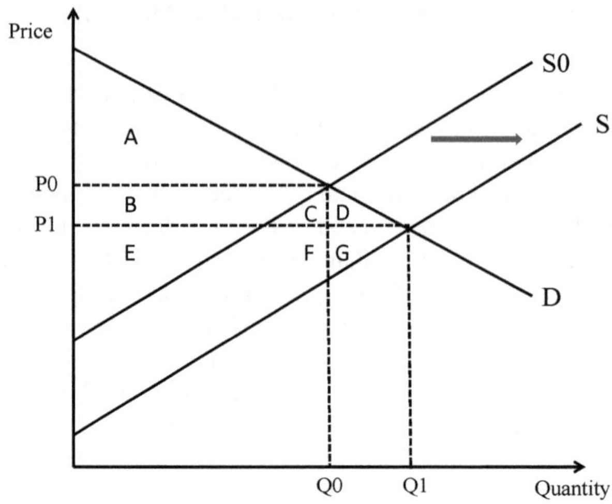

Fig. 4.1 Producer and consumer surplus changes due to supply shift. Notes: (1) Consumer surplus is the area below the demand curve and above the price. Producer surplus is the area above the supply curve and below the price. (2) Prior to the shift in supply from *S*0 to *S*1, producer surplus was areas $B+E$. After the supply shift, producer surplus is area $E+F+G$. Change in producer surplus is thus $F+G-B$. Consumer surplus before the shift was A. After the supply shift it is $A+B+C+D$. Change in consumer surplus is $B+C+D$

associated with increased output or decreased cost of production. Benefits to society are estimated as the sum of producer, consumer and innovator surplus and changes in deadweight losses. Figure 4.1 shows changes in producer and consumer surplus and deadweight losses due to a change in demand or supply.

This is a well-established methodology in economics literature and has been shown to provide valuable contributions to impact assessment efforts. When there are other impacts beyond the sector, researchers can use multi-market approaches. Impacts may be 'horizontal' in the sense that the adoption affects two or more commodity markets, or 'vertical', meaning that effects are transmitted among output and input markets.

Researchers initiate the estimation procedure of economic surplus by deciding on the type of model to use. A decision on the model type will also include decisions about the type of functions (linear vs curvilinear) and the type of supply or demand shift (parallel or pivotal). If there is a need to have preliminary estimations, procedures using a parallel shift and linear functions are handy. At this point, researchers must also consider whether there is data that is sufficient and of strong-enough quality for an econometric analysis.

Impact assessment practitioners choose from among several techniques to estimate economic surplus. The main advantage of econometric techniques is that these enable the researcher to test hypotheses about the parameters in the model with statistics. The mathematical programming method obviates the need for extensive data, but requires extensive knowledge about the processes and production characteristics of the innovation and has the disadvantage that it is hard to judge the robustness of the model.

The quasi-rent approach provides an expedient and simple means to measure economic surplus. However, it is significantly close to economic surplus only when the change in supply or demand is small, and the elasticity of supply is unitary. If the innovation under study departs from these narrow assumptions, quasi-rents and economic surplus will diverge. Equilibrium displacement models have a stronger theoretical basis, are relatively expedient to use and also enable the researcher to explore a range of policy questions. Minimal data is required and this is provided from the existing literature. However, because equilibrium displacement models are based on linear functions and parallel shifts, they may not be suitable for all problems encountered by practitioners.

General equilibrium analysis (GEA) is a technique used when the technology under study affects a large number of sectors in the economy, directly or indirectly. GEA is grounded on the principle that structural equations describe not only the stock of resources but also resource flows, processes and linkages among them. The impact of regulation can therefore be estimated by formulating the appropriate expected effects relative to the baseline model. Input/output models, linear programming models and computable general equilibrium models are GEA tools. For more details on partial and general equilibrium analysis, the reader is directed to Chap. 15 (Market Access and Trade).

4.2.3 Regulatory Impacts Assessment

Regulation can introduce costs and benefits that affect producers and society as a whole. Examples of studies in which these are measured, and the tools to measure them, are few. If the CBA (described above in general terms) is oriented towards estimating the regulation's impact on the private sector, or a representative firm, it is also known as a business impact methodology. If the CBA is based on examination of estimated or true budgets, then it is known as a budgetary analysis. Cost Effectiveness (CE) is a special case of CBA which can be used when benefits of a regulation cannot be easily estimated and when the regulatory policy sets specific target objectives. The CE of a regulatory policy is estimated by dividing the regulation's annual costs by the physical units described by the regulatory objective. For example, if the explicit objective of a biosafety regulation is to reach a pre-determined level of safety, then the cost of implementing the regulation is determined for this level of safety. The CE method does not identify an optimal regulatory level but can be used to compare alternative regulatory policy options. Additionally, CE does not preclude the selection of regulatory policy options exclusively on the basis of having 'the least-cost' option.[1]

Direct compliance cost (DCC) is the simplest means of assessing regulatory impacts. DCC equates the social cost of implementing a regulation to the sum of all

[1] This method is also used in the food security and medical literature where it may be inappropriate to quantify the value of human life. Typically, the CE method is the ratio of outcome (life saved, years of life saved) to the cost of achieving the outcome. Most common CE method is the Quality Life Adjusted Years (QALYs).

analytical procedures necessary to meet the regulation by a representative institution or private firms. In the case of private firms, DCC tends to overestimate the true value of the social cost of regulation because it ignores possible substitutions or changes in the production system. Moreover, analysts using DCC do not take into consideration the changes in capital or labour investments by private firms that result from implementing a biosafety regulation. This approach may approximate true social costs when expected changes in prices and quantities are small and there are few indirect effects.

4.3 Critical Assessment

Most practitioners agree that socio-economic impact assessment is not an end in itself but rather a way to identify and choose among innovation alternatives. Impact assessment can empower scientists and policy-makers by providing them with estimates of the potential costs and benefits associated with innovations, and in some cases, pinpointing ways to enhance overall benefits and reduce social costs.

However, these estimates should be considered with caution because they are by nature, speculative. Even in ex post analysis, some parameters and variables will be subject to some bias and measurement error, although not all error is systematic and bias may not alter qualitative findings (e.g. the relative rank of innovation options). In most cases, care should be taken in extrapolating or generalizing from one study context to another.

4.3.1 Cost/Benefit and Net Present Value Methods

Cost/Benefit and partial budget approaches are deceptively simple. In fact, considerable care must be used in applying them. For example, because of the costs of survey implementation, most early studies conducted in developing countries reported only gross margins. As compared to net margins, gross margins include the costs of intermediate inputs but exclude costs of labour and land. This was a shortcoming given that some crop-trait combinations, such as herbicide tolerance, are specifically designed to influence labour costs.

Partial budgets treat only one farm activity at a time. Even where farmers are fully commercialized, the net impact of adoption on whole-farm production and resource use cannot be deduced from a partial budget. Cross-activity impacts have rarely been systematically investigated in studies of GM crops, particularly in developing countries.

Partial budgets are also of limited utility in situations with missing or imperfect markets, such as in the case of subsistence food crops in many developing economies. In those cases, the effective prices that influence farmer decision-making depend on the characteristics of individual households and their access to markets, ranging between the consumer and producer price for the food crop. This can be addressed by employing sensitivity analysis based on the price range between the consumer and producer price.

Finally, risk and uncertainty have been considered explicitly in only a few of these studies. Stochastic analysis, Monte Carlo simulations and real option models can be utilized to introduce risk and uncertainty into CBA and partial budget analysis. Given the consistent evidence of outcome variability in the literature, examining the statistical distributions of impact variables (yields, costs, profits), in addition to average impacts, seems fundamental for future work.

4.3.2 Partial and General Equilibrium Analysis

The assumptions behind the partial and general equilibrium—including economic surplus approaches—most closely depict an industry with commercially oriented farmers who buy and sell in well-organized markets and grow their crops under relatively homogeneous conditions. This depiction is quite useful for portraying farmers in developed economies but unrealistic for most farmers in developing economies, and particularly those who produce staple food crops.

The quality of the underlying data is crucial for results validity. Generally, reliable cross-sectional, time-series data to support sector analyses of GM crops are not yet available in developing economies. At present, such data are probably too costly to assemble, maintain and disseminate publicly given the information infrastructure found in most of these countries.

One way researchers have compensated for the lack of large cross-sectional, time-series data has been to expand existing data from both primary and secondary sources using stochastic simulations and real option models, as mentioned above. These tools assume special significance when technologies are developed by farmers in heterogeneous production environments for uncertain markets, where location and year-specific effects on productivity can generate large coefficients of variation in model parameters, including farm profits, adoption rates and prices. If the number of input suppliers is small or markets must be segregated, risk and uncertainty in the market channel may be somewhat higher in the case of GM crops relative to other new crop varieties.

As the most used method to date is the economic surplus approach, additional information about the most common specific methods used under this approach is provided in Table 4.1 This table is useful when choosing between different options. For a more complete discussion of the different economic surplus methods, see Alston et al. (1995) and Alston and Pardey (2001).

The evaluation criteria to choose the appropriate methodology are expediency, available resources, data availability and quality, and the type of research or policy question to be answered. For example, the quasi-rent, standard surplus models and the equilibrium displacement models (EMDs) are very expedient, with little data required, and require relatively few resources to implement. However, the quasi-rent approach can only be thought of as a first approximation to check results from other economic surplus models. The EMDs can model policy implications readily; however, they require the researcher to have advanced knowledge of economic concepts and methods, as well as extensive knowledge of the policy implications in hand.

Table 4.1 Advantages and disadvantages of economic surplus estimation approaches

Estimation approach	Advantages	Disadvantages
Standard models	Very little data required Simple models Most data found in the literature	May be inflexible Linear models may not be appropriate for some production processes
Econometric	Ability to test statistically hypotheses about parameters Possibility of addressing some data problems	Large amount of good quality data required
Equilibrium Displacement Models (EDM)	Ability to model explicit economic and policy considerations Very little data required Data available in the literature	Linear models may not be appropriate for some production processes
Linear programming	Very little data required	Extensive production (engineering) knowledge required In most cases unable to statistically test hypotheses about parameters
Quasi-rent approaches	Expedient Relatively little data required	Converges to standard models of economic surplus only when cost or yield changes are small and when the elasticity of supply is unitary

4.3.3 Regulatory Impact Assessment

Studies about the impacts of regulations on research and development (R&D) and innovation in developed and developing economies have become more frequent. Given the salience of issues related to regulatory impact assessment issues such as supply channel performance, and industrial organization in the development and diffusion of GM crops, quantitative analyses of these issues are particularly needed. In addition, policy issues such as effects on health and the environment have not been adequately addressed. One explanation for the rarity of such studies is that these topics may be less amenable to analysis with conventional, applied research methods. Health and environmental analyses typically require interdisciplinary research design and analysis.

4.3.4 Overarching Issues

4.3.4.1 Developing Economies

Generally speaking, applied researchers must adapt their tools to the empirical context in which they conduct their analysis. This is also true when analyzing the impact of technology adoption on farmers in developing economies. As described in

Tripp (2000), such farmers typically work on small landholdings, and depend on a few key commodities, livestock or aquatic resources for food consumption and sale. Many of these are remote areas, where farmers have poor access to many goods and services, and particularly, modern farm inputs and related information. Smallholder farmers are often cash constrained unless they are linked to markets through farm credit or producer associations.

4.3.4.2 Data Problems

Data are the most critical ingredient in the above assessments. However, historical data series of prices and quantities supplied and demanded are often not available in developing economies. This limits the possibility of time-series econometric analysis. Differences in data collection procedures and definitions also limit the possibility of cross-national and regional comparisons. Survey data collected through interviews with farmers in developing countries, discussions related to various types of bias and ways to address them can be found in studies assembled by Smale and Falck-Zepeda (2012).

4.4 International Arena

The international experience with inclusion of socio-economic issues of benefits to producers and society in biosafety decision-making is relatively weak. Argentina (a non-party to the Cartagena Protocol on Biosafety) has a mandatory process that assesses economic consequence on exports and producer competitiveness. Brazil and Mexico allow the environmental and food/feed risk assessment to proceed. If any socio-economic issue is raised, a third party is then commissioned with a study. India and China do not have a formal requirement in their legislation/policy, but both have conducted socio-economic studies considering benefits to producers or society with different degrees of impact.

4.5 Administrative Consequences

Society can gain knowledge and information of technology impacts on producers and society. Inclusion of socio-economic considerations (SECs) focused on producers and society, especially as a result of thorough assessment studies, can help identify valuable technologies and discard those which are unacceptable for society. Assessment studies can help stakeholders take advantage of the benefits from such technologies by identifying limiting issues relevant to technology deployment and adoption. Furthermore, society can learn much about the multiple feedbacks and links between technology, science, R&D and the users' context in an innovation setting.

Inclusion of studies examining benefits to producers and society can increase regulatory costs of compliance with biosafety and/or technology approval regulations. Conducting socio-economic assessments will unequivocally increase implementation costs. In fact, the more complex and broader the portfolio of issues in an assessment, the more expensive a specific study will be to any developer and to government itself as it leads to a more complex regulatory process. An economic study examining impacts on biodiversity and long-term sociological and anthropological issues on producers will be far more complex (and expensive) to implement than one focused on a very specific topic (e.g. impact of exports/trade or impact on small-scale producers).

The likelihood of asymmetric impact on public sector R&D and public goods also increases. This scenario is particularly crucial to the public sector in developing countries that are likely to develop technologies of a public good nature. In fact, the introduction of additional regulatory hurdles can impact national and international public investments in R&D, which in turn may impact technology flows by causing a reduction in the number of potential technologies available to producers and society. This may be a result of additional regulatory complexity, cost implications and/or uncertainty. Bayer et al. (2010) have shown delays can have a significant impact on the net benefits from the potential adoption of GM crops in the Philippines. This research illustrated that even small delays of three years compared to the baseline can significantly decrease net benefits to producers. Increases in cost of compliance had a very small effect on net benefits. Inclusion of socio-economic studies will increase the cost of compliance and if not done in a careful manner and in close coordination with other assessments processes, they may increase the amount of time necessary to complete a biosafety regulatory assessment process.

Perhaps a more important impact of the inclusion of socio-economics into decision-making is the introduction of additional uncertainty into the process. A workable regulatory system can be defined as one where all elements of society are able to define, describe and trust the process and its outcome. Society actors can thus judge the system based on transparency, participation ability, predictability and robustness and cost and time efficiency. For example, developers (public or private sector) with predictable regulatory systems can attach a value on outcome by projecting potential gains by producers and consumers, and add the respective probabilities of success into their decision-making process. If developers cannot attach such probabilities due to an unpredictable or uncertain process and thus cannot attach success probabilities and calculate potential benefits, the likelihood is that they will not make investments in such jurisdiction.

4.6 Summary/Synthesis

- Sound estimates of the benefits and costs of innovations to producers, consumers and society as a whole are important because this information can be used by decision-makers to improve innovation systems.

- There is an array of approaches used by applied economists to estimate these costs and benefits; no approach is superior, each method has advantages and disadvantages, and the choice of approach depends on the nature of the innovation, the empirical context of the study, its feasibility, data requirements, the research budget, parameter uncertainty and relevance to the socio-economic context.
- A number of issues related to sampling, measurement and parameter variability should be carefully addressed by practitioners.
- Limitations and the applicability of methods are particularly acute when examining costs and benefits in developing agricultural economies. Approaches have been recommended in the literature.
- While there is no 'best' practice, applied researchers should attempt to follow good practice. Practitioners are encouraged to describe the challenges met during study implementation if these may affect study outcomes, and report how they were addressed.

References

Alston JM, Norton GW, Pardey PG (1995) Science under scarcity: principles and practice for agricultural research evaluation and priority setting. Cornell University Press, Ithaca

Alston JM, Pardey PG (2001) Attribution and other problems in assessing the returns to agricultural R & D. Agric Econ 25:141–152

Areal FJ, Riesgo L, Rodriguez-Cerezo E (2012) Economic and agronomic impact of commercialized GM crops: a meta-analysis. J Agric Sci. CJO2012 doi:10.1017/S0021859612000111. Accessed Dec 2012

Bayer, J. C., G. W. Norton, and J. B. Falck-Zepeda. (2010). Cost of compliance with biotechnology regulation in the Philippines: Implications for developing countries. *AgBioForum* 13(1): 53-62. Available online at http://www.agbioforum.org/v13n1/v13n1a04-norton.htm

Falck-Zepeda JB, Traxler G, Nelson RG (2000) Surplus distribution from the introduction of a biotechnology innovation. Am J Agric Econ 82(May 2000):360–369

Demont M, Tollens E (2001) Uncertainties of estimating the welfare effects of agricultural biotechnologies in the European Union. Working Paper Number 58, Department of Agricultural and Environmental Economics, Katholieke Universiteit, Leuven

Falck-Zepeda J, Sanders A, Rogelio Trabanino C, Batallas-Huacon R (2012) Caught between scylla and charybdis: impact estimation issues from the early adoption of GM maize in Honduras. AgBioForum 15(2):138–151. http://www.agbioforum.org

Hofs JL, Fok M, Vaissayre M (2006) Impact of Bt cotton adoption on pesticide use by smallholders: a 2-year survey in Makhatini Flats (South Africa). Crop Prot 25(9):984–988

Huang J, Hu R, Fan C et al (2002) Bt cotton benefits, costs, and impacts in China. AgBioForum 5(4):153–166. http://www.agbioforum.org

National Academy of Sciences of the United States (2000) Transgenic plants and world agriculture. National Academy Press, Washington, DC

Pemsl D, Waibel H, Orphal J (2004) A methodology to assess the profitability of Bt-cotton: case study results from the state of Karnataka, India. Crop Prot 23(12):1249–1257

Pemsl D, Waibel H, Gutierrez AP (2005) Why do some Bt-cotton farmers in China continue to use high levels of pesticides? Int J Agric Sustain 3(1):44–56

Pontifical Academy of Sciences (2001) Science and the future of mankind: science for man and man for science. http://www.casinapioiv.va/content/dam/accademia/pdf/sv99.pdf. Accessed Dec 2012

Potrykus I, Ammann K (eds) (2010) Proceedings of a Study Week of the Pontifical Academy of Sciences New Biotechnology 27(5):445–718
Qaim M (2009) The economics of genetically modified crops. Ann Rev Resour Econ 1:665–694
Smale M, Falck-Zepeda J (eds) (2012) Farmers and researchers discovering biotech crops: experiences measuring economic impacts among new adopters. AgBioForum Special Issue 151(1):7–33
Smale M, Zambrano P, Gruère G et al (2009) Impacts of transgenic crops in developing countries during the first decade: approaches, findings, and future directions. IFPRI Food Policy Review 10. International Food Policy Research Institute, Washington, DC
Smyth SJ, Phillips PWB, Castle D (eds) (forthcoming) Handbook on agriculture, biotechnology and development. Edward Elgar Publishing, Cheltenham
Tripp R (2000) Can biotechnology reach the poor? The adequacy of seed and information delivery. Paper presented at the 4th International Conference on the Economics of Agricultural Biotechnology, Ravello, Italy, 24–28 August 2000

Chapter 5
Consumer Choice

Vivian Moses and Siglinde Fischer

5.1 Introduction

In the closing years of the twentieth century, Europe witnessed a series of disturbing food-related crises. They encompassed cases of deliberate and illicit adulteration, contamination with noxious chemicals from industrial effluents and the involvement of animal diseases, including bacterial and viral infections, and bovine spongiform encephalopathy ('mad cow disease'). In some countries, this generated growing scepticism about information, particularly assurances about food safety, deriving from industry as well as from governmental and other official sources. These food problems were the precursors of the genetically modified (GM) food debate which remains partly unresolved to this day.

The 'debate' about GM foods and products has raged in Europe for 15 years. European consumers (and European institutions) are widely regarded as the very epitome of anti-GM sentiment. Is that really true or is it the widely held perception of anti-GM attitudes rather than the reality which prevails? Moreover, how are the views of the public to be evaluated: by their statements or by their actions?

In spring 2004, customers in a German city were offered 'pretend' GM bread in a bakery and French fries at a lunch counter. The products, labelled as containing GM ingredients (although they did not), were offered at reduced prices alongside 'non-GM' equivalents (which were, of course, identical). Four times more of the cheaper 'GM' loaves and more than 20 times more 'GM' fries were sold compared with the 'conventional' variety (Südwestrundfunk 2004). An experiment with asparagus revealed similar results. In the UK, an experiment showed that 28% of customers were willing to buy GM breakfast cereals at equal or at lower prices compared with conventional counterparts (Moon and Balasubramanian 2004). Is price thus a (or the) determining factor?

V. Moses (✉)
Division of Nutritional Sciences, King's College, London SE1 9NH, UK
e-mail: V.Moses@qmul.ac.uk

S. Fischer
BioAlliance AG, 60313, Frankfurt, Germany

K. Ludlow et al. (eds.), *Socio-Economic Considerations in Biotechnology Regulation*, Natural Resource Management and Policy 37, DOI 10.1007/978-1-4614-9440-9_5,

When, in April 2004, the EU Regulation 1829/2003 legislation (European Union 2003)—requiring the labelling of more than a 0.9% content of adventitious GM-derived material in any ingredient in a foodstuff—came into force a number of such products were already on sale in the grocery stores of several Member States. There was indeed a widespread expectation (including pronouncements by some government ministers (Deutsche Welle 2004)) that more and more GM products would be offered for sale in the supermarkets and grocery shops of Europe (GMO Compass 2007). That accordingly seemed to be the right time to observe the appearance in European food stores of GM products together with consumers' reactions to them and accordingly prompted the CONSUMERCHOICE proposal which directly asked the question 'Do European consumers buy GM foods?' (CONSUMERCHOICE 2008).

It soon became clear that while some predictions were fulfilled, others were not. Before the enactment of Regulation 1829/2003 there was no obligation for food manufacturers and suppliers to indicate the presence of GM ingredients on packaging and very few had done so. Some retailers had announced specifically that they were removing GM ingredients from own-label products. It was left to private organisations (non-governmental organisations (NGOs) and others) inclined to do so to attempt to identify products containing the GM ingredients and make the facts known to the public.

The Eurobarometer polls and other reports showed widespread scepticism to GM food (Durant et al. 1998; Gaskell et al. 2000, 2003; Grove-White et al. 1997, 2000; Marris et al. 2002; Wagner et al. 2001). The arguments and underlying premises of popular viewpoints have been investigated in qualitative studies using in-depth interviews and focus groups. Such studies have shown that attitudes to GM food (and its labelling) are linked to moral, existential and epistemological issues about trust and people's sense of agency. Lay scepticism about GM foods may be influenced by a lack of trust in the institutions and the actors responsible for the new technology (Grove-White et al. 1997; Marris et al. 2002; Wagner et al. 2001), or by a lack of a sense of agency (Gaskell et al. 2000; Wagner et al. 2001; Wibeck 2002). In addition, GM food is sometimes perceived as 'unnatural', challenging traditional perceptions of nature, agriculture and humanity's place in nature, which may bring about moral objections (Gaskell et al. 2000, 2003).

Over recent years there has been a gradual decline in antipathy to GM foods and crops, more so in some countries than in others. Thus, a recent UK study showed a decline in concerns about GM foods from 25% in 2006 to 20% in 2007 (Food Standards Agency, UK 2008), a trend which is continuing (Food Standards Agency, UK 2009). Swedish consumer opinion polls point to a relatively negative public opinion to GM foods (Gaskell et al. 2006). For 2005, the number of opponents in Sweden was markedly higher than the total percentage of opponents in Europe generally, which amounted to 58%. There were, however, indications that the Swedish negative opinion was not absolute (The Consumer Organisation 2007). It is plausible that under certain circumstances, for example if environmental benefits could be proved, there would be some willingness among Swedish consumers to buy GM foodstuffs.

Various interesting examples of consumer reactions have been observed. In Sweden, a beer is brewed containing GM maize grown in Germany (Øresund Food Excellence 2004). It was for a time sold in a large Swedish retailer chain but was withdrawn due to consumer protests. It is now offered in some restaurants and in southern Sweden through the Swedish state-owned liquor monopoly Systembolaget (Independent on Sunday 2004). This Kenth beer was available for tasting at the Food and Drink Expo 2004 exhibition in Birmingham in March 2004; passers-by and visitors to the booth were invited to sample it. Of about 2,000 people so invited, only 12 refused on the grounds that it contained a GM ingredient. In that same exhibition, visitors were asked, as they had been in 2002, to predict when they expected to see GM products in the stores. The period has become shorter, with the overwhelming proportion of consumers expecting GM foods to become part of normal existence in the next 2–5 years. That has turned out to be rather optimistic.

While some reports continue of consumer antipathy to GM (Muschel 2011), others perceive changes in attitude. Some not unreasonably see consumers more willing to consume GM if they can perceive personal advantage (Halliday 2010): Nutritional or health benefits (Health Canal 2011) seem more attractive than price advantages since the latter will often be small and not necessarily readily perceptible by shoppers faced with constant variability in food prices. Other studies take a more general view, asking what the prevailing consumer attitudes are and trying to understand the reasons for them. The IGD in the UK updated their study of consumer attitudes in 2010 (IGD 2010) (following earlier studies in 2003 (Aerni et al. 2011) and 2008 (IGD 2008)).

A 2011 experimental study (Aerni et al. 2011) offered three clearly labelled types of corn bread at five market stands across the French- and German-speaking regions of Switzerland: one made with organic corn flour, one with conventional and one with GM maize. In addition, the consistency was tested between purchasing decisions at the market stand and the previous voting decisions on GMOs in 2005 by means of an ex post questionnaire; note that in 2005, Swiss citizens had expressed their negative attitude towards genetic engineering in agriculture by voting in favour of a ban against using GM crops in their domestic agriculture. The results of this discrete choice analysis showed that Swiss consumers treat GM foods just like any other type of novel food. A similar earlier experiment (2007) involving six countries (Belgium, France, Germany, Sweden, the UK and New Zealand) involved setting up real roadside fruit stalls based on a choice modelling experimental design. At each site, conventional fruit was placed on sale labelled as ‘organic’, ‘spray-free genetically modified’ or ‘conventional’ (or appropriate translations of same in the prevalent local language) and offered at varying price levels. The price for each fruit category was set at one of three levels: the median market price in that locality, median plus 15 % or median minus 15 %. A parsimonious main effects balanced fractional factorial design was used to generate nine price and fruit offerings. Research assistants fluent in the local language operated the stalls. The experimenters commented (Knight et al. 2007) that their research revealed a significant (and in some markets, surprisingly high) percentage of consumers in European countries were apparently willing to choose GM food provided there is a price advantage coupled with a consumer benefit (in this case 'spray-free' status). But it might be an uphill job in some places. While awareness in some European countries might be growing, in others it still seems to be at a remarkably low level. Thus, a report from Poland (Gentle 2012) noted a growing understanding but from a low base: Only 3 % of Poles knew correctly what 'GMO' meant while two-thirds of those asked were unable to answer the question.

5.2 Methodologies

In asking 'Do European consumers buy GM foods?' it will never be the case that any conceivable positive answer could or would apply to all the hundreds of millions of European consumers. The response to the CONSUMERCHOICE study question could be no more than that some people did purchase such foods as a matter of choice or disinterest in the GM issue; investigating in depth their motivations was beyond the resources of the study. Similarly, some consumers buy kosher, halal or vegetarian products while others do not. Retailers sell the products as a matter of marketing, leaving individuals to decide whether to buy. There is no public consultation on ethics or whether to do so; it is seen as a normal commercial transaction. But with GM products there are additional factors: objections (sometimes vigorous to the extent of vandalism) of special interest groups, policies of retailers, as they put it, 'catering (albeit selectively) to customer demand' and the influence of the media and public policy in the different Member States which may actively encourage or discourage a favourable view of GM products.

With only a few GM products actually available to European consumers (most probably unlabelled in the earlier period), most public opinion assessments have been undertaken using various types of opinion polls. Such polls usually take the form of questionnaires to be answered either in person or in writing. There might be many or just one question; if just one it might be as leading as 'Would you eat Frankenstein food?'. But other polls ask a number of questions, sometimes fairly balanced, at other times less so, and accordingly derive more interesting and relevant information. Thus the Food Standards Agency, UK (2009), in its quarterly evaluation of (UK) public attitudes found that only 4 % spontaneously expressed concern about GM food but around 20 % did so when prompted (i.e. asked specifically whether they had concerns). Exactly similar effects were obtained with all the other questions asked, ranging from food poisoning and the amount of salt in food to GM food and BSE (the two lowest concerns recorded). In the summer of 2012, The Independent (Grice 2012) reported that '[a]sked whether the Government should encourage experiments on GM crops so that farmers can reduce the amount of pesticides they use, 64 per cent of the public agreed and 27 per cent disagreed, while 9 per cent replied "don't know"'; that seems to have justified the headline 'Dramatic change as two-thirds now support GM crop testing.'

What is it that such questionnaires and polls reveal? At the very least they show how people respond to such questions in the circumstances in which they are asked.

But do they predict how people will *behave* when presented with a real choice situation in the familiar environment of their normal activities (Friese et al. 2012)? Our data (see below) suggest perhaps not.

Nevertheless, as noted above, there have been some focus group studies and a few 'experiments' staged in which the reactions of consumers supposedly offered a choice of GM or non-GM items are recorded. To what extent the participants in such activities realised, at least in Europe, that the 'GM items' were not GM at all, is not clear; perhaps some of them did not and thought they had a real choice. Most, but not all, evidence of the published opinion polls have all along offered a variable picture of European consumer attitudes—and still do. By and large, they have revealed a small fraction of consumers (perhaps 10% or so) vehemently against GM, a similar proportion enthusiastic and the rest indifferent to varying degrees. The evidence of the 'experiments'—and, indeed, of supermarket sales—suggested and suggests that when offered products really or falsely labelled 'GM' at a favourable price, or they see GM products in the familiar environment of the supermarket, consumers tend to buy. Not all consumers, of course, but then there is probably no single product that all consumers want. The discrepancy between what consumers say in questionnaires and polls, and what they do when they shop casts doubt on whether opinion polls actually provide reliable indications of how consumers would or do behave when presented with real rather than theoretical choices.

With the EU labelling regulations coming into force, consumers would have the information reasonably required to decide whether they wished to consume products containing or made from GMOs. As the new products appeared on the shelves, an opportunity existed of observing what shoppers actually bought when faced with this new choice, rather than what they might have said they would purchase; this prompted the development of the CONSUMERCHOICE project. A possible discrepancy between the public opinion polls about GM foods and the actual behaviour of customers when faced with real choice had not previously been extensively explored. The objective of CONSUMERCHOICE was thus to determine what consumers actually do when buying (labelled) food as distinct from what they say they will do as reported in polls and questionnaires.

The experiments commented upon earlier sought to explore consumer behaviour with respect to choosing or rejecting GM products. Such studies have value in seeking to understand consumer decision-making when presented with choices 'clearly in the context of an experiment'. The problem from the perspective of the real conditions of daily life in non-experimental environments is whether consumers going about their normal business of buying food in food stores behave in the same way (Rousu et al. 2007; McCluskey et al. 2003; Lusk et al. 2004; Colson et al. 2011). We had our doubts, justified as it turned out.

Rather than concentrating on what consumers said they might do with respect to buying GM foods, or how they behaved in experiments, CONSUMERCHOICE explored as far as possible what consumers actually bought in those countries where such foods were on sale. The variety and presence of labelled products on the shelves, sales data, the recorded purchasing behaviour of customers and published material were combined with the project's own surveys of opinion using polls and focus groups to offer a view of the real attitudes and behaviour of consumers in a number of countries towards foods containing ingredients derived from GM sources. In Member States with no or only a few such products on the shelves, attention was directed to consumer responses to 'GM-free' labels.

An ideal exploration would have involved taking up station in the appropriate sections of food stores, waiting for customers to choose GM products and then asking why they had done so. For all sorts of reasons, that was impracticable: It is not clear that food store managers would have agreed and the resources to do anything of the sort in several countries—on a scale which would have been statistically plausible—were simply not available.

The CONSUMERCHOICE project (2006–2008) was accordingly configured to investigate as far as practical:

- What products labelled 'GM', 'non-GM' or 'GM-free' were on sale in the participating Member States, and where
- What evidence there was for ongoing sales of such products: whether shoppers actually bought GM products when they were present on the shelves
- How both purchasers and non-purchasers responded to questions about their attitudes towards buying these products
- What consumers said they would do if they had the opportunity to purchase freely

In addition, there were limited inquiries into what Europeans did when they resided in or visited North America where many GM foods are on sale, none of them labelled. All this information was set against the background of the public mood as expressed by the media and in polls in each country together with major decisions made by national governments and their agencies.

Several GM products were consistently offered for sale in the Czech Republic, Estonia, the Netherlands, Poland, Spain and the UK, but not in Greece or Slovenia. In Sweden there was just one product (a beer containing GM maize) on sale only under very restricted circumstances. In Germany GM products have appeared in supermarket chains over the years only to be withdrawn soon after.

In most cases the product was oil or lecithin derived from GM soya, packaged as cooking or salad oil, or incorporated into other products, including margarines. A few maize products were found, including popcorn in the Czech Republic and maize oil and chips/crisps in the Netherlands.

Generally, the GM-labelled foods were sited on the shelves next to related products; in most cases there were non-GM equivalents to the GM products readily available in the same store. There was no special shelf labelling or other overt indication that the product was of GM origin aside from the container label. In the Czech Republic, there appeared to be a consistent price differential between the less expensive GM versus the more expensive non-GM cooking oil.

That supermarkets and grocery stores in some countries have continued for years to offer GM food items indicates that they must be selling. Store shelf space is valuable; products are not displayed for long periods if they do not sell. We therefore conclude that some consumers have for those long periods been buying foodstuffs labelled as containing GM ingredients, just as those pioneers once did with tomato purée in the UK in the mid- to late-1990s (National Centre for Biotechnology Education 2006).

An essential consideration in deciding whether or not European consumers bought GM-labelled food was to find out what they actually did when given the opportunity rather than what they said they might—or would do—should such an opportunity arise.

It is clearly not possible to observe purchasing behaviour if there are no relevant products. Asking people what they would do about purchasing GM-labelled products brings consideration of that issue to the forefront of their minds. Were they actually just buying their groceries as usual, they might pay no attention at all to whether or not the items contained GM ingredients. One can set up trial experiments, as some people have done on occasion, pretending to offer consumers GM products, perhaps at a reduced price, in order to test how they react (Moon and Balasubramanian 2004). But it is not clear that consumers really accept this ruse, particularly as it will be an unusual event for them (very unusual indeed; it would be virtually unique!); they may realise what is going on and frame their responses accordingly.

The closest it was possible to get to reality in a European context was to compare actual purchases of GM products made by consumers with their subsequent responses to questions, without focussing or otherwise bringing to their attention the fact that they might have already bought such items.

GfK, a market research company, maintains panels of shoppers in a number of countries. Provided with a bar code reader, the panellists scan each and every item they have purchased so that, for each of those individuals, GfK has an exact record of their purchases. In collaboration with GfK, CONSUMERCHOICE submitted—'after their grocery shopping was complete and the barcode data recorded'—a list of ten questions to individual consumers in four countries (the Czech Republic, the Netherlands, Poland and Spain) asking for attitudes towards GM-food purchases. Without, of course, their personal identities being disclosed, it was then possible to compare individuals' authentic purchases of food products as reported via their bar code readers with their responses to the specific questions.

Interestingly, almost no difference was found between buyers and non-buyers of GM-labelled products in the four countries, all of which have a range of foods containing GM ingredients available in the shops. When they were asked whether they 'bought food labelled as containing GM ingredients', half of all respondents said they did not buy such products.

Just more than 21 % and 23 % of consumers who actually had or had not chosen GM foods, respectively, said they believed they had bought GM products. Even more remarkable is the fact that of the people who did buy GM-labelled foods, 48 % said they would not buy such products. Did this represent confusion on their part about what they were buying or a lack of awareness that GM foods were what they had already bought? Did they not read the labels? If so, was that because they had no interest in doing so? Or had they read the labels but not fully understood them?

The 22.9 % willingness to buy GM-labelled foods on the part of non-buyers might suggest there is a potentially significant market for GM-labelled products which has not so far been satisfied because products of interest to them were not available with GM content. This conclusion is supported by the finding that only one person in five is careful to avoid choosing GM foods. It might well be the case that the non-buyers did not buy GM-labelled foods because they had no interest in the particular items on offer which did carry GM labels, so neither a GM nor a GM-free label would have meaning for them in that context. Had their favoured products been available with GM ingredients, they might well have bought them.

The views expressed in the focus groups during the CONSUMERCHOICE inquiry showed that concerns about harm from the use of GMOs were voiced when people are questioned directly, with differences between those who buy GM and those who do not. In most countries participating in this project (Sweden was a notable exception), a majority of consumers claimed they were aware of the requirement for GM products to be labelled but only half of the respondents agreed that they read the labels. On the other hand, there was a significant difference between the answers from buyers and non-buyers of GM-free-labelled products, suggesting that buying GM-free is a more conscious choice.

There are however contradictions. Many people say it matters to them that the products they buy do not contain GM ingredients. With this in mind one would expect the shoppers to be very careful when buying their food products. Yet when asked how careful they are to make sure to avoid GM products, many say they are not careful. In the UK, for example, 48 % replied that they cared whether their food contained GM ingredients yet the only food so labelled which is relatively widely on sale is soya cooking oil: Do consumers really take care not to buy GM-labelled products when there is actually only one on sale? And how great is the demand specifically for cooking oil derived from soya compared with oils from other sources?

5.3 Critical Assessment

It was and is, of course, impossible to predict in advance how in real life consumers will respond to food labelled as containing GM ingredients, hence the dilemma for retailers, manufacturers and farmers. Some retailers claim no philosophical objection to offering GM products but are clearly worried about the effect on their sales or protests by activists, especially if they become the first locally to do so. However, providing products for a minority of consumers with incompatible special requirements presents few problems for retailers: They already do so for patrons with religious requirements or wishing to avoid animal products, while offering other products in the same stores for the bulk of their customers. If they so decided, it could be done in the same way for GM products.

It is important to bear in mind that questionnaires, interviews and focus group discussions on GM food have dealt mainly with hypothetical products and scenarios, since clearly labelled GM products have rarely and mostly only comparatively recently been available on the European market. Together with extensive anecdotal observations, the polls often generate uncertain and conflicting conclusions. Consumers generally may not be so antagonistic as some retailers fear (Øresund Food Excellence 2004; Independent on Sunday 2004; Muschel 2011) and not all food suppliers focussing on non-GM foodstuffs are necessarily successful. For example, sales of specifically non-GM pork (from animals fed non-GM fodder) by a Danish meat-producing group fell far short of expectation (Politiken 2004). But, so far, few food manufacturers and retailers have withstood pressures (mainly from NGOs) to withdraw GM-labelled products from their shelves (The Times 2004; Langelüddeke and Deichmann 2004). Many, perhaps most, large retailers have somewhere on their websites a statement about their policies regarding GM products, although those pages are often difficult to find and are, indeed, becoming rarer; some retailers appear to have deleted them.

Lay persons' expressed attitudes to GM food products may well differ from their actual choices when such products are available in the stores. Moreover, whatever form consumer reaction takes to GM foods in the stores, public reaction is conducted in the context of government decisions, media news items, discussions, articles and presentations, as well as a range of activities by scientific, civic and industrial bodies, and by NGOs. No matter its ultimate origin, most members of the public acquire their information on GMO topics from the media; what the newspapers and magazines print—and the broadcasters say—is obviously important.

5.4 International Arena

There are no internationally agreed rules on labelling of GM food or products, although national or regional labelling regulations exist. The *Codex Alimentarius* Commission (Codex) develops international food standards that, amongst other things, identify a processed food product and its essential composition and quality factors, provide labelling requirements and establish the scientific procedures used to sample and analyse product. Each member country is encouraged to incorporate Codex standards into relevant domestic regulation, but they may unilaterally impose more stringent food safety regulations for consumer protection, provided the different standards are scientifically justifiable. Jackson and Jansen (2010) provide a detailed discussion of the science-based risk assessment process for food safety and its relationship to World Trade Organization (WTO) dispute cases.

Codex plays an important role in the agri-food trade because its standards, guidelines and recommendations are acknowledged in the Sanitary and Phytosanitary and Technical Barriers to Trade (TBT) Agreements of the WTO. There are currently no Codex standards for products of biotechnology, although there has been significant effort by Codex to develop a standard for the labelling of food products derived from biotechnology. The Codex Committee on Food Labelling was tasked in 1993 to initiate work on the development of a standard on the labelling of GM-derived foods and for nearly 20 years the Committee's efforts were gridlocked. In 2011, the US relented on its opposition to the labelling of GM food products and in 2012, Codex adopted principles for a risk analysis of foods derived from biotechnology, which establishes that if a risk is identified, labelling is an appropriate management strategy. Codex stresses that any risk analysis of biotechnology derived foods has to be science-based and that these principles do not address 'environmental, ethical, moral and socio-economic aspects…' (Codex Alimentarius Commission 2012, p. 1). It is important to note that this is a Codex principle on risk analysis of foods derived from biotechnology and not the standard on the labelling of GM foods that the Committee was tasked with 20 years ago.

5.5 Administrative Consequences

To introduce regulations on GM food labelling, a number of policy decisions would need to be made. All of these decisions would need to be guided by clear information on the purpose of the labelling. Gathering real data on consumer and retailer concerns requires further research and cannot rely only on polls conducted in an artificial 'what if' scenario. Policy decisions that would need to be made include clearly defining which products were to be labelled: what is 'GM' for these purposes, how far down the food chain would the label be required (only at the final retail level, only packaged food or food in restaurants etc.), what would the label be required to tell consumers, how and where should the label appear on packaging and so on. A further, and more difficult, question is how much 'GM' content a food could contain before labelling requirements apply. No food is 100% pure—and even if not intentional, adventitious presence is unavoidable. A feasible threshold needs to be set. The additional costs of labelling (and the segregation and contractual obligations required to enable accurate labelling) would need to be assessed and recognised. Issues with markets outside the particular domestic market having different thresholds would also need to be considered in making these decisions. Further, to be worthwhile, labelling regulations would need to be enforced by monitoring food products available in the marketplace. These costs would also need to be assessed and recognised.

5.6 Summary/Synthesis

- Assessing consumer opinion and the reasons for those opinions is difficult.
- Asking consumers what they would do under ill-defined circumstances is not necessarily a guide to what they actually do in real life.
- Consumers choose labelled GM products although they say they have not. This may indicate that the sensitivity food retailers fear among consumers is exaggerated.
- Past consumer behaviour shows customers do not necessarily desert stores selling GM products. Desertion did not happen when ASDA's and Morrisons' egg suppliers in the UK introduced GM feed for their chickens; outside the GM debate, consumer groups with strong attitudes to certain food types continue

to shop in stores offering products they actively avoid, provided the products wanted by them are labelled according to their needs and clearly separated from forbidden items. Fear of disruptive demonstrations by anti-GM activists may also have a chilling effect on retailers although it is unknown how extensive demonstrations would now be, if they occurred.
- Without understanding the reasons for consumers' opinions and motivations for reluctance by retailers, devising labelling regulations to address these matters is difficult.

References

Aerni P, Scholderer J, Ermen D (2011) How would Swiss consumers decide if they had freedom of choice? Evidence from a field study with organic, conventional and GM corn bread. Food Policy 36:6:830–838

Bayer A (2011) What countries have banned GMO crops? The Examiner, 18 June. http://www.examiner.com/article/what-countries-have-banned-gmo-crops. Accessed 28 Feb 2013

Codex Alimentarius Commission (2012) Principles for the risk analysis of foods derived from modern biotechnology. http://www.codexalimentarius.org/standards/list-of-standards/en/?no_cache=1. Accessed 28 Feb 2013

Colson G, Huffman WE, Rousu M (2011) Will consumers pay more for product enhanced attributes: evidence from food experiments. J Agric Resour Econ 36:343–364

CONSUMERCHOICE (2008) Do European consumers buy GM foods? European Commission (November). http://www.kcl.ac.uk/medicine/research/divisions/dns/projects/consumerchoice/index.aspx. Accessed 28 Feb 2013

Deutsche Welle (2004) Germany to greenlight GM Food labelling. http://www.dw-world.de/dw/article/0,,1084846,00.html. Accessed 28 Feb 2013

Durant J, Bauer M, Gaskell G (eds) (1998) Biotechnology in the public sphere. A European sourcebook. Science Museum, London

European Union (2003) Regulation (EC) No 1829/2003 of the European Parliament and of the Council of 22 September 2003 on genetically modified food and feed. http://ec.europa.eu/food/food/animalnutrition/labelling/Reg_1829_2003_en.pdf. Accessed 16 Oct 2013

Finz S (2012) Prop. 37: Genetic food labels defeated. San Francisco Chronicle http://www.sfgate.com/news/article/Prop-37-Genetic-food-labels-loses-4014669.php. Accessed 28 Feb 2013

Food Standards Agency, UK (2008) Consumer Attitudes to Food Standards: Wave 8. UK Report Final. http://products.ihs.com/Ohsis-SEO/906566.html. Accessed 28 Feb 2013

Food Standards Agency, UK (2009) Quarterly public attitudes tracker: December. http://www.food.gov.uk/multimedia/pdfs/trackerdecem09.pdf. Accessed 28 Feb 2013

Friese M, Tucker Smith C, Plischke T et al (2012) Do implicit attitudes predict actual voting behavior particularly for undecided voters? PLOS One, 7(8) e44130. doi:10.1371/journal.pone.0044130

Gaskell G et al (2000) Biotechnology and the European public. Nat Biotechnol 18:935–938

Gaskell G, Allum N, Stares S (2003) Europeans and biotechnology in 2002; Eurobarometer 58.0 (2nd edn). http://ec.europa.eu/public_opinion/archives/ebs/ebs_177_en.pdf. Accessed 28 Feb 2013

Gaskell G et al (2006) Europeans and biotechnology in 2005: patterns and trends. Eurobarometer 64.3. http://www.ec.europa.eu/research/press/2006/pdf/pr1906_eb_64_3_final_report-may2006_en.pdf. Accessed 28 Feb 2013

Gentle P (2012) Awareness on genetically modified crops low 'but growing' in Poland. News PL. http://www.thenews.pl/1/12/Artykul/93306,Awareness-on-genetically-modified-crops-low-but-growing-in-Poland. Accessed 28 Feb 2013
GMO Compass (2007) Labelled goods hard to find. http://www.gmo-compass.org/eng/regulation/labelling/92.gmo_labelling_labelled_goods.html. Accessed 28 Feb 2013
Grice A (2012) Dramatic change as two-thirds now support GM crop testing. The Independent. http://www.independent.co.uk/news/uk/politics/dramatic-change-as-twothirds-now-support-gm-crop-testing-7973432.html?origin=internalSearch. Accessed 28 Feb 2013
Grove-White R, Macnaghten P, Mayer S et al (1997) Uncertain world. genetically modified organisms, food and public attitudes in Britain. The Centre for the Study of Environmental Change. Lancaster University, Lancaster. http://www.ncbe.reading.ac.uk/ncbe/gmfood/publications.html. Accessed 28 Feb 2013
Grove-White R, Macnaghten P, Wynne B (2000) Wising up. The public and new technologies. The Centre for the Study of Environmental Change. Lancaster University, Lancaster. http://csec.lancs.ac.uk/docs/wising_upmacnaghten.pdf. Accessed 28 Feb 2013
Halliday J (2010) GM fruit may be more acceptable if consumers see personal gain. Food Navigator. http://www.foodnavigator.com/Science-Nutrition/GM-fruit-may-be-more-acceptable-if-consumers-see-personal-gain/?c=gTN1p4qDILWA3fyyM5OfLw=&utm_source=newsletter_daily&utm_medium=email&utm_campaign=Newsletter+Daily. Accessed 28 Feb 2013
Health Canal (2011) Consumers willing to pay premium for healthier genetically modified foods: ISU study. http://www.healthcanal.com/public-health-safety/20716-Consumers-willing-pay-premium-for-healthier-genetically-modified-foods-ISU-study.html. Accessed 28 Feb 2013
IGD (2008) Genetically modified foods—consumer research. http://www.igd.com/index.asp?id=1&fid=1&sid=8&tid=33&cid=548; link no longer available
IGD (2010) Consumer attitudes towards genetically modified (GM) foods. http://www.igd.com/print.asp?pid=1&pflid=1&plid=5&pcid=1487 (link no longer available)
Independent on Sunday (2004) Europe prepares for pull of GM lager. http://enjoyment.independent.co.uk/food_and_drink/news/story.jsp?story=539985 (this document is no longer available)
Jackson LA, Jansen M (2010) Risk assessment in the International Food Safety Policy arena. Can the multilateral institutions encourage unbiased outcomes? Food Policy 35:538–547
Knight JG, Mather DW, Holdsworth DK et al (2007) Acceptance of GM food—an experiment in six countries. Nat Biotechnol 25:507
Langelüddeke P, Deichmann T (2004) Was Konsumenten wollen müssen. Novo-Magazin. http://www.novo-magazin.de/71/novo7124.htm. Accessed 28 Feb 2013
Lusk JL, House LO, Valli C et al (2004) Effect of Information about benefits of biotechnology on consumer acceptance of genetically modified food: evidence from experimental auctions in United States, England, and France. Eur Rev Agric Econ 31:179–204
Marris C, Wynne B, Simmons P et al (2002) Public perceptions of agricultural biotechnologies in Europe. Final report of the PABE research project. http://csec.lancs.ac.uk/archive/pabe/docs/summary.doc. Accessed 28 Feb 2013
McCluskey JJ, Grimsrud KM, Ouchi H et al (2003) Consumer response to genetically modified food products in Japan. Agric Resour Econ Rev 32:2:222–231
Moon W, Balasubramanian SK (2004) Is there a market for genetically modified foods in Europe? Contingent valuation of GM and non-GM breakfast cereals in the United Kingdom. AgBioForum 6:3:article 6. http://www.agbioforum.org/v6n3/v6n3a06-moon.htm. Accessed 28 Feb 2013
Muschel R (2011) Die Verbraucher wollen keine Gentechnik. Südwest Press Online. http://www.swp.de/ulm/nachrichten/suedwestumschau/Die-Verbraucher-wollen-keine-Gentechnik;art4319,1106953. Accessed 28 Feb 2013
National Centre for Biotechnology Education (2006) Genetically modified food. Case studies: tomato puree. http://www.ncbe.reading.ac.uk/ncbe/gmfood/tomato.html. Accessed 28 Feb 2013
Office of the Gene Technology Regulator. Australia (no date) About the OGTR. Australian Government Department of Health and Aging. http://www.ogtr.gov.au/internet/ogtr/publishing.nsf/content/about-index-1. Accessed 28 Feb 2013

Øresund Food Excellence (2004) Première for GM foods in Sweden. http://www.foodoresund.com/composite-347.htm (this link is no longer available)
Politiken (2004) Danish Crown stopper GMO-frit kød. http://politiken.dk/erhverv/ECE89429/danish-crown-stopper-gmo-frit-koed/. Accessed 28 Feb 2013
Rousu M, Huffman WE, Shogre JF et al (2007) Effects and value of verifiable information in a controversial market: evidence from lab auctions of genetically modified food. Econ Inquiry 45:409–432
Südwestrundfunk (2004) Gene food—a question of the price? http://www.swr.de/plusminus/beitrag/04_04_13/beitrag4.html (this link is no longer available)
The Consumer Organisation Stockholm (2007) Attitude surveys on gene technology and food. http://www.konsumentforeningenstockholm.se/upload/Konsumentfrågor/SKOP_genteknik_sep2007_rapport.pdf (this link is no longer available)
The Times (2004) Sainsbury's cowed into non-GM milk. http://www.lobbywatch.org/archive2.asp?arcid=3637. Accessed 16 October 2013
Wagner W et al (2001) Nature in disorder: the troubled public of biotechnology. In: Gaskell G, Bauer M (eds) Biotechnology 1996–2000: the years of controversy. Science Museum, London
Wibeck V (2002) Genmat i fokus. Analyser av fokusgruppssamtal om genförändrade livsmedel (Genetically Modified Food in Focus. Analyses of focus group discussions). Diss. Linköping: Linköping Studies in Arts and Science, 260. http://www.worldcat.org/title/genmat-i-fokus-analyser-av-fokusgruppssamtal-om-genforandrade-livsmedel/oclc/186131791. Accessed 28 Feb 2013

Chapter 6
Environmental Impacts

Justus Wesseler and Richard Smart

6.1 Introduction

Concerns about the introduction of genetically modified organisms (GMOs) have been driven by impacts on: (1) human and animal health; and (2) the environment (see, e.g., EU Council for the Environment 1999). Rules and regulations have been implemented for GMOs in response to these concerns to: govern their approval process; cultivation; international trade; and products derived from them.

Implementing regulations incurs administration costs. Delays caused by this process entail costs of foregone benefits for companies, which have to be justified by the benefits of compliance. However, knowing this in advance is always unclear. Companies only introduce a new product if there is a market for it. A potential market exists if the new product is superior to what is currently available or more competitively priced. If users expect this to be the case, they adopt the new product with the allied expectation of increasing their net benefits, and society benefits as more goods can be produced with the same input of resources, or the same amount of goods with fewer resources.

Producing and consuming GMOs may have negative impacts on human health and/or the environment. Including these negative impacts in the net-benefit assessment at the user level may warrant GMOs being either restricted or banned. Conversely, if the impacts are included and result in positive net gains, additional constraints on GMO use or a ban may be unjustified from a cost-benefit perspective. Hence, it is unclear if introducing a GMO warrants additional use restrictions or a ban merely because its use has a negative environmental impact. Despite all forms of production in agriculture impacting the environment, the impact of GMOs may be smaller than that of the non-GMOs they replace.

J. Wesseler (✉) · R. Smart
Research Department of Agriculture Economics, Center of Life and Food Sciences Weihenstephan, Technische Universität München, Freising-Weihenstephan, Germany
e-mail: justus.wesseler@wzw.tum.de

K. Ludlow et al. (eds.), *Socio-Economic Considerations in Biotechnology Regulation*, Natural Resource Management and Policy 37, DOI 10.1007/978-1-4614-9440-9_6,

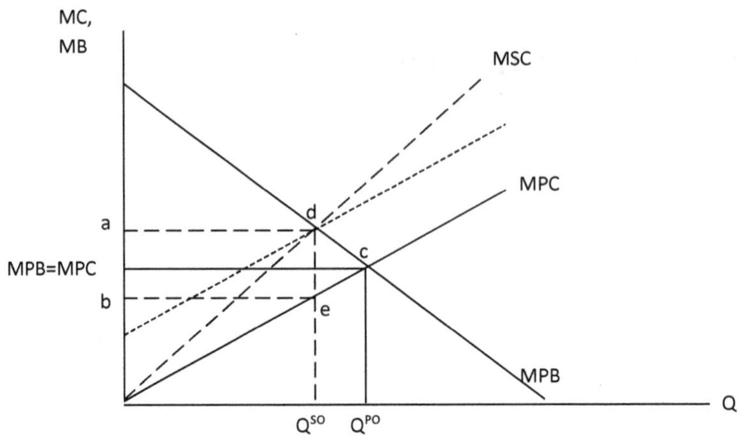

Fig. 6.1 Internalizing external effects of agriculture production

6.2 Methodologies

A framework for assessing the environmental benefits and costs of GMOs is presented in Fig. 6.1. The x-axis depicts the quantity, Q, produced, either for a single crop or a portfolio of crops on a specific plot, farm, or region. The y-axis represents marginal benefits (MBs) and costs (MCs) of producing quantity, Q. The marginal private benefits (MPBs) decrease with an increase in the production of Q while the marginal private costs (MPCs) increase with production: that is, as the production of Q increases, private costs increase and private benefits decrease due to the inverse relationship between benefits and costs. The optimal quantity, Q, produced is at point c, where MPBs equal MPCs, and reflects the optimal level of production for society if no additional benefits or costs need to be considered.

Producing Q may bear additional costs unconsidered under private costs, which if added to MPCs, give us marginal social costs (MSCs), which are greater than MPCs, and the societal optimal level of production, Q, decreases from Q^{PO} to Q^{SO}. Q could be reduced by taxing its production. The optimal tax rate—the Pigouvian Tax—should increase private costs so that MPCs intersect at point d with MPBs. The external effects of production are then internalized, and the polluter pays for the extra environmental damage caused, equivalent to a minus b. Figure 6.1's important message is: despite Q's production causing external environmental damage, reducing its production to zero is suboptimal. A ban is unjustified from a benefit-cost perspective by merely observing that producing an agricultural crop causes environmental impacts.

The Pigouvian view presented on regulating environmental externalities has been criticized, most prominently by Coase (1960), who argued that observing externalities does not necessarily justify government intervention via, for example, a Pigouvian Tax. Stakeholders should have an incentive to reduce environmental

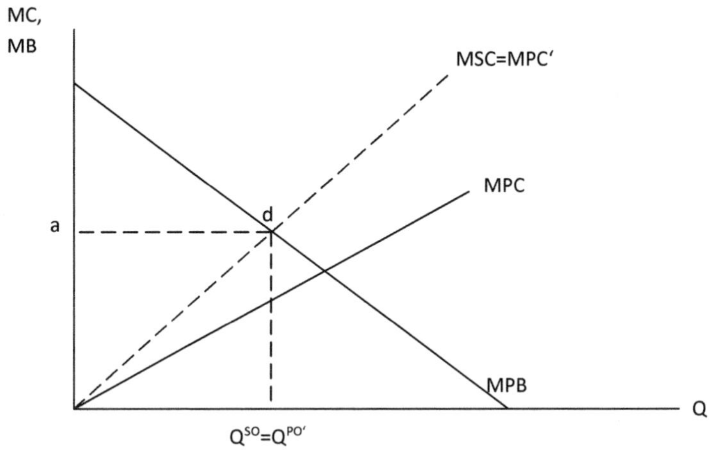

Fig. 6.2 Internalized external effects of agriculture production where MSC=MPC'

pollution. An investigation is necessary to determine if government intervention can improve the current situation of observed environmental pollution; all institutional arrangements available to address the problem should be considered. As a reference, Coase suggests comparing the outcome of alternative institutional arrangements with the existing situation. An intervention is warranted if a different institutional arrangement improves the outcome, as presented in Fig. 6.2, where the MPC has been adjusted by internalizing the external effects of production so that MPCs are equivalent to MSCs, indicated by MSC=MPC'.

Coase's view is challenged by libertarians; the question of government intervention, however, depends on property rights. They argue externalities should be settled by the courts. "We have concluded that everyone should be able to do what he likes, except if he commits an overt act of aggression against the person and property of another. Only this act should be illegal, and it should be prosecutable only in the courts under tort law, with the victim or his heirs and assigns pressing the case against the legal aggressor" (Rothbard 1997, p. 169). While ex post liability can address a number of environmental externalities, this does not per se exclude the use of ex ante regulations, even under the libertarian view, if, for example, "everyone" freely decides to work together in a group to implement regulations imposed on members of the group. Farmers may voluntarily form a group and decide about their own production standards. Further, implementing ex post liability has its own problems (Shleifer 2010). However, Ludlow and Smyth (2011) have argued that innovations are vital to society and hence, the future of an entire innovation should not be decided by the courts, but rather by governments.

However, the libertarian view does not necessarily contradict the situation shown in Fig. 6.2. The expected ex post liability costs increase MPCs. Further, ex post liability costs provide incentives for implementing ex ante measures to reduce

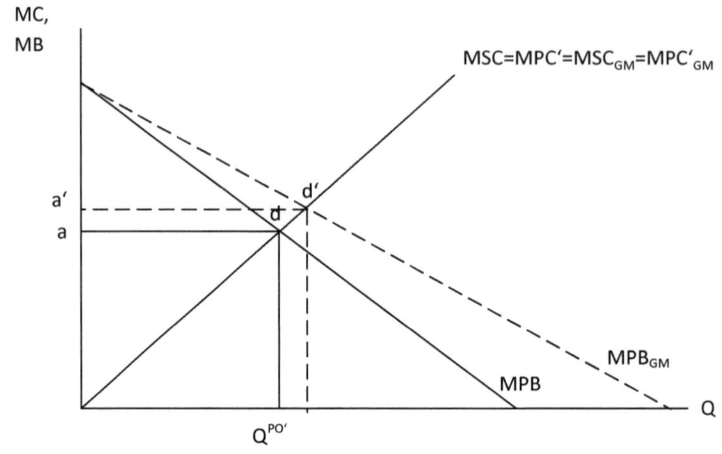

Fig. 6.3 Equilibrium with positive marginal private producer benefits of GMOs and no changes in marginal social costs

ex post liability, hence increasing the MPC compared to a situation where this possibility is absent, as discussed, for example, by Beckmann and Wesseler (2007) for coexistence.

In conclusion, externalities bear additional costs under the Pigouvian, Coasian, and libertarian views; and views on measuring the costs and appropriate responses differ. These views reach the same conclusion: the mere existence of externalities per se does not justify a ban.

Figure 6.3 illustrates the case where the introduction of a GM crop increases MPBs compared to the second best alternative, but the MSC function remains unaffected. An increase in MPB is necessary otherwise no market for the GM crop exists. Here, the quantity of agricultural production and hence MSCs increase; but the overall change in social welfare is always positive.

The situation differs if the introduction of the GM crop changes the MSC function. In Fig. 6.4, the MSC = MPC' function steepens, and it is unclear if the introduction of the GM crop is desirable from a social welfare perspective. Private benefits increase, but so do social costs. Because it is assumed external costs are internalized, private costs also increase. In the end, if $MPB_{GM} - MPC'_{GM} > MPB - MPC'$, farmers will adopt the GM crop. Conversely, if $MPB_{GM} - MPC'_{GM} < MPB - MPC'$, farmers will fail to adopt the GM crop. If regulations are not adopted and the MSC function steepens, more will be produced than is socially optimal.

Figure 6.5 illustrates the case where the MSC curve flattens; the GM crop reduces the environmental impact, and farmers plus other members of society benefit from its introduction. If, in this case, the regulations for crop production remain unadjusted and the MSC function remains unchanged (does not flatten), overall welfare will only be reduced at a production level corresponding to point c.

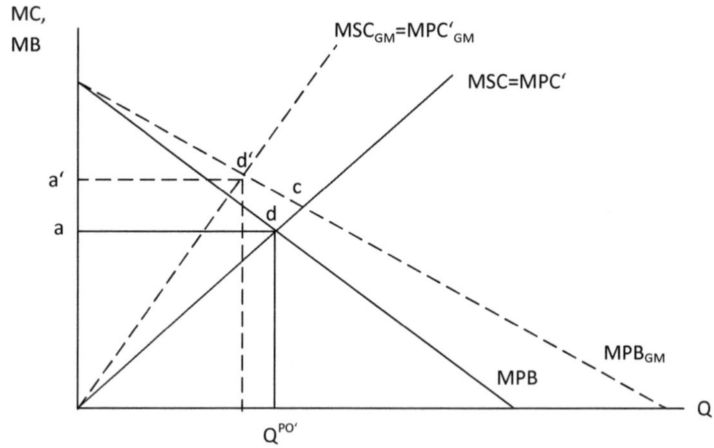

Fig. 6.4 Equilibrium with positive marginal private producer benefits of GMOs and an increase in marginal social costs

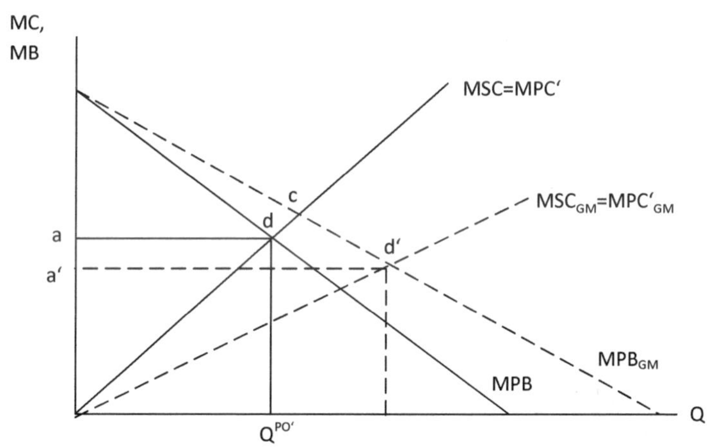

Fig. 6.5 Equilibrium with positive marginal private producer benefits of GMOs and a decrease in marginal social costs

6.2.1 The Precautionary Principle

The previous discussion failed to differentiate between different types of external costs. Environmental impacts may be irreversible and/or catastrophic, which is one of the reasons why the precautionary principle is mentioned in many regulations on GMOs (such as the Cartagena Protocol) or regulations on release within the European Union (EC 2001).

The precautionary principle is diversely interpreted; the most widely held is the prospect that harmful effects of a new technology take precedence over the prospect of beneficial effects. As harmful effects are potentially catastrophic, and this possibility cannot be excluded, and "the infinite costs of a possible catastrophic outcome necessarily outweigh even the slightest probability of its occurrence" (Van den Belt 2003, p. 1123), the result would be a ban of GMOs and all other new technologies (e.g., nanotechnology, cellular telephones). Many disagree with this view, and line of reasoning as pointed out in 2003 by philosopher Henk van den Belt.

For GMO approval, where catastrophic effects cannot be excluded, this interpretation of the precautionary principle is unhelpful. Van den Belt recommends comparing the benefits and costs of possible errors as a guideline for approval, which corresponds with recommendations by leading economists who state: "…regulate until the incremental benefits from regulation are just off-set by the incremental costs. In practice, however, the problem is much more difficult, in large part because of inherent problems in measuring marginal benefits and costs" (Arrow et al. 1996, p. 221).

A method of addressing potential environmental impacts in line with the precautionary principle, and in particular considering uncertainties and irreversible damage, is by performing an extended benefit-cost analysis suggested, amongst others, by Wesseler et al. (2007). They propose modeling the uncertainty of future net benefits using a stochastic process. The economic literature suggests if a policy includes irreversible costs, the net benefits arising from the policy have to be larger than otherwise. The additional net benefits needed to compensate for irreversible costs are calculated by using real-option models.

6.2.2 The Environmental Impact Quotient

Measuring the direct impact on the environment of using GMOs remains unsolved. However, the Environmental Impact Quotient (EIQ) is an indicator, which includes the impact of pesticides on the ecology, farm workers, and consumers (Kovach et al. 1992) and is computed by summing the calculated impacts of a given pesticide on the aforementioned parameters (Nillesen et al. 2006). Some of the EIQ's shortcomings, addressed by Kleter and Kuiper (2005), are that temporal aspects—important for measuring the effect on water reservoirs of a continuous use of glyphosate on herbicide resistant (HR) crops or changes in insecticide use caused by pest resistance—are ignored. Such long-term effects pose problems for environmental risk assessment in general (European Food Safety Authority (EFSA) Panel on Genetically Modified Organisms 2010).

6.3 Critical Assessment

The environmental impacts of GM crops arise at different levels: direct effects are due to changes in input (pesticide, fuel) use and indirect effects are due to changes in yield, land use, and pesticide use.

6.3.1 Yield Increase

Most GM food and feed crops have shown relative increases in yield by reducing deviations from the yield frontier caused by biotic and abiotic stress. Higher yields reduce pressure on land use. Ex post calculations by Brookes and Barfoot (2012) indicate substantial yield increases of GM crops versus non-GM crops. Therefore, GM crops can potentially contribute towards reducing pressure on natural habitats from agricultural land uses—the existence and magnitude of which is yet to be addressed.

6.3.2 Pesticide Use

Pesticides are used to control pests and nontarget plants. A direct benefit of GM crop cultivation is the reduction of pesticide applications, incidentally reducing farmers' exposure to chemicals (Hossain et al. 2004; Huang et al. 2005), and lowering pesticide residues in crops. Also, quantities of chemicals released into the environment are lower, potentially increasing on-farm biodiversity (Nickson 2005). Additionally, the level of mycotoxins in crops can be reduced through genetic engineering (Wu 2006).

Decreased pesticide use is one of the most important direct impacts of GM crops on the environment (Kleter and Kuiper 2005), but how to measure this impact remains unsolved. However, applying the EIQ to HR soybeans indicates an overall positive environmental impact versus non-HR soybeans. Kleter et al. (2007) calculated that for pesticide applications on conventional versus GM oilseed rape in the USA, applications of pesticide-active ingredients; total ecological impact per hectare; ecological impact; and farmer impact were 30, 42, 39, and 54% lower, respectively.

Brookes and Barfoot (2008) found that for conventional and GM crops at the national level, EIQ values decreased by 15.4%, and that for HR canola in North America, the amount of active chemical ingredients applied decreased by 7.9 million kg (12.6%). Smyth et al. (2011a) noted that the results of this study, and hence the net overall benefit, may be biased as they assumed the highest application rate was used in all instances.

Gusta et al. (2011) and Smyth et al. (2011a, b) show the adoption of HR canola has changed weed control practices in Canada, shifting from soil-incorporated to foliar-applied post-emergent herbicides. More than 60% of the respondents reported a simplification of weed management. As a result, the environmental impact of canola production—based on a modified EIQ—dropped by 59% between 1995 and 2006.

There are concerns about the control of volunteer canola (e.g., Ellstrand 2001). According to Smyth et al. (2011b), 8% of the HR canola-growing farmers mentioned volunteer canola as a problem, of which 35% reported that additional effort to control volunteers was needed, thus supporting results by Beckie et al. (2006) and Serecon Management Consulting Inc (2005). Control costs for volunteer canola are less than Canadian $ 3 per hectare (Gusta et al 2011).

The selection pressure caused by glyphosate use on nontarget flora is of concern. Glyphosate resistance was reported for *Amaranthus palmeri* (Gaines et al. 2010). Up until 2007, 13 glyphosate-resistant (GR) weeds were reported worldwide (Service 2007) posing a medium- to long-term threat on HR technology (Bonny 2008). However, Frisvold and Reeves (2010) contend the emergence of weed and pest resistance could be addressed by appropriate crop management strategies.

Since the late 1990s, there has been a two to sevenfold suppression of the cotton bollworm population in areas where Bt cotton was introduced in China (Wu et al. 2008), and less damage to non-Bt cotton and other affected crops, plus a host-preference change for the cotton bollworm (Jongsma et al. 2010). Kuosmanen et al. (2006) report that Chinese farmers growing Bt cotton may continue using insecticides, as is the case in India, because of inferior seed quality (Herring 2009), or the problem of secondary pests (Pemsl et al. 2008; Lu et al. 2010). The implications of target-pest suppression for resistance and pest management are poorly understood and need further investigation (Jongsma et al. 2010; Carrière et al. 2003).

Gusta et al. (2011) quantified spillover benefits of HR canola cultivation in Canada on herbicide applications of the succeeding crop to be worth Canadian $ 37 per hectare (54 % of the respondents reported a benefit) in 2007.

6.3.3 Fertilizer Use

Fertilizers are used to increase both crop yield and quality by providing plants with additional nutrients. Quantitatively, the most important fertilizers are: nitrogen (N); phosphorus (P); and potassium (K).

The contribution of biotechnology to improvements in N use efficiency (NUE; in agronomic terms, the ratio of crop yield to the N fertilizer supplied (Moose and Below 2008)) is exclusively indirect via yield-improving traits (pest and/or herbicide resistance); for example, reduced damage to the root system of GM corn resistant to corn rootworm can lead to greater N uptake. Trigo and Cap (2003) note that GR soybean adoption has increased the area under no-till practices exponentially in Argentina, with a potential positive effect on NUE (Rao and Dao 1996). Conversely, the adoption of GR soybeans increases the use of glyphosate, which is toxic to the N-fixing symbiont *Bradyrhizobium japonicum*—important for supplying soybeans with N; Zablotowicz and Reddy (2007) found slight negative effects in N fixation at label use rates, but a consistent reduction at excess rates of glyphosate use.

Concerns exist about the impacts of GM crops on soil microbes and hence nutrient cycling, but empirical evidence is lacking (Motavalli et al. 2004; Al-Deeb et al. 2003; Saxena and Stotzky 2001).

From the above, it is inconclusive that yield gains due to genetic modification directly result in NUE improvements. Studies investigating effects of GM crops consider biotechnology to be neutral in terms of fertilizer use (see, e.g., Qaim and Traxler 2005).

6.3.4 Environmental Safety Issues

The transfer of pest- and herbicide-resistant traits to weedy species and the persistence of feral crop plants carrying these traits raise issues about their impacts on the environment. Environmental safety issues focus on the effects—direct or indirect—of GM crops on nontarget organisms and the transfer of GM traits to populations of wild plants (Food and Agriculture Organization 2003).

Gene flow via pollen transfers GM traits from crops to wild relatives and non-GM crops. Seed escaping during harvest, transportation, or processing contributes to this process (Dunwell and Ford 2005). If the GM trait confers a selective advantage over wild plants, then persistence and introgression of this trait into wild or weedy populations is more likely (Jenczewski et al. 2003). If the trait confers a physiological disadvantage, the transgene will be competed out of the population by selection pressure. Therefore, GM traits have the potential to cause different environmental and agronomic impacts (Dunwell and Ford 2005).

Movement of herbicide-tolerance traits into wild populations will only confer an advantage where herbicides are applied. The physiological effort of sustaining the trait in the absence of herbicide selection may be costly in the longer term (Snow et al. 1999; Gueritaine et al. 2002) resulting in selection against these plants. Essentially, processes of intrapopulation genetic drift determine the fate of traits that confer no benefit (Pilson and Prendeville 2004).

The movement of GM traits into wild plant populations has the potential to reduce the number and/or diversity of wild plants, thereby altering the ecological structure of communities. Wild relatives may become extinct as a result of swamping by competitive plants and repeated hybridization. Traits like drought and salt tolerance may enable plants to invade new habitats leading to unwanted ecological change (Dunwell and Ford 2005). However, Van de Wiel et al. (2005) point out the high variations in the results of gene flow studies make it difficult to gain a consistent view about their implications for the environment. It seems regional aspects are important in quantifying the magnitude of gene flow (Scatasta 2005).

6.3.5 Tillage and Greenhouse Gas Emission Effects

Estimates show greenhouse gas (GHG) emissions from agriculture contribute to 15 % of annual emissions globally. Interestingly, adopting GM crops can contribute towards reducing these emissions (Hassan 2005) via an increase in reduced- or zero-tillage systems (Frisvold et al. 2009; Ward et al. 2002).

In the USA, estimated savings for diesel use under reduced-tillage systems were 37 L/ha (Griffith and Parsons 1980), and 1.43 L/ha for HR sugar beets in Europe from savings in pesticide applications (Demont et al. 2004).

Fields planted with HR crops require less tillage between crop plantings to manage weeds, possibly contributing to reduced soil erosion (Fawcett and Towery 2003; Nickson 2005). In the USA, no-tillage and conservation-tillage reduced soil erosion by up to 90% (Baker and Johnson 1979). Reduced-tillage contributes to increases in soil organic matter, and improvements in soil structure and water-holding capacity (Karlen and Sharpley 1994), and pesticide runoff can be reduced significantly (Baker 1990; Waibel and Fleischer 1998). Improved water-holding capacity reduces soil-nutrient losses (Blevins et al. 1983; Karlen 1995), and improvements in above- and below-ground water quality have been reported because of reduced N emissions (Wheatley et al. 1995). A survey among HR canola-growing farmers in Canada (Smyth et al. 2011a) confirms higher soil moisture content and fewer erosion problems.

Reduced-tillage positively affects the biodiversity of soil microorganisms and above-ground fauna (see, e.g., Castrale 1985; Best 1985; Basore et al. 1986), and populations of small mammals (Basore et al. 1986) and benthic invertebrate communities (Barton and Farmer 1997). While reduced-tillage provides important environmental benefits, these might be reduced through the use of glyphosate or other broad-spectrum herbicides in reduced-tillage systems. Nevertheless, the US Environmental Protection Agency considers glyphosate to have a minimal toxicity towards mammals, fish, and invertebrates (United States Environmental Protection Agency (US-EPA) 1986).

Phillips (2003), Beckie et al. (2006) and Kleter et al. (2007) found correlations between adopting HR oilseed rape and zero-tillage systems. According to Smyth et al. (2011a), tillage operations among farmers growing HR canola in Canada dropped by more than 70% (2.73 down to 0.74 passes).

Noteworthy is the contribution HR crops have made to the adoption of reduced- and zero-tillage systems (Frisvold et al. 2009), and the indirect overall net environmental benefits of the change in tillage systems induced by their adoption.

6.4 International Arena

Introducing GMOs into a country has been perceived to be a threat to biodiversity, human health, and the economy of rural communities, hence threatening sustainable development. In parallel, scientists, especially in developing countries, fear being bypassed by the new technology (Wesseler 2010).

The Cartagena Protocol on Biosafety is an international agreement protecting biological diversity from the potential risks posed by GMOs. Countries may ban GMO imports if they feel there is insufficient scientific evidence on the product's safety, and exporters are required to label shipments containing GMOs (Secretariat of the Convention on Biological Diversity 2000).

The World Trade Organization (WTO) Agreement on the Application of Sanitary and Phytosanitary Measures allows its members to impose trade measures protecting human health, animal or plant life, and health. For food products, the WTO uses the Codex Alimentarius Commission's scientifically based standards and definitions for risk assessments (Hobbs 2010).

Consumers and/or environmentalists have sought protection from GMOs, especially in the EU where, in 2003, the GMO approval process resulted in separate

WTO complaints by Argentina, Canada, and the USA. The WTO ruled in 2006 that the EU's GMO policies from 1984 to 2004 were, in effect, a ban on GMOs and illegal. Restricted market access remains a concern despite continued negotiations; GMOs have nevertheless been approved for import since 2004 (Wesseler 2010).

6.5 Administrative Consequences

Empirical evidence indicates substantial environmental benefits being provided by the cultivation of GM crops, especially reductions in GHG emissions and pesticide use. Documenting these effects should continue and be institutionalized to maintain current information.

Relatively little information about yield increases on land conservation is available, and should be addressed at the international level. To date, information on the impact of yield increases on land use has been poorly investigated.

Problems related to pest and weed resistance have recently emerged. Resistance problems can be reduced, if not avoided, by appropriate site-specific management plans (requiring monitoring of pest populations and their properties), the costs of which need to be balanced against their benefits; local farming conditions, and the economic environment also need to be considered.

Administrative bodies involved in the approval process of GMOs should address potential risks, and weigh those risks against the potential benefit/s of the specific GMOs under consideration. The benefits of adopting GM crops, especially on the environment are often ignored. As agricultural production always impacts the environment, an environmental impact assessment should be done, comparing the GMO under consideration with the second best alternative.

6.6 Summary/Synthesis

- Substantial positive environmental impacts of GM crop cultivation have been reported, especially on reductions in pesticide use and emissions of GHGs, and do not justify a ban on environmental grounds.
- Bans and delays in approval of herbicide and pest-resistant traits result in unnecessary damage to the environment.
- Pest resistance has emerged and needs to be addressed to maintain the environmental benefits already generated by GM crops.
- All agricultural activities have environmental impacts—environmental impact assessments therefore need to be done to draw comparisons with the second best alternative.
- A comprehensive environmental impact assessment should account for reversible and irreversible benefits and costs, and differentiate between private and external benefits and costs in relation to the existing regulatory environment.

References

Al-Deeb MA, Wilde GE, Blair JM et al (2003) Effect of Bt corn for corn rootworm control on non-target soil micro arthropods and nematodes. Environ Entomol 32(4):859–865

Arrow KJ, Cropper ML, Eads GC et al (1996) Is there a role for benefit-cost analysis in environmental, health, and safety regulation? Science 272:221–222

Baker DB (1990) Herbicide concentrations in Ohio's drinking water supplies: a quantitative assessment. In: Weigmann DL (ed) Pesticides in the next decade: the challenges ahead. Virginia Water Resources Center and VPI, Blacksburg

Baker JL, Johnson HP (1979) The effect of tillage system on pesticides in runoff from small watersheds. Trans Am Soc Agric Eng 22:554–559

Barton DR, Farmer MED (1997) The effects of conservation tillage practices on benthic invertebrate communities in headwater streams in southwestern Ontario, Canada. Environ Pollut 96(2):207–215

Basore NS, Best LB, Wooley JB Jr (1986) Bird nesting in Iowa no-tillage and tilled cropland. J Wildl Manag 50:19–28

Beckie HJ, Harker KN, Hall LM et al (2006) A decade of herbicide-resistant crops in Canada. Can J Plant Sci 86(4):1243–1264

Beckmann V, Wesseler J (2007) Spatial dimension of externalities and the coase theorem: implications for coexistence of transgenic crops. In: Heijman W (ed) Regional externalities. Springer, Berlin, pp 215–234

Best LB (1985) Impacts of tillage practices on wildlife habitat and populations. In: Farming for a better environment. Soil and Water Conservation Society, Ankeny

Blevins RL, Thomas GW, Smith MS et al (1983) Changes in soil properties after 10 years of continuous no-tilled and conventionally tilled corn. Soil Tillage Res 3:135–146

Bonny S (2008) Genetically modified glyphosate-tolerant soybean in the USA: adoption factors, impacts and prospects. A review. Agron Sustain Dev 28(1):21–32

Brookes G, Barfoot P (2008) Global impact of biotech crops: Socio-economic and environmental effects, 1996–2006. AgBioForum 11(1):21–38

Brookes G, Barfoot P (2012) GM crops: global socio-economic and environmental impacts 1996–2010. PG Economics Ltd, Dorchester

Carrière Y, Ellers-Kirk C, Sisterson M et al (2003) Long-term regional suppression of pink bollworm by *Bacillus thuringiensis* cotton. Proc Natl Acad Sci 100(4):1519–1523

Castrale JS (1985) Responses of wildlife to various tillage conditions. Trans North Am Wildl Nat Resour Conf 50:142–156

Coase R (1960) The problem of social cost. J Law Econ 3:1–44

Demont M, Wesseler J, Tollens E (2004) Biodiversity versus transgenic sugar beets—the one Euro question. Eur Rev Agric Econ 31(1):1–18

Dunwell JM, Ford CS (2005) Technologies for biological containment of GM and non-GM crops. Defra Contract CPEC 47. DEFRA, London

EC (2001) Directive 2001/18/EC of the European Parliament and of the Council of March 2001 on the deliberate release into the environment of genetically modified organisms and repealing Council Directive 90/220/EC. Off J Eur Communities L 106:1–38 (17.4.2001)

Ellstrand NC (2001) When transgenes wander, should we worry? Plant Physiol 125:1543–1545

EU Council for the Environment (1999) 2194th Council meeting, Luxembourg, 24/25 June 1999. C/99/203/. http://europa.eu/rapid/pressReleasesAction.do?reference=PRES/99/203&format=HTML&aged=0&lg=fi&guiLanguage=en. Accessed 27 Nov 2012

European Food Safety Authority (EFSA) Panel on Genetically Modified Organisms (2010) Guidance on the environmental risk assessment of genetically modified plants. EFSA J 8(11):1879

Fawcett R, Towery D (2003) Conservation tillage and plant biotechnology: how new technologies can improve the environment by reducing the need to plow. Purdue University: Conservation Technology Information Center (CTIC), West Lafayette

Food and Agriculture Organization (2003) Report of the FAO expert consultation on the environmental effects of genetically modified crops 16–18 June 2003. Food and Agriculture Organization of the United Nations, Rome

Frisvold GB, Reeves JM (2010) Resistance management and sustainable use of agricultural biotechnology. AgBioForum 13(4):343–359

Frisvold GB, Boor A, Reeves JM (2009) Simultaneous diffusion of herbicide resistant cotton and conservation tillage. AgBioForum 12(3–4):249–257

Gaines TA, Zhang W, Wang D et al (2010) Gene amplification confers glyphosate resistance in *Amaranthus palmeri*. Proc Natl Acad Sci 107(3):1029–1034

Griffith DR, Parsons SD (1980) Energy requirements for various tillage—planting systems. NCR Extension Publication NCR-202-W. Purdue University, West Lafayette

Gueritaine G, Sester M, Eber F et al (2002) Fitness of backcross six hybrids of hybrids between transgenic oilseed rape (*Brassica napus*) and wild radish (*Raphanus raphanistrum*). Mol Ecol 11:1419–1426

Gusta M, Smyth SJ, Belcher K et al (2011) Economic benefits of genetically modified herbicide tolerant canola for producers. AgBioForum 14(1):1–13

Hassan R (2005) Millennium ecosystem assessment. In: Ecosystems and human well-being, Vol. 1 Current state and trends. Island Press, Washington DC

Herring R (2009) Persistent narratives: why is the "Failure of Bt Cotton in India" story still with Us? AgBioForum 12(1):14–22

Hobbs JE (2010) Public and private standards for food safety and quality: international trade implications. http://www.esteyjournal.com/j_pdfs/hobbs11-1.pdf. Accessed 10 Jan 2013

Hossain F, Pray CE, Lu Y et al (2004) Genetically modified cotton and farmers' health in China. Int J Occup Environ Health 10:296–303

Huang J, Hu R, Rozelle S et al (2005) Insect-resistance GM rice in farmers' fields: assessing productivity and health effects in China. Science 308:688–690

Jenczewski E, Ronfort J, Chevre AM (2003) Crop-to-wild gene flow, introgression and possible fitness effects of transgenes. Environ Biosaf Res 2:9–24

Jongsma MA, Gould F, Legros M et al (2010) Insect oviposition behavior affects the evolution of adaptation to Bt crops: consequences for refuge policies. Evol Ecol 24(5):1017–1030

Karlen DL (1995) Tillage effects on soil fertility. In: Farming for a better environment. Soil and Water Conservation Society, Ankeny

Karlen DL, Sharpley AN (1994) Management strategies for sustainable soil fertility. In: Hatfield JL, Karlen DL (eds) Sustainable agricultural systems. Lewis Publishing, Boca Raton

Kleter G, Kuiper H (2005) Assessing the environmental impact of changes in pesticide use on transgenic crops. In: Wesseler J (ed) Environmental costs and benefits of transgenic crops. Wageningen UR Frontis Series Vol. 7. Springer, Dordrecht

Kleter GA, Bhula R, Bodnaruk K et al (2007) Altered pesticide use on transgenic crops and the associated general impact from an environmental perspective. Pest Manag Sci 63:1107–1115

Kovach J, Petzoldt C, Degni J et al (1992) A method to measure the environmental impact of pesticides. New York's Food and Life Sciences Bulletin no 139. Cornell University, Ithaca

Kuosmanen T, Pemsl D, Wesseler J (2006) Specification and estimation of production functions involving damage control inputs: a two-stage, semi-parametric approach. Am J Agric Econ 88(2):499–511

Lu Y, Wu K, Jiang Y et al (2010) Mirid bug outbreaks in multiple crops correlated with wide-scale adoption of Bt cotton in China. Science 328:1151–1153

Ludlow K, Smyth SJ (2011) The quandary of agricultural biotechnology, pure economic loss and non-adopters: comparing Australia, Canada, and the United States. Jurimetrics 52:7–41

Moose S, Below FE (2008) Biotechnology approaches to improving maize nitrogen use efficiency. In: Kritz AL, Larkins BA (eds) Biotechnology in agriculture and forestry: molecular genetic approaches to maize improvement. Springer, Berlin

Motavalli PP, Kremer RJ, Fang M et al (2004) Impact of genetically modified crops and their management on soil microbially mediated plant nutrient transformations. J Environ Qual 33(3):816–824

Nickson TE (2005) Crop biotechnology—the state of play. In: Poppy GM, Wilkinson MJ (eds) Gene flow from GM plants. Blackwell, Oxford

Nillesen E, Scatasta S, Wesseler J (2006) Do environmental impacts differ for Bt, Ht and conventional corn with respect to pesticide use in Europe? An empirical assessment using the Environmental Impact Quotient. IOBC/wprs Bull 29(5):109–118

Pemsl D, Gutierrez AP, Waibel H (2008) The economics of biotechnology under ecosystem disruption. Ecol Econ 66:177–183

Phillips PWB (2003) The economic impact of herbicide tolerant canola in Canada. In: Kalaitzandonakes N (ed) The economic and environmental impacts of agbiotech: a global perspective. Kleuwer Academic Publishers, New York

Pilson D, Prendeville HR (2004) Ecological effects of transgenic crops and the escape of transgenes into wild populations. Annu Rev Ecol Evol Syst 35:149–174

Qaim M, Traxler G (2005) Roundup ready soybeans in Argentina: farm level and aggregate welfare effects. Agric Econ 32:73–86

Rao SC, Dao TH (1996) Nitrogen placement and tillage effects on dry matter and nitrogen accumulation and redistribution in winter wheat. Agron J 88:365–371

Rothbard M (1997) The logic of action two. Edward Elgar, Cheltenham

Saxena D, Stotzky G (2001) *Bacillus thuringiensis* (Bt) toxin released from root exudates and biomass of Bt corn has no apparent effect on earthworms, nematodes, protozoa, bacteria and fungi in soil. Soil Biol Biochem 33:1225–1230

Scatasta S (2005) Comment on Van de Wiel, Groot and Den Nijs: gene flow from crop to wild plants and its population-ecological consequences in the context of GM-crop biosafety, including some recent experiences from lettuce. In: Wesseler J (ed) Environmental costs and benefits of transgenic crops. Wageningen UR Frontis Series Vol. 7. Springer, Dordrecht

Secretariat of the Convention on Biological Diversity (2000) Cartagena Protocol on Biosafety to the Convention on Biological Diversity: text and annexes. Secretariat of the Convention on Biological Diversity, Montreal

Serecon Management Consulting Inc (2005) Herbicide tolerant volunteer canola management in subsequent crops. Canola Council of Canada (CCC), Winnipeg

Service RF (2007) A growing threat down on the farm. Science 316:114–117

Shleifer A (2010) Efficient regulation. In: Kessler DP (ed) Regulation vs. litigation: perspectives from economics and law. University of Chicago Press, Chicago. http://www.nber.org/books/kess09-1. Accessed 27 Nov 2012

Smyth SJ, Gusta M, Belcher K et al (2011a) Environmental impacts from herbicide tolerant canola production in western Canada. Agric Syst 104:403–410

Smyth SJ, Gusta M, Belcher K et al (2011b) Changes in herbicide use following the adoption of herbicide tolerant canola in western Canada. Weed Technol 25:492–500

Snow AA, Andersen B, Jorgensen RB (1999) Costs of transgenic herbicide resistance introgressed from *Brassica napus* into weedy *B. rapa*. Mol Ecol 8:605–615

Trigo EJ, Cap EJ (2003) The impact of the introduction of transgenic crops in Argentinean agriculture. AgBioForum 6:87–94

United States Environmental Protection Agency (US-EPA) (1986) Pesticide Fact Sheet: Glyphosate. No. 173. EPA 540/FS-88-124. Office of Pesticide Programs, Washington DC

Van de Wiel C, Groot M, den Nijs H (2005) Gene flow from crop to wild plants and its population—ecological consequences in the context of GM-crop biosafety, including some recent experiences from lettuce. In: Wesseler J (ed) Environmental costs and benefits of transgenic crops. Wageningen UR Frontis Series Vol. 7, Springer, Dordrecht

Van den Belt H (2003) Debating the precautionary principle: "Guilty until Proven Innocent" or "Innocent until Proven Guilty"? Plant Physiol 132:1122–1126

Waibel H, Fleischer G (1998) Kosten und Nutzen des chemischen Pflanzenschutzes in der deutschen Landwirtschaft aus gesamtwirtschaftlicher Sicht. Vauk, Kiel

Ward C, Flanders A, Isengildina O et al (2002) Efficiency of alternative technologies and cultural practices for cotton in Georgia. AgBioForum 5(1):10–13

Wesseler J (2010) Genetically modified organisms (GMOs) and the agriculture and food policy. In: Oskam A, Meesters G, Silvis H (eds) EU Policy for agriculture, food and rural areas. Wageningen Academic Publishers, Wageningen

Wesseler J, Scatasta S, Nillesen E (2007) The maximum incremental social tolerable irreversible costs (MISTICs) and other benefits and costs of introducing transgenic maize in the EU-15. Pedobiologia 51(3):261–269

Wheatley DM, Macleod DA, Jessop RS (1995) Influence of tillage treatments on N_2 fixation of soybean. Soil Biol Biochem 27(4/5):571–574

Wu F (2006) Mycotoxin reduction in Bt corn: potential economic, health, and regulatory impacts. Transgenic Res 15:277–289

Wu K, Lu Y, Feng H et al (2008) Suppressing of cotton bollworm in multiple crops in China in areas with Bt toxin-containing cotton. Science 321:1676–1678

Zablotowicz RM, Reddy KN (2007) Nitrogenase activity, nitrogen content, and yield responses to glyphosate in glyphosate-resistant soybean. Crop Prot 26(3):370–376

[illegible]

Chapter 7
Ethics and Equity

Paul B. Thompson

7.1 Introduction

Perspectives and views regarding rights are often developed as an attempt to specify the basic freedoms or capabilities that are necessary for human flourishing. Although views on rights acknowledge that natural scarcities constrain the potential for flourishing, they insist that when the action of one human limits the potential of another, this is *the* paradigmatic case calling for ethical critique. Rights are intended to protect human beings from oppression by other human beings; they are not to be understood as entitlements against the natural world. An alternative starting point is that of "values." In either case, what is needed is an articulation of the ethical theories under the identified rights or values.

In a broad sense, ethics is simply "doing the right thing." To consider criteria for acting ethically is to ask "How should we act?" In principle, *all* considerations relevant to answering this question are of potential relevance to any assessment of ethics. *Any* element in a technology assessment is therefore also a component in ethical assessment, given this broad definition. For practical purposes, however, criteria become categorized as "ethical" as distinct from environmental, economic, or social criteria for one of three reasons.

First, the criteria may concern domains of activity or types of impact that have been omitted from an assessment focused on environmental, safety, or economic outcomes. Outcomes that do not affect key decision-makers (or their client groups) directly are frequently and sometimes willfully overlooked. When someone classifies an omitted element as having ethical significance, it can simply be a way of asserting that the assessment or decision procedure *should have* taken it into consideration.

P. B. Thompson (✉)
Department of Philosophy, Department of Community Sustainability and
Department of Agricultural, Food and Resource Economics, Michigan State University,
East Lansing, MI 48824 USA
e-mail: thomp@anr.msu.edu

K. Ludlow et al. (eds.), *Socio-Economic Considerations in Biotechnology Regulation*,
Natural Resource Management and Policy 37, DOI 10.1007/978-1-4614-9440-9_7,

Second, some criteria for evaluating an emerging technology are routine while others are contested or controversial. The fact that people disagree about the relevance or significance of a given feature is liable to result in that feature being relegated to the category of ethics. In general, criteria that bear directly on human health are noncontroversial. However, they do involve a value judgment. Nature places no particular value on the survival or flourishing of human beings, so the fact that emphasis on health-related outcomes is utterly uncontroversial should not be viewed as a "value-free" assessment. In contrast, there is no unanimity about claims that an environmental impact is "intrinsically valuable" (meaning that it has significance apart from its impact on human beings), or claims about the spiritual value of an impact. There are also deep and longstanding controversies about when and whether impact on an individual or group's economic well-being should be reflected in a technology assessment. In each of these cases, categorizing an impact or outcome as "ethical" may be a way of acknowledging that there is controversy and lack of agreement over the value that should be attributed to it.

Finally, procedural characteristics of the assessment or decision process itself are likely to be characterized as ethical. Who is involved in making judgments about which outcomes are to be included, or how is the relative value of outcomes to be weighted? When there is controversy about the value that should be attributed to an outcome, there is likely to be controversy about the way that an assessment process is structured and carried out. This is especially the case when a process for evaluating action or policy can be shown to have systematic or characteristic weaknesses and omissions. Hence, inattention to ethical dimensions (as defined in the first two senses) may support the judgment that the procedure itself is ethically deficient. A structural bias in the assessment procedure toward outcomes that can be measured using biological or medical methods may be categorized as an ethical deficiency in the assessment procedure.

Although a focus on these three domains of ethics narrows the scope for this chapter considerably, the range of potential topics for inclusion under the heading of ethics in the case of agricultural biotechnology is still broad. This chapter will emphasize criteria that have been characterized in terms of fairness, equity, and equality.

7.2 Methodologies

Neoliberal thought arguably blends elements of utilitarian and libertarian ethics with the theory of economic growth. Resistance to neoliberalism plausibly relies upon ethical critiques of utilitarianism and libertarianism that will be summarized below. These critiques have circulated for over 200 years and there is every reason to suspect that such ideas motivate many opponents of genetic modification, as well. There is, however, very little empirical work that links resistance and philosophical ideals of any kind. What is more, sociological analyses of neoliberalism and globalization do not cite philosophical discussions of the role that utilitarianism and libertarianism might play in the development and evaluation of GM crops. As such, the intriguing possibility of linking these sociological and philosophical analyses remains undeveloped.

Fairness or fair play is a commonsense notion that may have a basis in the fundamental way that human beings experience interpersonal relationships. Research in moral psychology indicates that children develop an innate sense of fairness very early in their cognitive development (Fehr and Fischbacher 2004). Theoretical characterizations of fairness that associate the idea with equity are often grounded in the tradition of economic and policy analysis known as social choice. Here, society is viewed *as if* the relative wealth and well-being of all persons, their respective relationships of power, interdependence, and vulnerability, and also their freedom and capabilities for future action were the result of a decision or choice. Doing so motivates a way of evaluating factors that give rise to alternative states of affairs in a manner analogous to the way that an individual would evaluate a decision to pursue one course of action rather than another (Sen 2009).

The social choice perspective can be deployed both as an abstract exercise in the comparative evaluation of alternative states of society, as well as in a more practical assessment of specific actions, initiatives, or policies that will alter the status quo in more or less predictable ways. In the latter case, an analysis of the outcomes that an initiative or policy is expected to produce can be utilized as a decision aide for real-world decision-makers, including managers, administrative officials, legislators, and in some cases, judges. Social choice is a tradition of normative (or ethical) analysis closely associated with the utilitarian philosophy articulated by Jeremy Bentham and John Stuart Mill, who argued that the utilitarian maxim—act so as to produce the greatest good for the greatest number—provides a general rule to be followed in selecting which action, initiative, or policy should be undertaken in any given situation.

For public officials contemplating a policy or a piece of legislation, the utilitarian approach to social choice requires first an assessment of all effects or impacts—the total utility—that the action (as well as its alternatives) has on individuals within society. Bentham characterized this assessment as an inquiry into the pain and pleasure that the action will produce. Mill, and most subsequent utilitarians, have understood utility in terms of whether a given individual is more or less satisfied in each of the possible states of society under consideration, but have followed Bentham in insisting that the utility experienced by all parties must be included in the assessment of competing alternatives. This formulation of social choice has influenced welfare economics, where theorists have proposed methods to characterize the satisfaction of preferences in the language of benefit and cost, and have proposed a succession of principles to more carefully express the intent of the utilitarian maxim. The Kaldor-Hicks rule—total benefits should be large enough to compensate for total costs—has been an especially influential rule (Thompson et al 1994).

Classical economic theory argued that no reallocation of resources could be made without making someone worse off. A nineteenth century economist, Vilfredo Pareto, advocated that resources could be reallocated such that at least one individual would be better off, without making anyone worse off. This has come to be known as a Pareto improvement. When there are no further allocations possible without making one individual worse off, this is known to be Pareto efficient. GM-crop commercialization that adversely affects nonadopters would establish that nonadopters have been made worse off because of negative externalities, such as lower prices for comingled commodities. The Kaldor-Hicks criteria, however, can address this situation. In an attempt to propose a means of allowing innovation that would result in the nonadopters not being made worse off, Kaldor and Hicks proposed an improvement that would be Pareto efficient, provided two criteria were followed. The Kaldor-Hicks criteria hold that: first, if the "winners" of an innovation are able to compensate the "losers" then the innovation is a Pareto improvement; and second, if the "losers" are unable to bribe the "winners" to prevent the commercialization of the innovation it is also an improvement. Provided the compensation is at least equal to the losses suffered by the nonadopter, they are, in terms of Pareto efficiency, no worse off.

Whether one focuses on Bentham's original idea of maximizing social utility or one adopts the more feasible (but also less ambitious) Kaldor-Hicks criteria, this family of approaches in social choice has always been seen to have an ethical weakness in its apparent insensitivity to the distribution of costs and benefits. By focusing on the total or net utility for all affected parties, utilitarian decision rules allocate no special significance to outcomes that are experienced by a particular subgroup. Bentham and Mill *intended* utilitarian social choice theory as a critique of social policies that tended to benefit a small subgroup consisting of the wealthy elite. But so long as net utility is positive, there is nothing in the utilitarian tradition that prevents a persistent pattern of "winners"—people who gain benefits—at the expense of "losers"—people who bear costs. Thus, the general problem of distributive justice can be conceptualized as one of articulating a principle or criterion that would either constrain the set of options available to a decision-maker to those that can be considered "fair" or that would at least provide a basis for comparing alternatives in terms of their distributive impact.

Perhaps the most straightforward way to do this is to stipulate that benefits and costs should be shared equally: no one person or subgroup should reap a disproportionate share of benefit or bear a disproportionate share of costs. This is *the principle of equality*. Equality and the feeling of injustice that arise when distributions are unequal may have a basis in the cognitive structure of human experience. There may be a root ability to perceive a distribution of rewards as "equal" or "unequal" and to associate fairness or unfairness with this perception that develops relatively early in childhood and transcends cultural variation. If so, there is a profound sense in which the principle of equality is a primitive building block of cognition. There is, however, little doubt that the adult human ability to recognize a given distribution as equal or unequal in the ethical relevant sense is subject to numerous cultural, historical, and experiential modifications. Thus, for example, adult males may feel fully justified in claiming the largest share of a household's food supply in a given social context, and other household members may be fully reconciled to this pattern. In a different social context, children may be favored. Such differences may be supported by extended belief systems, say that adult males need strength and energy for hunting or other physical work, in the first case, or that the unique developmental needs of children take precedence, in the latter case.

Economists have developed a variety of tests to measure differential impacts on winners and losers, and even to estimate the degree to which these impacts can be attributed to economic growth or to specific technological innovations (Acemoglu 2000; Barro 2000). Many of these tests attempt to determine the value of "losses" to the "losers." The redistribution of wealth has taxed many a society and government. There is a variety of wealth redistribution programs and mechanisms and according to Alston et al. (1998), many economists assumed that the cost of wealth redistribution, regardless of the economic sector, was equal to the revenue that was redirected. These original assumptions held that the redistribution of money by government was costless. This is obviously not the case and research provides estimates on the marginal social welfare cost that ranges from US$ 1.20 to 1.50 for every dollar distributed (Alston and Hurd 1990). In a review of this literature, Fullerton (1991) reconciled the results and suggests that the marginal cost of taxation (in the USA) is considerably lower than first thought, ranging from US$ 1.07 to 1.25. In essence, the redistribution of every $ 1 of government revenue will cost between 7 and 25 %, thus reducing the amount of revenue that is available to be redistributed.

These tests offer insight into the sense in which any given cluster of technological innovations—in the present case new seeds or other input technologies—can be viewed with respect to distributive impact. However, applying these tests to support an ethical judgment about the equity of an emerging technology presupposes the resolution of key interpretive ambiguities that reside within the social choice approach. In important respects, it is the *process* of arriving at a consistent conceptual framework for conceptualizing the impacts that lies at the heart of ethics.

One important pattern emphasizes the idea that one's share of the rewards from a productive activity should be proportional to one's contribution to the productive process. Under this interpretation, an unequal share of rewards would occur when one person who has contributed relatively little to the production process receives more than someone who has contributed a great deal. The most dramatic instance of inequality involves the failure to recognize a given contribution entirely. Critics of GM crops have argued that biotechnology companies fail to recognize the contribution that farmers have made to the germplasm of crops they sell. The criticism extends a longstanding claim against breeders in the developed world, who collect seeds often from poor farmers in developing countries, and then utilize the germplasm to develop economically valuable varieties (Juma 1989). Historically, farmers, who may have made extensive contributions to the quality of the germplasm over centuries of trial-and-error seed selection, have been excluded from any share in the economic benefit or consequential intellectual property.

Unequal power represents an important shift in the way that a principle of equality is interpreted. Equality is no longer being interpreted in terms of whether the benefits are being shared equitably. The key idea here is *procedural equality*, or what is often referred to as "equality under the law." This notion of equality has a lengthy history, developing in the European political tradition. It descends from doctrines such as Magna Carta, by which absolute monarchs recognized a domain of law in which their authority would be limited. At first all nobles, then later all citizens, were said to be equally subject to the authority and procedures of law. Rank or status would accord no favorable treatment when questions of law were at hand. The idea of procedural equality then enjoyed a considerable expansion with the development of neoclassical economic theory, which postulated conditions in which all parties freely agreed to have transactions under conditions of complete information.

Economists postulated these conditions as prerequisites of rational exchange. Transactions in which one party is *not* free to say yes or no or in which one party has information that the other lacks do not qualify as instances of rational choice. But "rationality" in this economic sense conforms quite closely to the conditions for an ethically justifiable or defensible exchange. A transaction or pattern of transactions in which a weaker party is forced to accept terms reluctantly exposes unequal power relationships. There is thus an important sense in which the rationality and hence the legitimacy of economic exchanges depends on whether parties are truly free to accept or decline the terms of trade. Poverty can produce procedural inequalities in that poor people may be forced to accept terms of trade that better-off trading partners would regard as unacceptable (Sen 1999). In such cases, inequality in terms of overall wealth can create conditions in which less well-off people can be forced to accept low wages or high prices because the alternative is starvation. The morally relevant inequality is grounded in the coercive nature of the situation, violating conditions of procedural equality, and only incidentally related to overall wealth.

7.3 Critical Assessment

Scientific risk assessment has played a crucial and contested role in the evaluation of agricultural biotechnologies. What is crucial from an ethics perspective is that assessments presuppose an ability to recognize a given outcome as *harmful*, and this is a value judgment. The working conception of fair and equitable distribution of the benefits and burdens of new technology developed so far presupposes that people will not be *coerced* into enduring harmful impacts from new technology, but the responsibility for ensuring that this is indeed the case is delegated to agencies that rely heavily on scientific expertise. Some impacts (death or disease among human beings) are noncontroversially recognized as ethically significant harms, and scientists may be able to identify otherwise poorly understood mechanisms that lead to these noncontroversial harms. However, science cannot exclude the possibility that some unknown mechanism exists. Hence, one source of controversy arises when nonscientists engage in speculation about unknown mechanisms. But there are also forms of harm that are themselves subject to debate. Some would assert that simply being placed in a situation where they consume something unknowingly constitutes a form of harm. People who observe certain religious dietary prohibitions or practice vegetarianism may think this way, for example (Thompson 2002). Thus, while scientific risk assessment provides an indispensable element in the ethics of emerging technology, it does not provide an infallible or uncontestable way to assure the ethical legitimacy of technological innovations.

But how do these rather formal conceptions of procedural equality apply to emerging technologies? New technologies that succeed in the sense of being widely and voluntarily adopted can be presumed to be benefiting both those who adopt them as well as those who profit from their development, manufacture, and sale. To the extent that greater efficiencies lead to lower prices for consumers—a feature common in the case of agricultural technologies—benefits are significantly multiplied. New technologies are also associated with upheaval and change in the workplace, with the bankruptcy of firms that are slow to innovate, and the loss of employment from those firms. It is thus plausible to view technological innovation as a process that is amenable to evaluation using the general framework of social choice. The innovation (or cluster of innovations) can be evaluated according to a Kaldor-Hicks test: if benefits outweigh costs or losses, the technology can be viewed as ethically justified (Wise 1978).

But the inequality between winners and losers due to the implementation of new technology provides a particularly challenging puzzle from the perspective of social choice. If there is a persistent pattern among the "losers" (e.g., if it is always the low-wage workers and those too poor to afford the new technology who lose) it is possible to see technological innovations that pass a Kaldor-Hicks test by a large margin as troubling from the perspective of distributive justice. Yet some of the losers can be thought of as victims of their own bad decision-making or bad luck just as readily as they can be thought of as victims of a social decision to implement a new set of technological means.

Should technological innovation even be evaluated from the social choice perspective? Recall that in social choice theory society is viewed *as if* the relative wealth and well-being of all persons, respective power relationships, and their capabilities for future action are the result of a decision or choice. It is possible to apply this framework to events that are unambiguously not the result of decision-making. One can evaluate the costs and benefits of a natural disaster, such as a hurricane, from the perspective of social choice. Although a hurricane is unlikely to generate benefits that outweigh costs, such an analysis might indeed provide interesting insight into the way that benefits and costs are distributed. But in this case, there is clearly no "decision" to launch a hurricane. It is possible that technological innovation is more like a hurricane than a social choice.

The development and dissemination of new technology require a number of clear decision points that distinguish it from a natural event, but these decision points are quite diverse: they accrue to different individuals or groups existing at different times and places, and occupying very different social roles. Some decisions are specifically focused on the research and development of the technology, while others involve funding the development process through capital investment or grants. Other decisions concern the regulatory environment in which the technology will operate, while still others concern the rules for patenting, licensing, and transferring the technology. Other decisions focus on liability for losses. Furthermore, although these decisions may have profound implications for the likely benefits and costs of technological innovation, many of the key legal and policy choices are not even focused on the particular technology in question, at all. These legal and policy decisions may be structured as social choices about the structure for a wide array of technological innovations that will occur in fields such as information technology or health care, and their impact on agricultural production may not be taken into consideration by decision-makers at all.

The sheer complexity of the decision-making processes in question means that technological innovation is, in some key respects, more like a hurricane than a singular activity or action that would clearly be amenable to evaluation from a social choice perspective. In cases where individuals or groups do make clear decisions, they face uncertainties about the disposition of other human decision-makers (not to mention natural events) that sharply limits their ability to execute an informed analysis of costs and benefits to all affected parties. Thus, one school of thought within the tradition of social choice holds that the managerial or planning perspective implied by utilitarian ethics is simply incapable of delivering the result implied by any formulation of the utilitarian maxim: Decisions about whether or not to innovate or to adopt technology *should not* be based on a prediction of expected utility. Instead, public policy should be confined to rules that protect individual liberty and stipulate conditions for the utilization and exchange of private property (Nozick 1974; De Gregory 2001). This *libertarian* perspective on social choice holds that however unfortunate distributive inequalities may be for the individuals who end up on the losing end of technological innovations, it is inappropriate for public authorities to do anything to relieve or reverse them.

This is not to say that libertarians see technological innovation as equivalent to natural disasters. One important difference is that unlike a hurricane, technological change comes about because human agents undertake planned activities with some expectation of personal benefit. Thus while innovations initiate patterns of change that defy *social* planning, they are unlike natural disasters in that post hoc attempts to compensate losers have the effect of interfering with the planning and intentional action of *individuals.* A second important difference relies on utilitarian reasoning about the relative net value of outcomes. While hurricanes do produce winners (people in the salvage or construction industry see a pickup in business), the overall net social value is overwhelmingly negative. The history of technological innovation, however, is one of improvements in the efficiency of production processes and public health, and the creation of entirely new social goods, especially in the arts, entertainment, and recreation. A blend of these views produces an ethic that ties a bias in favor of innovation with the view that losers are not entitled to compensation.

Although the libertarian position removes considerations of equity from the domain of considerations relevant to public authorities, it does not preclude actions to address inequality by private individuals. Individuals may make gifts or donate to charitable organizations that are dedicated toward helping the poor without violating the libertarian's master principle. The upshot of this twist in the social choice perspective is a shift in the meaning of "ethics." It is perfectly appropriate to encourage individuals to redress inequalities through charitable acts, and to claim that they have an ethical obligation to do so. However, investment decisions should be steered by a principle of profit maximization, which is believed to be a more reliable guide to promoting the public good than attempting to predict the complex interaction of forces that cause social inequalities, in the first place. "Ethics" then comes to be seen as a domain of private charity, while "capital" indicates the investment decisions that should steer technology (Nozick 1974). One conceptual weakness of this position is that it provides no guidance to individuals (or private firms) as to how much of their wealth they should allocate to ethics as compared to capital.

A further difficulty with models suggested to assess ethics is the metric used. Measures of economic growth or material progress are being used as proxies for "benefit," which could ultimately be understood as well-being or happiness, rather than economic well-being (Sparrow 2007). Further, benefits can accrue to individuals although their individual wealth or that of their community does not change.

7.4 International Arena

The CPB itself is the most pertinent international agreement relevant to the inclusion of an assessment of ethical considerations. However, any international agreement or protocol arguably articulates the international community's values and demonstrates what "rights" exist (Australian Gene Technology Ethics and Community Consultative Committee, GTECCC). Thus, for example, the international community's response to "sterile" conifers demonstrates a particular attitude to such developments and is perhaps the evidence of embedded values and motivations.

7.5 Administrative Consequences

Agricultural biotechnology raises potential ethical concerns with regard to the inherent characteristics of the technology itself, its products, and the consequences of the use of such products, and therefore any administrative body involved in addressing these concerns will need to adopt a broader approach. Further, as in the context of any developing technology, any model used to measure ethical concerns will need continual updating to reflect the technology's development, particularly to remain relevant regarding what its actual or likely consequences are (Weckert 2007). Societal values change rapidly and are difficult to identify accurately (Australian GTECCC) but regular measures of these will also be needed for an assessment process that includes ethical and equity concerns.

7.6 Summary/Synthesis

- Equity can be measured by reference to the distribution of benefits and costs occurring as the outcome of technological innovation. This utilitarian approach includes noncontroversial criteria (such as safety risks) and changes in SECs such as wealth, well-being, power, or social status as outcomes.
- An alternative equity measure focuses on interference with personal freedoms and political and economic activity of others. This libertarian approach values equal rights but not necessarily equal benefits and loss. Financial losses and changes in SECs are treated as the result of market forces and not directly relevant to equity.
- An alternative focus is agency and autonomy whereby an individual or social group's ability to undertake planned activity and affect the course of events is emphasized. When situations are consistently dominated by the same relatively few actors or where one group or class persistently lacks agency and autonomy, they are unjust.
- Different measures will be appropriate for developed as opposed to developing countries. For example, in developing countries, an assessment of farmers' perception of the new technology and also the social conditions under which it is being offered as being consistent (or not) with their ability to have power or influence over the conditions in which they live and farm would be relevant. A test for equity emphasizing whether farmers' decisions to adopt biotechnology was made on self-interested economic grounds might be wholly inappropriate in such a context.
- Decisions about whether or not to innovate or adopt technology *should not* be based on a prediction of expected utility. Instead, public policy should be confined to rules that protect individual liberty and stipulate conditions for the utilization and exchange of private property.

References

Acemoglu D (2000) Technical change, inequality, and the labor market. National bureau of economic research working paper series 7800

Alston JM, Hurd BH (1990) Some neglected social costs of government spending in farm programs. Am J Agric Econ 72:149–156

Alston JM, Norton G W, Pardey PG (eds.) (1998) Science under scarcity: principles and practices for agricultural research and priority setting. CABI Publishing, Wallingford

Australian GTECCC (Gene Technology Ethics and Community Consultative Committee) (undated) Discussion paper: environmental ethics as it relates to gene technology in Australia

Barro R (2000) Inequality and growth in a panel of countries. J Econ Growth 5(1):5–32

De Gregory TR (2001) Agriculture and modern technology: a defense. Iowa State University Press, Ames

Fehr E, Fischbacher U (2004) Social norms and human cooperation. Trends Cogn Sci 8(4):185–190

Fullerton D (1991) Reconciling recent estimates of the marginal welfare cost of taxation. Am Econ Rev 81(1):302–308
Juma C (1989) The gene hunters: biotechnology and the scramble for seeds. Zed Books, London
Nozick R (1974) Anarchy, State, and Utopia. Basic Books, New York
Sen A (1999) Development as freedom. Anchor Books, New York
Sen AK (2009) The idea of justice. Harvard University Press, Cambridge
Sparrow R (2007) Negotiating the nanodivides. In: Hodge G, Bowman D and Ludlow K (eds) New global frontiers in regulation: the age of nanotechnology. Edward Elgar, UK
Thompson PB (2002) Why food biotechnology needs an opt out. In: Bailey B, Lappé M (eds) Engineering the farm: ethical and social aspects of agricultural biotechnology. Island Press, Washington, DC, pp 27–44
Thompson PB, Matthews RJ et al (1994) Ethics, public policy, and agriculture. Macmillan, New York
Weckert J (2007) An approach to nanoethics. In: Hodge G, Bowman D and Ludlow K (eds) New global frontiers in regulation: the age of nanotechnology. Edward Elgar, UK
Wise WS (1978) The economic analysis of agricultural research. R & D Manag 8(3):185–190

Chapter 8
Food Security and Safety

Debra M. Strauss

8.1 Introduction

Times have changed since the days of royal tasters. As our concerns extend to the rest of the kingdom and beyond national borders, we face the critical need to develop increasingly complex policies to ensure the safety of the mainstream food supply. Incidents involving food contamination, particularly salmonella and *E. coli* in eggs, peanuts, and produce have been numerous and widespread. Tainted foods have caused illnesses and deaths that perhaps could have been prevented by more rigorous and proactive policies. Recognition has emerged that consumers need greater protection before these outbreaks occur, through more stringent requirements and better enforcement of food safety standards, including inspections. Moreover, traceability and recall mechanisms are necessary to resolve the problems that do arise. Food safety is important for all foods, regardless of the process to produce them. These concerns are heightened in the area of genetically modified organisms (GMOs), where scientific uncertainty compounds the issues in the effort to determine and evaluate the risks of harm to human health and the environment as essential elements in developing food safety regulation.

A study commissioned by the World Health Organization (WHO) identified several risks presented by GMOs and GM foods for human health as part of its safety assessment, including: direct health effects (toxicity); tendencies to provoke allergic reaction (allergenicity); specific components with toxic properties; the stability of the inserted gene; nutritional impact; and any unintended effects that could result from genetic modification (WHO Study 2005; see also Strauss 2006). Of particular concern is gene transfer, whereby genes from GM foods could transfer to bacteria in the gastrointestinal tract or to cells of the body and cause negative health effects (WHO Food Safety 2012). Recognizing that the long-term effects are unknown, the WHO Study (2005: iii) cited additional risks to the environment as "unintended

D. M. Strauss (✉)
Fairfield University, Fairfield, CT, USA
e-mail: dstrauss@fairfield.edu

K. Ludlow et al. (eds.), *Socio-Economic Considerations in Biotechnology Regulation*, Natural Resource Management and Policy 37, DOI 10.1007/978-1-4614-9440-9_8,

effects on non-target organisms, ecosystems, and biodiversity." For instance, outcrossing—the spread of transgenes in the natural environment through cross-pollination—has since been documented to occur and threatens to make GM crops the dominant species in the ecosystem; herbicide-resistant "superweeds" have also arisen (NRC 2000; NRC 2010; Bratspies 2003; Strauss 2010). Long-term scientific studies in humans have not yet been done, but some initial findings in animals and several case examples suggesting hazards have caused significant concern (Losey et al 1999[1]; Pusztai 2001[2]; Strauss 2006; Bt and Monarch Butterflies 2012; Angelo 2007). Moreover, a 2011 Canadian study reported that "the blood of 93 % of pregnant women and 80 % of their umbilical-cord blood samples contained a pesticide implanted in GMO corn by Monsanto, though digestion was supposed to remove it from the body." The researchers concluded, "given the potential toxicity of these environmental pollutants and the fragility of the fetus, more studies are needed." (Aziz and Leblanc 2011, p. 532).

Ironically, the issues raised are no longer merely matters of science and answers no longer hinge on scientific knowledge, claimed to be inadequate in this area both for the purposes of performing risk assessment and for determining unknown future effects (Strauss 2009). Instead, policy makers should use a broader perspective to examine the important implications for the international community and reshape matters of international trade and economics in line with long-term public interest. Implementing more international regulations on domestic issues, for purposes such as food security and safety may detract significantly from state sovereignty, limiting the nation-states' abilities to act in the best interests of their citizens, follow their cultural norms, or adhere to previously established international agreements (Strauss and Strauss 2009). Sensitivity to this position could help strike a balance between the global governance approach fueled by international trade aspects and the states' abilities to protect the food security rights of their citizens. Food safety in the regulation of GMOs raises key questions for policy makers, including:

- What is the appropriate role of risk assessment in the face of scientific uncertainty?
- Who should bear the risk of these products—consumers or biotech companies?
- What is the responsibility of government and regulatory bodies to ensure the safety of the food supply?
- Are there conflicts of interest behind the scenes affecting food policies? If so, they must be clearly identified, cautiously approached, and explicitly balanced, thus making transparency critical to the process.

[1] The journal PNAS (Proceedings of the National Academy of Science) published a volume in 2001 (http://www.pnas.org/content/98/21.toc) containing 6 articles that refute Losey's study. These studies posit that the conditions created by Losey in the laboratory would never be created in nature.

[2] The Royal Society in the UK released a report that found that there was no evidence of adverse effects from feeding GM potatoes to rats as reported by Pusztai. See: http://royalsociety.org/uploadedFiles/Royal_Society_Content/policy/publications/1999/10092.pdf.

- What are some of the cultural differences in perception of risk, how have these views shaped regulatory structures, and how can differences be accommodated on a global level while still respecting national sovereignty?
- How do GMOs fit with broader food safety issues and policies, and how can these policies be made consistent (e.g., a proactive, preventative regime versus a proof-of-harm approach)?

8.2 Methodologies

GMO regulation reflects cultural differences in consumers' degree of trust in their government as connected with their aversions to risk given scientific uncertainty and the government's decision as to whose perceptions of risk should determine public policy in view of possibly competing stakeholder interests—subgroups of consumers, consumer and trade groups, biotech companies, industry suppliers, farmers using GM crops, and those with conventional or organic techniques (see Chap. 18 for greater detail on cultural aspects of biotechnology). Moreover, regulators have varying levels of responsiveness to public opinion due in part to public demand and accountability corresponding to the level of trust in the government. For example, perhaps because of food scares that hit Europe before the USA, US consumers have more trust in their government and the US approach has been laissez faire, in contrast to the strict regulatory regime of the EU (Strauss 2006). However, recent food contamination issues have raised American awareness and resulted in new food safety legislation embodied in the FDA Food Safety Modernization Act (FSMA), and it is possible that this proactive mandate will trigger greater regulation of GM foods (Strauss 2011a, 2012; Lin 2012).

The divergent legal approaches reflect this cultural difference in consumer attitudes toward genetic engineering technology through the level to which scientific uncertainty is factored into risk assessment as part of the regulatory process. One approach would be to treat GM foods the same as other foods and focus on the end product's properties in assessing food safety. As an illustration, the USA has chosen to fit GM crops and food products into existing laws. In the USA, these crops are not subject to special regulatory scrutiny despite their novel properties and there is no acknowledgement of any risks inherent in the technology through which they were derived. US consumers appear less aware of potential risks and more trusting of regulatory agencies; whereas Europeans are more risk averse to possible human health and safety issues associated with GM food products. Consequently, Europe, Japan, and most of the international community give greater weight to this uncertainty than does the US government.

US regulations do not mandate labeling of GM foods, instead recommending voluntary labeling of GM foods and requesting that companies notify the Food and Drug Administration (FDA) of their intent to market GM foods at least 120 days before launch (FDA 2001). The inquiry focuses on whether the GM foods are *substantially equivalent* to their parent crops.[3] If so, only the general labeling requirements for all foods apply. The absence of mandatory labeling and monitoring, as well as a premarket approval process, stands in stark contrast to the approach in some other jurisdictions.

Another way to treat the same scientific uncertainty as to the unknown effects of this new technology would be to adopt a precautionary and proactive approach and special regulatory regime. For instance, EU regulators treat biotechnology as "a novel process requiring novel regulatory provisions," and have launched EU and international initiatives that take into account a wider range of known and unknown risks to human health and the environment (Strauss 2006, p. 176). Most significant is EU Directive 2001/18/EC, regulating and restricting the distribution of GMOs and foods containing GM ingredients (EU Council Directive 2001; see also BINAS Online 2006). Recognizing that the effects may be irreversible, the Directive mandates that "due attention be given to controlling risks from the deliberate release into the environment of (GMOs)" (cls 4 and 5). This Directive provides a notification procedure before a GM product is placed on the market, a period of public comment, an assessment report, principles for environmental risk assessment, field testing, and a gradual scale of release with proper evaluation at each step (cls 24 and 25).

The Directive also has specific provisions for labeling and packaging, including a requirement that "this product contains genetically modified organisms" appear either on a label or in an accompanying document (Art 13, § 2, f.). For products with unavoidable traces of authorized GMOs, minimum threshold levels for the labeling requirement shall be established (Art 21, § 2). Postmarket monitoring by the industry is required, as well as notification of authorities of new information and taking immediate measures necessary to protect human health and the environment (Arts 19, § 2, f. and 20). The Directive allows a temporary ban of GM products if there is evidence exposing risks to human health or the environment, a measure that led to a trade dispute initiated by the USA, Canada, and Argentina at the WTO (EC-Biotech Reports 2006; Strauss 2008).

More recently, the EU enacted regulations establishing a stricter framework for monitoring GMOs, strengthening existing labeling rules for GM food, implementing labeling of GM feed, and dictating an authorization procedure for deliberate release into the environment (EC Regulation 1830/2003 and 1829/2003; see also European Commission, Questions and Answers). A tolerance level of 0.9% is set for non-GM foods, feed, and processed products, as allowable "adventitious presence" or unintended low-level presence of an EU-approved GM substance. All products containing more than this level must be labeled as containing GMOs. The practical effect of these laws is to impose a "zero tolerance" standard for non-EU-approved GM crops. Traceability provisions require segregation of GM crops at all stages of

[3] The FDA asks the industry to compare the compositions of GM and non-GM crops; when they are not significantly different the two are regarded as 'substantially equivalent', and no additional labeling or animal testing is required. This concept has been disfavored in Europe where the capability to classify a novel food as being substantially equivalent no longer justifies a lack of safety assessments.

production, handling, storage, shipment, processing, and marketing, including the prevention of pollen drift to non-GM fields.

In response to a series of food safety failures, in May 2003, the EU created the European Food Safety Authority (EFSA) to serve as a "food safety watchdog" (Podger 2004; Strauss 2006, pp. 180–181). EFSA deals only with the science of risk assessment (determining what risks exist), while the FDA, its US counterpart, also handles the policy decisions involved in risk management (determining whether those risks are considered acceptable). The European Parliament wanted an organization that gave "genuinely objective, independent and public advice", leaving policy judgments to the European Commission (EC) (Podger 2004; Strauss 2006, pp. 180–181). Another goal is to instill "a much clearer degree of scientific input into the risk management measures adopted by the EU" by taking care not to avoid difficult scientific issues of risk assessment for fear of unpopularity (Podger 2004). At the same time, in light of European sensitivity to food issues and past food scares, EFSA seeks to achieve more transparency and restore public confidence. The new regulations further strengthened its charge, requiring EFSA to provide more detailed justification in its opinions on individual applications for permits, particularly as to overruling scientific objections raised by the national authorities; and asking applicants as well as EFSA in their risk assessments to address potential long-term effects and biodiversity issues (Strauss 2008).

These alternative regulatory regimes illustrate differences in the weight given to considerations of food security, primarily in their view of the level of risk the government is willing to accept and the comfort level of its citizens in being chosen to bear those risks. Commentators have asserted that: "Risk analysis examines the distribution of risks *and all distribution questions are inherently political*" (Olufs 2012, p. 16 (emphasis in the original)). These decisions are not transparent or subject to democratic scrutiny when made by bureaucratic organizations; moreover, "consumers and food producers often have radically different views of risk" (p. 16). US rules favor large food companies rather than seeking and giving more weight to the public and consumer organizations in an open dialogue. In the US government's view, there is no scientific basis to presuppose that biotech foods are more risky or substantially different from other food products. In contrast, by establishing a separate food safety agency that addresses problems with risk assessment in the face of scientific uncertainty, the EU has made an effort to separate science from politics and policy, increase transparency, and promote traceability. These differences reflect divergent views on the role of science, decisions on who bears the risk of uncertainty, the duties of the government and agencies, and the involvement of other organizations.

8.3 Critical Assessment

The potential safety issues from unintended and unknown risks and scientific uncertainty have sparked calls for a more effective approach to risk assessment in the USA, and resistance to the pressure to extend this laissez-faire approach internationally. The ultimate question is who bears the burden of proving that these substances are not hazardous—the companies before approval or the consumers. Recognizing that science is fallible, who should bear the risk of as-yet undetected hazards? If rigorous preapproval processes are not required, this consideration would at a minimum necessitate that the food should be labeled to enable consumers to be informed and individually make this choice as to the level of risk they are willing to undertake. Surveys of Americans and Europeans consistently show that consumers, when prompted, are concerned about food safety and want labeling in order to exercise their right to choose whether they eat GM food (European Commission 2001; Pew Initiative 2005; Institute for Agriculture and Trade Policy 2002). However, when consumers are asked generally about food labeling, only 3 % mention that they would like additional information about biotechnology on the label (IFIC 2012). Ultimately, only through labeling, segregation, and monitoring can there be some chance of removal from the market if, through the development of scientific assessment techniques and long-term studies, problems are discovered. Under an ideal approach that most values food safety, these substances would not be allowed to enter the food supply unless their safety can be established with enough reliable certainty in advance.

Countries with mandatory labeling legislation for GM foods allow their consumers a choice in selecting foods according to their comfort level. Transparency can be ensured only by requiring labeling and traceability of food products derived from GM plants at all stages of production and distribution. It has been suggested that countries tailor their regulations to minimize harm to trade while also responding to consumer concerns. Mandatory labeling of internationally traded GM foods would help address cultural differences and risk factors, while still allowing free trade and economic markets to control the process. Industry and trade associations have also begun to respond to the perceived risks of biotechnology in food. Fearful of losing buyers, large food producers have underscored their acceptance of consumer demands for labeling and have asked suppliers to segregate fields, grain bins, and storage elevators, with some even paying a premium for non-GM crops (e.g., Strauss 2006).

The international community should pay close attention to the problems engendered by the USA due to its failure to set up segregation mechanisms. Of significant concern to US producers is "the fact that U.S. farm, grain storage, and transportation systems are not designed to segregate bulk, untagged, biotechnology agricultural products, on a large scale and with precision, from conventional varieties" (Stamps 2002, p. 7). As a result, "the U.S. [g]overnment [cannot] certify that certain varieties are completely absent from export channels" (Stamps 2002, p. 7, quoting US Dept of State 2001). These changes in storage and transportation structure would place added costs on the US farm sector but may become necessary due to the concerns of international trade, economic loss for traditional and organic farmers, and potential future policy changes under a new proactive safety mandate. Most significantly, such an ineffective system highlights the dangers of unintended cross-contamination—that biotechnology crops will crossbreed with other plants resulting in unintended harmful breeds—the consequences for biodiversity are far more severe than simple economic costs such as labeling. If these genetic modifications cannot be monitored effectively, a more extreme remedy such as a ban may be necessary.

Critics of the US approach have also observed that food policy and its implementation are too fractionated and that "'[t]his fragmented federal system makes communication, efficiency and uniformity almost impossible during an emergency.' As a consequence, the need for better coordination is paramount" (Strauss 2012, pp. 312–313; see also Benson 2010; Notes 2007; Taylor 2004). There have been calls for a single federal agency dedicated to food safety; however, in an area of such complexity with overlapping and intersecting spheres of expertise (e.g., food and components, plants, environmental hazards), perhaps the effort would be better spent on improving coordination, communication, and management under a common mandate rather than creating yet another administrative agency (Strauss 2012). Such a proactive mandate for food security may now be embodied in the FSMA, which as a preventative approach for food safety that gives increased authority to the FDA including expansion of overseas inspections and restrictions on imports, should be extended in the future to other agencies such as the USDA as well as GM foods (Strauss 2011a). Currently each agency focuses only on its own narrow charge without viewing the broader scope of food safety, and they lack the scientific expertise to comprehend the real potential impact of GM crops on the environment or even raise the most relevant health and safety concerns (Strauss 2012; Aoki 2011). Moreover, the systemic problem is compounded by agency reliance on information provided by the companies they regulate and conflicts of interest in that research.

These limitations could be ameliorated by the use of expert groups and advisory panels with transparent process and balanced representation, similar to those consulted by the WTO in international trade disputes (Strauss 2011b, 2008). Other stakeholders, particularly trade associations and suppliers, organic and conventional farmers and consumers, will also need to remain vigilant in their demands for a proactive regulatory regime. In an area of such increasing scientific and socioeconomic complexity, a unified multilateral approach is clearly warranted. Governmental units and experts must work together to study the long-term human health and environmental effects of GMOs and prevent further contamination and extinction of non-GM crops. Food security includes biodiversity, which may be lost if GM crops become the dominant species and when they spur the development of "superweeds"; some of these effects have already been documented (e.g., NRC 2010; Kaskey 2011). Until more of these effects are known, mandatory labeling, monitoring, and segregation of crops are the most prudent approach to protect the integrity and security of the food supply (Strauss 2012). Similar concerns apply on the international level (Strauss 2009).

8.4 International Arena

Several international agreements hold special relevance for food security, most significantly embodying the precautionary principle that reflects and reinforces the risk-aversion approach of the EU and international community. The CBD recognizes that "biological diversity is about more than plants, animals and microorganisms and their ecosystems—it is about people and our need for food security, medicines, fresh air and water, shelter, and a clean and healthy environment in which to live" (CBD, About the Convention 2013). As the only international regulatory instrument established to protect biological diversity from the risks of biotechnology, the Cartagena Protocol expressly focuses on the potential adverse effects of living modified organisms (LMOs) on the environment, while taking into account the risks to human health as a secondary consideration (Art 1).

According to the WHO Study (2005), the Cartagena Protocol is only the first step in the international regulation of GM foods. The Protocol's scope does not consider GM foods that do not meet the definition of an LMO. GM foods are within the scope of the Cartagena Protocol only if they contain LMOs capable of transferring or replicating genetic material. Moreover, the primary focus on biodiversity limits its consideration of human health issues. As a further limitation, the leading GM exporters—Argentina, Canada, and the USA—have not signed on and are not bound by its terms. In the EC-Biotech case, the WTO declined to apply it to these nonsignatories nor to extend its precautionary principle to the level of customary international law (EC-Biotech Reports 2006; Strauss 2008).

To provide international consistency in GM food assessment, the *Codex Alimentarius* Commission—an international standard setting body for food safety jointly administered by two UN agencies, the Food and Agricultural Organization (FAO) and WHO—adopted principles setting a uniform standard for assessing food safety for foods derived from modern biotechnology (FAO/WHO 2003). The Codex principles set forth a premarket assessment, implemented on a case-by-case basis, including an evaluation of direct effects from the inserted gene and unintended effects that may arise. The safety assessment principles for GM foods require an investigation of the risks previously identified, namely, toxicity, allergenicity, specific components having nutritional or toxic properties, the stability of the inserted gene, the nutritional effects of the specific gene modification, and any unintended effects from the gene insertion. The WTO Agreements such as the SPS Agreement embrace the Codex principles, although the Codex does not in itself have a binding effect on national legislation. These principles are further referred to as a standard in cases of trade disputes (Strauss 2006). However, there has been criticism that the standards determined by the Codex give preferentiality to supporting trade and biotechnology, over protecting consumer interests and safety (Morse 2007; Strauss and Strauss 2009; Olufs 2012). Biotech-labeling standards, which the Codex was tasked with in 1993, have still not been developed.

In the international trade of food, the SPS Agreement acknowledges that states can protect domestic food supplies for scientific, well-documented health and safety risks, provided this does not unfairly discriminate against a particular industry or country for unfounded reasons. Instead of banning all products, a state could implement less costly or less restrictive alternatives, such as mandatory labeling or quality testing, to sustain free trade while protecting its citizens (Strauss and Strauss 2009). Under the SPS Agreement (Art 5.1), risk assessments must be used to justify such a ban (Strauss 2008). Article 5.7 provides an exception to the risk assessment requirement where relevant scientific evidence is insufficient.

However, the WTO in the EC-Biotech dispute rejected attempts by several EU member states to ban GM food under this exception because of their failure to do a scientific risk assessment, despite their contention that there was substantial scientific uncertainty as to the risks and that the insufficiency of scientific evidence made it impossible to conduct risk assessments (EC-Biotech Reports 2006; Strauss 2008). Despite its narrow interpretation, the WTO ruling does not preclude the possibility of utilizing other WTO agreements, such as the TBT Agreement. The TBT Agreement would be a particularly appropriate vehicle in view of recent proposals for the labeling of GM foods as a form of "least restrictive trade" measure (Strauss 2008, pp. 820–821).

Developing countries have a particular stake in the international agreements under interpretation in these trade disputes. In attacking the EU's ban, the USA was sending a message to developing countries not to use their rights under the Cartagena Protocol. In the EC-Biotech dispute, the WTO refused to embrace the precautionary principle as a key principle in environmental governance. For developing countries, the protection of transnational corporate interests at the expense of their citizens may also fall within the rubric of the International Covenant on Economic, Social, and Cultural Rights (CESCR 1966), which (in Art 2(1)) mandates that its members work to ensure the right to the highest attainable standard of health care and other socio-economic rights, including the right to food (Straub 2006; Strauss 2009). The vulnerability of developing countries as to their own food security highlights the need to reaffirm their rights and augment them if necessary.

Notwithstanding these international agreements that provide a foundation for further regulations on LMOs, it will be difficult to develop a global consensus on standardized policies due to "the inherent limits to regulatory harmonization in this policy area" (Falkner 2007, p. 108).

8.5 Administrative Consequences

Although ultimately it may be determined that food security issues involving GMOs should be controlled by national agencies, according to the food policies of each country, this discussion explores administrative consequences at the international level to consider which international bodies might be the most appropriate venues for handling GMOs in the global food supply. The key questions for consideration would be whether to:

- Establish a separate agency for food safety?
- Separate science and policy makers?
- Involve advisors from scientific and university community with expertise?
- Include nongovernmental organizations (NGOs), transnational corporations (TNCs), and/or other organizations to shape food policy?

Past disputes such as the EC-Biotech case illustrate that the WTO is not the appropriate body to make determinations about issues involving biotechnology and global food security due to its exclusive focus on international trade. As a result, the WTO has been unsuccessful in considering scientific matters such as risks to human health and the environment and has repeatedly failed to acknowledge the importance of the precautionary principle and environmental treaties (Lee-Muramoto 2012; Strauss 2008; UNU-IAS Report 2005). Moreover, by failing to take into account cultural values and excluding consideration of public health and environmental issues, the WTO has arguably legislated beyond the regulatory scope of international trade and undermined its credibility in the international community (Strauss 2008; Zurek 2007).

Because genetic engineering is a new technology, there are uncertainties as to where GMOs fit within the international regulatory framework. While the standards set forth by Codex have been a useful reference point in WTO disputes, this very link to the trade organization and its process of principle making may limit its appropriateness as a purely scientific body (Strauss 2009; Post 2006).[4] The WHO or the FAO could be more appropriate bodies, because they take scientific evidence into consideration when determining international policies. The WHO has recognized the importance of giving consideration to human health issues, traditional knowledge protection, and equitable benefit sharing (WHO Traditional Medicine Workshop 2000). However, the WHO is not an ideal candidate due to a lack of resources and effective enforcement mechanisms. Although the WHO has been instrumental in studying the risks of GMOs to human health and the environment, the organization has had limited capacity beyond reporting its findings (WHO Study 2005).

Perhaps an international body such as the FAO would be an appropriate administrator in this area. The FAO currently serves as the world forum for discussing the "use, control, and conservation of germplasm" (Strauss 2009, p. 314). The FAO Commission on Plant Genetic Resources adopted the International Undertaking on Plant Genetic Resources (FAO Undertaking), a nonbinding declaration that aims to remedy the inequity in exchanges of plant genetic resources.[5] It stated as its objective "to ensure that plant genetic resources of economic and/or social interest, particularly for agriculture, will be explored, preserved, evaluated and made available for plant breeding and scientific purposes" (FAO Undertaking, Art 1). As the culmination of this process, the International Treaty on Plant Genetic Resources for

---

[4] On the other hand, as an international organization Codex has been successful in openly including relevant stakeholders in its process and thereby achieving consensus. See Post 2006.

[5] Over 100 countries signed the FAO Undertaking, but the USA did not.

Food and Agriculture (ITPGR 2001) further clarified the level of intellectual property protection available for banked seeds subsequently modified by the recipient (Strauss 2009; ITPGR 2001). As the primary international organization in the area of food and agriculture, the FAO has supported a cautious approach that legitimizes the role of member states in controlling the safety of their food supply.

Sharing an initiative under the WHO Study, the Organisation for Economic Co-operation and Development (OECD 2012) established the Internal Coordination Group on Biotechnology in 1993 to aid international coordination in the areas of agriculture, technology, and trade. The OECD BioTrack provides a clearinghouse of information on biotechnology products and field trials, as well as Consensus Documents for the Work on Harmonisation of Regulatory Oversight in Biotechnology (OECD Consensus Documents 2006). This OECD effort seeks to promote international harmonization in the safety assessment and regulation of biotechnology food products, including conforming food labeling practices, which otherwise have a potential to impede international trade in food products as nontariff trade barriers. Although this organization's mission includes consideration of the environmental implications of economic and social development, its primary charge is to promote economic growth and financial stability (OECD). Most recently, the OECD and FAO jointly produced a report indicating that increased productivity and a more sustainable food system will improve global food security (OECD-FAO 2012).

If an existing international organization is not fully appropriate, responsibility for GM foods could span multiple international bodies each with specific spheres of authority. A combination of international organizations might be able to overcome the inherent limits of each. For example, the Codex, FAO, and WHO could work together to establish labeling guidelines that would serve as risk management tools and provide consumers with more information on the presence of GM ingredients along with warnings about potential allergens (Strauss 2009; Codex 2006). If such a collaboration is insufficient, perhaps a new transnational regulatory body could be established as a scientific and policy-making entity to focus more specifically on the global food supply. While such an entity should incorporate the cultural, economic, and scientific aspects of GMOs into its policies, there must also be one or more independent scientific bodies that could focus on safety and efficacy apart from policy, akin to how the EFSA functions. As discussed above, this component in shaping food policy could include the use of advisory panels of scientific experts and a transparent representation of all stakeholders.

8.6 Summary/Synthesis

- In an area of increasing scientific and socioeconomic complexity, a unified multilateral approach is warranted, engaging in the dialogue between governmental units, international organizations, and other stakeholders, including trade associations and suppliers, organic and conventional farmers, and academics and consumers.

- Government units and international experts must avoid conflicts of interest and work together to study long-term human health and environmental effects of GMOs and prevent further contamination and extinction of non-GM crops, while developing more appropriate means of risk assessment in the face of scientific uncertainty.
- Until these effects are known, mandatory labeling, traceability requirements, and monitoring and segregation of crops would be the most prudent approach to protect the integrity and security of the food supply; such an approach would also help address cultural differences and risk aversions, respect national sovereignty and special concerns of developing countries, and facilitate free trade and global economic markets.
- The Codex, FAO, and WHO could work together to establish labeling guidelines that would serve as a risk management tool and provide consumers with more information on the presence of GM ingredients and warnings about potential allergens.
- If the collaboration of already established international organizations is insufficient, a new transnational regulatory body could be established to incorporate the cultural, economic, and scientific aspects of GMOs into its policies, through consultation with one or more independent scientific bodies with the ability to focus exclusively on safety and efficacy.

References

Angelo MJ (2007) Regulating evolution for sale: an evolutionary biology model for regulating the unnatural selection of genetically modified organisms. Wake Forest L Rev 42:93–165

Aoki K (2011) Food forethought: intergenerational equity and global food supply—past, present, and future. Wis L Rev 2011:399–478

Aziz A, Leblanc S (2011) Maternal and fetal exposure to pesticides associated to genetically modified foods in eastern townships of Quebec, Canada. Repro Toxicol 41:528–533

Benson SM (2010) Guidance for improving the federal response to foodborne illness outbreaks associated with fresh produce. Food & Drug LJ 65:503–524

BINAS Online (2006) Facts on GMOS in the EU. http://binas.unido.org/binas/regs.php. Accessed 16 Mar 2006

Bobo JA (2007) The role of international agreements in achieving food security: how many lawyers does it take to feed a village? Vand J Transnat'l L 40:937–947

Bratspies RM (2003) Myths of voluntary compliance: lessons from the starlink corn fiasco. Wm & Mary Envtl L & Pol'y Rev 27:593–649

Bt and Monarch Butterflies (2012). http://www.biotech-info.net/butterflies_btcorn.html. Accessed 19 Aug 2012

CBD (2013) About the Convention. http://www.biodiv.org/convention/default.shtml. Accessed 17 Mar 2013

Codex Alimentarius Commission FAO & World Health Organization (WHO) (2006) Report of the 34th session of the Codex Committee on Food Labelling, Ottawa, Canada, 1–5 May

Comisión Nacional de Recursos Fitogenéticos (2012) Frequently asked questions about the Cartagena Protocol on Biosafety. http://www.conarefi.ucr.ac.cr/Bioseguridad1.htm. Accessed 19 Aug 2012

EC-Biotech Reports (2006) WTO, Reports of the Panel, European Communities—measures affecting the approval and marketing of biotech products, WT/DS291, WT/DS292 and WT/DS293 (29 September)
EC Regulation No 1829/2003 (2003) of the European Parliament and Council on Genetically Modified Food and Feed enacted 22 September 22 2003 and implemented 18 April 2004, 2003 OJ (L268) 25
EC Regulation No 1830/2003 (2003) of the European Parliament and Council on the Traceability and Labeling of Genetically Modified Organisms enacted 22 September 22 2003 and implemented 18 April 2004, 2003 OJ (L268) 24
European Commission Research Directorate-General (2001) Europeans, science and technology. http://ec.europa.eu/public_opinion/archives/ebs/ebs_154_en.pdf. Accessed 2 Jan 2013
European Commission (2012) Questions and answers on the regulation of GMOs in the European Community. http://ec.europa.eu/food/food/biotechnology/gmfood/qanda_en.pdf. Accessed 19 Aug 2012
EU (2001) Council Directive 2001/18/EC, 2001 O.J. (L106). http://eur-lex.europa.eu/LexUriServ/LexUriServ.do?uri = OJ:L:2001:106:0001:0001:EN:PDF. Accessed 2 Jan 2013
Falkner R (2007) The global biotech food fight: why the United States got it so wrong. Brown J World Aff 14:99–110
FAO (1983) Res. 8/83 (FAO Undertaking) International Undertaking on Plant Genetic Resources, UN Food and Agriculture Organization, 22nd Sess., Annex 1, Agenda Item 16, U.N. Doc. C/83/REP (1983)
FAO/WHO Food Standards (2003) Codex Alimentarius Commission, Principles for the Risk Analysis of Foods Derived from Modern Biotechnology, CAC/GL 44-2003. http://www.codexalimentarius.net/download/standards/10007/CXG_044e.pdf
FAO (2012) Biotechnology in food and agriculture, FAO statement on biotechnology. http://www.fao.org/biotech/fao-statement-on-biotechnology/en/. Accessed 19 Aug 2012
FDA (2001) Premarket notice concerning bioengineered foods, 21 C.F.R. Parts 192 and 592. http://www.fda.gov/OHRMS/DOCKETS/98fr/cf0128.pdf. Accessed 2 Jan 2013
Institute for Agriculture and Trade Policy (2002) American public opinion polls on GE foods (citing an ABCNews.com poll, June 2001). http://www.iatp.org/news/american-public-opinion-polls-on-ge-foods. Accessed 3 Jan 2013
International Covenant on Economic, Social, and Cultural Rights (CESCR) (1966) 16 December 1966, 6 I.L.M. 360. http://www2.ohchr.org/english/law/cescr.htm. Accessed 2 Jan 2013
International Food Information Council (2012) Food safety poll. Online at: http://www.foodinsight.org/Default.aspx
International Treaty on Plant Genetic Resources for Food and Agriculture (ITPGR), art. 12.3(d), Nov. 3, 2001. ftp://ftp.fao.org/docrep/fao/011/i0510e/i0510e.pdf. Accessed 13 Jan 2013
Kaskey (2011) Attack of the superweed. Bloomberg Businessweek. http://www.businessweek.com/magazine/attack-of-the-superweed-09082011.html. Accessed 13 Jan 2013
Lee-Muramoto MR (2012) Reforming the "Uncoordinated" framework for regulation of biotechnology. Drake J Agr L 17:311–367
Lin CF (2012) SPS-PLUS and bilateral treaty network: a "Global" solution to the global food-safety problem? Wis Int'l LJ 29:694–734
Losey JE et al (1999) Transgenic pollen harms monarch larvae. Nature 399:214
Morse EA (2007) Sound science and trade barriers: democracy, autonomy, and the limits of the SPS Agreement. http://ssrn.com/abstract=982626. Accessed 2 Jan 2013
National Research Council (2000) Genetically Modified Pest-Protected Plants: Science and Regulation. http://books.nap.edu/catalog.php?record_id=9795. Accessed 2 Jan 2013
National Research Council (2010) Impact of genetically engineered crops on farm sustainability in the United States. http://www.nap.edu/openbook.php?record_id=12804 & page=1. Accessed 12 Jan 2013
Notes (2007) Reforming the food safety system: what if consolidation isn't enough? Harvard L Rev 120:1345–1366

OECD (2012) What we do and how. http://www.oecd.org/about/whatwedoandhow/. Accessed 17 Aug 2012
OECD-FAO (2012) OECD-FAO Agricultural Outlook 2012–2021. OECD Publishing and FAO. http://dx.doi.org/10.1787/agr_outlook-2012-en
OECD Consensus documents for the work on harmonisation of regulatory oversight in biotechnology. http://www.oecd.org/document/51/0,2340,en_2649_37437_1889395_1_1_1_37437,00.html. Accessed 13 May 2006
Olufs S (2012) The Politics of food safety: lessons for a theory of organization. Paper prepared for annual meeting of the Western Political Science Association, Portland, Oregon, 22–24 March. http://wpsa.research.pdx.edu/meet/2012/olufs.pdf. Accessed 2 Jan 2013
Pew Initiative on Food and Biotechnology, Public Sentiment About Genetically Modified Food (2005), http://www.pewtrusts.org/uploadedFiles/wwwpewtrustsorg/Public_Opinion/Food_and_Biotechnology/2005summary.pdf. Accessed 2 Jan 2013
Podger G (2004) European Food Safety Authority will focus on science. Eur Aff 5 http://www.europeanaffairs.org/20041202309/Winter-2004/european-food-safety-authority-will-focus-on-science.html. Accessed 2 Jan 2013
Post DL (2006) The precautionary principle and risk assessment in international food safety: how the World Trade Organization influences standards. Risk Anal 26:1259–1273
Pusztai A (2001) Genetically modified foods: are they a risk to human/animal health? ActionBioscience Org. http://www.actionbioscience.org/biotech/pusztai.html. Accessed 2 Jan 2013
Stamps J (2002) Trade in biotechnology food products. Int'l Econ Rev 5–14
Straub P (2006) Farmers in the IP Wrench—how patents on gene-modified crops violate the right to food in developing countries. Hastings Intl'l & Comp L Rev 29:187–213
Strauss DM (2006) The international regulation of genetically modified organisms: importing caution into the U.S. food supply. Food & Drug LJ 61:167–196
Strauss DM (2008) Feast or famine: the impact of the WTO decision favoring the U.S. biotechnology industry in the EU ban of genetically modified foods. Am Bus LJ 45:775–826
Strauss DM (2009) The application of TRIPS to GMOs: international intellectual property rights and biotechnology. 45 Stan J Int'l L 45:287–320
Strauss DM (2010) We reap what we sow: the legal liability risks of genetically modified food. J Legal Stud Bus 16:149–177
Strauss DM (2011a) An analysis of the FDA Food Safety Modernization Act: protection for consumers and boon for business. Food & Drug LJ 66:353–376
Strauss DM (2011b) Achieving the food safety mandate: bringing the USDA to the table. Hamline J Pub L & Pol'y 33:1–47
Strauss DM (2012) The role of courts, agencies, and congress in GMOs: a multilateral approach to ensuring the safety of the food supply. Symposium edition: genetically modified organisms: trade and law in a global market. Idaho L Rev 48:267–319
Strauss DM, Strauss MC (2009) Globalization and national sovereignty: controlling the international food supply in the age of biotechnology. J Legal Stud Bus 15:75–94
Taylor MR (2004) Lead or react? A game plan for modernizing the food safety system in the United States. Food & Drug LJ 59:399–403
UN Conference on Environment and Development (1992) June 3–14 Rio Declaration on Environment and Development. 6 UN Doc A/CONF.151/26/Rev. 1 (1 Jan 1993)
UN Convention on Biological Diversity (CBD) (1992) 5 June 1992, 31 I.L.M. 818 http://www.cbd.int/convention/convention.shtml
UNU-IAS Report (2005) Trading precaution: the precautionary principle and the WTO http://www.ias.unu.edu/binaries2/Precautionary%20Principle%20and%20WTO.pdf. Accessed 13 Jan 2013
US Department of State (2001) Fact sheet, frequently asked questions about biotechnology. http://www.state.gov/e/eb/rls/fs/1142pf.hm
World Health Organization (WHO) (2005) Modern food biotechnology, human health and development: an evidence-based study. http://www.who.int/foodsafety/publications/biotech/biotech_en.pdf. Accessed 3 Jan 2013

World Health Organization (WHO) (2012) Food Safety, 20 questions on genetically modified (GM) foods. http://www.who.int/foodsafety/publications/biotech/20questions/en/index.html. Accessed 19 Aug 2012

WHO Traditional Medicine Workshop (2000) Report of the Interregional Workshop on Intellectual Property Rights in the Context of Traditional Medicine. WHO/EDM/TRM/2001.1 (2000). Bangkok, Thailand, 6–8 December

WTO Agreement on the Application of Sanitary and Phytosanitary Measures (SPS Agreement) (1994). http://www.wto.org/english/tratop_e/sps_e/spsagr_e.htm

WTO Agreement on Technical Barriers to Trade (TBT Agreement) (1994), available at http://www.wto.org/english/docs_e/legal_e/17-tbt_e.htm

Zurek L (2007) The European Communities biotech dispute: how the WTO fails to consider cultural factors in the genetically modified food debate. Tex Int'l LJ 42:345–368

Chapter 9
Health Impacts

Martina Newell-McGloughlin

9.1 Introduction

From a public health perspective, an important group of the coming generations of genetically modified (GM) crop plants and livestock are those with the value-added output traits of improved nutrition and food functionality. Continuing improvements in molecular and genomic technologies contribute to the acceleration of development of these products. Newell-McGloughlin (2008) presents examples of crops that have been genetically modified with macronutrient and micronutrient traits that may provide benefits to consumers and domestic animals. These new products and new approaches require a reassessment of appropriate criteria to assess benefits for human and animal health and well-being, and manage potential risks, while ensuring that the development of innovative technologies and processes is encouraged to provide value-added commodities for the consumer.

At a fundamental level, food is a source of nutrition to meet daily requirements at a minimum in order to survive but with an ever greater focus on the desire to thrive. In the latter instance, there is ever-growing interest in the functionality of food. "Functional foods" are modified foods or food ingredients that may provide a health benefit beyond the traditional nutrients they contain (Bidlack and Rodriguez 2011; Goldberg 1994).

From the basic nutrition perspective, there is a clear dichotomy in the demonstrated need between different regions and socioeconomic groups, the starkest being overconsumption in the developed world and undernourishment in less developed countries. Dramatic increases in the occurrence of obesity and related ailments in developed countries are in sharp contrast to the chronic malnutrition in many less developed countries. Both problems require a modified food supply, and the tools of biotechnology have a part to play.

Worldwide, plant-based products comprise the vast majority of human food intake, irrespective of location or financial status (Mathers 2006). In some cultures,

M. Newell-McGloughlin (✉)
University of California, Davis, USA
e-mail: mmmcgloughlin@ucdavis.edu

K. Ludlow et al. (eds.), *Socio-Economic Considerations in Biotechnology Regulation*, Natural Resource Management and Policy 37, DOI 10.1007/978-1-4614-9440-9_9,

either by design or default, plant-based nutrition comprises 100% of the diet. Therefore, it is to be expected that nutritional improvement can be achieved via modifications of staple crops. While the correlative link between food and health is still open to debate, a growing body of evidence indicates that food components can influence physiological processes at all stages of life. Functional foods are of increasing interest in the prevention and/or treatment of at least four of the leading causes of death in the USA: cancer, diabetes, cardiovascular disease, and hypertension. The US National Cancer Institute estimates that one in three cancer deaths are diet related and that eight of ten cancers have a nutrition/diet component (Block et al. 1992; Potter and Steinmetz 1996). Inverse relationships have been observed between carotenoid-rich foods and certain cancers (Botella-Pavı´a and Rodriguez-Conceptı´on (2006). Other nutrient-related correlations link dietary fat and fiber to the prevention of colon cancer, folate to the prevention of neural tube defects, calcium to the prevention of osteoporosis, psyllium to the lowering of blood lipid levels, and antioxidant nutrients to the scavenging of reactive oxidant species and protection against oxidative damage of cells that may lead to chronic disease, to list just a few (Mutch et al. 2005; Mathers 2006). Further, many food components are known to influence the expression of both structural genes and transcription factors (Tfs) in humans (Go et al. 2005; Mazzatti et al. 2007). Examples of these phytochemicals can be found in Newell-McGloughlin (2008, Table 9.1).

On the functionality side, there is a mirror component from the perspective of the genetic makeup of the individual doing the consumption. This field of personal response to nutrients is further divided into two thematic subsets. Nutrigenomics is the prospective analysis of differences among nutrients in the regulation of gene expression, while nutrigenetics is the analysis of genetic variations among individuals with respect to the interaction between diet and disease. These spheres of enquiry are designed to provide nutritional recommendations for personalized or individualized nutrition (Marti et al. 2010; Brigelius-Flohe and Joost 2006). Haplotyping studies are beginning to indicate gender- and ethnicity-specific polymorphisms that are implicated in susceptibilities to polygenic disorders such as diabetes, cardiovascular disease, and some cancers (Corthe´sy-Theulaz et al. 2005; Mutch et al. 2005; Brigelius-Flohe and Joost 2006).

It is estimated that plants produce up to 200,000 phytochemicals across their many and diverse members (Oksman-Caldenty and Inze´ 2004); obviously, a smaller number is available on our food plate, with approximately 25,000 different phytochemicals in food plants (Go et al. 2005). The quality of crop plants, nutritionally or otherwise, is a direct function of this metabolite content (Memelink 2004). From a health perspective, plant components of dietary interest can be broadly divided into four main categories, the first two to be enhanced and the latter two to be limited or removed: macronutrients (proteins, carbohydrates, lipids (oils), fiber); micronutrients (vitamins, minerals, functional metabolites); anti-nutrients (substances such as phytate that limit the bioavailability of nutrients); and allergens (intolerances and toxins).

The flagship of GM-improved nutritional varieties is the beta-carotene-enhanced rice commonly referred to as Golden Rice. Despite being under consideration and

Table 9.1 Examples of crops in research and/or development with nutritionally improved traits intended to provide health benefits for consumers and animals. (Modified from ILSI 2004, 2004a)

Trait	Crop (trait detail)	Reference
Protein and amino acids		
Protein quality and level	Bahiagrass (protein↑)	Luciani et al. 2005
	Canola (amino acid composition)	Roesler et al. 1997
	Maize (amino acid composition; protein↑)	Cromwell 1967, 1969;Yang et al. 2002; O'Quinn et al. 2000; Young et al. 2004
	Potato (amino acid composition; protein↑)	Chakraborty et al. 2000; Li et al. 2001; Yu and Ao 1997; Atanassov et al. 2004
	Rice (protein↑; amino acid)	Katsube et al. 1999
	Soybean (amino acid balance)	Rapp 2002; Dinkins et al. 2001
	Sweet Potato (protein↑)	Prakash et al. 2000
	Wheat (protein↑)	Uauy et al. 2006
Essential amino acids	Canola (lysine↑)	Falco et al. 1995
	Lupin (methionine↑)	White et al. 2001
	Maize (lysine↑; methionine↑)	Agbios 2006; Lai and Messing 2002
	Potato (methionine↑)	Zeh et al. 2001
	Sorghum (lysine↑)	Zhao et al. 2003
	Soybean (lysine↑; tryptophan↑)	Falco et al. 1995; Galili et al. 2002
Oils and Fatty Acids		
	Canola (lauric acid↑; γ-linolenic acid↑; + ω-3 fatty acids; 8:0 and 10:0 fatty acids↑; lauric + myristic acid↑; oleic acid↑)	Del Vecchio 1996; Froman and Ursin 2002; James et al. 2003; Ursin 2003, Dehesh et al. 1996; Agbios 2006; Roesler et al. 1997
	Cotton (oleic acid↑; oleic acid + stearic acid↑)	Chapman et al. 2001; Liu et al. 2002
	Linseed (+ ω-3 and -6 fatty acids)	Abbadi et al. 2004
	Maize (oil↑)	Young et al. 2004
	Oil Palm (oleic acid↑ or stearic acid↑; oleic acid↑ + palmitic acid↓)	Parveez 2003; Jalani et al. 1997
	Rice (α-linolenic acid↑)	Anai et al. 2003
	Soybean (oleic acid↑; γ-linolenic acid↑)	Kinney and Knowlton 1998; Reddy and Thomas 1996
	Safflower (γ-linoleic acid GLA↑)	Arcadia 2008
Carbohydrates		
Fructans	Chicory, (fructan↑; fructan modification)	Smeekens 1997; Sprenger et al. 1997 Sévenier et al. 1998
	Maize (fructan↑)	Caimi et al. 1996
	Potato (fructan↑)	Hellwege et al. 1997
	Sugar beet (fructan↑)	Smeekens 1997
Fructose, raffinose, stachyose	Soybean	Hartwig et al. 1997

Table 9.1 (continued)

Trait	Crop (trait detail)	Reference
Inulin	Potato (inulin↑)	Hellwege et al. 2000
Starch	Rice (amylase ↑)	Chiang et al. 2005; Schwall 2000
Micronutrients and functional Metabolites		
Vitamins and carotenoids	Canola (vitamin E↑)	Shintani and DellaPenna 1998
	Maize (vitamin E↑; vitamin C↑; folate↑; lycopene)	Rocheford et al. 2002; Cahoon et al. 2003; Chen et al. 2003; Bekaert 2008; Naqvi et al. 2009; Harjes 2010
	Cassava (+β-carotene)	Welsch R et al. 2010
	Mustard (+β-carotene)	Shewmaker et al. 1999
	Potato (+β-carotene and lutein↑)	Ducreux et al. 2005; Diretto et al. 2010
	Rice (+ β-carotene, folate↑)	Ye et al. 2000; Storozhenko et al. 2007
	Strawberry (vitamin C↑)	Agius et al. 2003
	Tomato (folate↑; phytoene and β-carotene↑; lycopene↑; provitamin A↑)	Della Penna 2007, Díaz de la Garza et al. 2004; Enfissi et al. 2005; Mehta et al. 2002; Fraser et al. 2001; Rosati 2000; Sun et al. 2012; Klee et al. 2012
Functional 2ndrymetabolites	Apple (+ stilbenes)	Szanowski et al. 2003
	Alfalfa (+ resveratrol)	Hipskind and Paiva 2000
	Kiwi (+ resveratrol)	Kobayashi et al. 2000
	Maize (flavonoids↑)	Yu et al. 2000
	Potato (anthocyanin and alkaloid glycoside↓; solanin↓)	Lukaszewicz et al. 2004
	Rice (flavonoids↑; + resveratrol)	Shin et al. 2006; Stark-Lorenzen 1997
	Soybean (flavonoids↑)	Yu et al. 2003
	Tomato (+ resveratrol; chlorogenic acid↑;flavonoids↑; stilbene↑anthocynanins↑)	Giovinazzo et al. 2005; Niggeweg et al. 2004; Muir et al. 2001; Rosati 2000; Gonzali et al. 2009
	Wheat (caffeic and ferulic acids↑; + resveratrol)	UPI 2002
Mineral availabilities	Alfalfa (phytase↑)	Austin-Phillips et al. 1999
	Lettuce (iron↑)	Goto et al. 2000
	Rice (iron↑)	Lucca et al. 2002
	Maize(phytase↑, ferritin↑)	Drakakaki 2005, Han 2009
	Soybean (phytase↑)	Denbow et al. 1998
	Wheat (phytase↑)	Brinch-Pedersen et al. 2000, 2006

Excludes protein/starch functionality, shelf life, taste/aesthetics, fiber quality, and allergen/toxin reduction traits

risk assessments since the late nineties, it was finally approved in the Philippines only in 2013. Ingo Potrykus, the developer, says an unreasonable amount of testing has been required, without scientific justification. In a 2010 *Nature* article, he lays the blame solely at the door of the regulatory process which he considers excessive, stating "I therefore hold the regulation of genetic engineering responsible for the death and blindness of thousands of children and young mothers" (Potrykus 2010, p 561). Scientific, civic, and religious opinion leaders from all over the world have expressed support for the value of this technology. Florence Wambugu of Kenya (1999) states that the great potential of biotechnology to increase agriculture in Africa lies in its "packaged technology in the seed," which ensures technology benefits without changing local cultural practices. Potrykus' Golden Rice is a seminal example. Incorporation of beta carotene into rice cultivars and widespread distribution of this "packaged technology in the seed" could prevent one to two million deaths each year. Wambugu (1999) observes that in the past, many foreign donors funded high-input projects, which have not been sustainable because they failed to address social and economic issues such as changes in cultural practice. Since the trait is incorporated into the seed this will obviate the need to learn novel cultural practices in order to avail of the nutritional improvement.

While the correlative link between food and health, beyond meeting basic nutrition requirements, has only been unequivocally proven in a number of cases, a growing body of evidence indicates that food components can influence physiological processes at all stages of life. Nutrition intervention from a functionality perspective has a personal dimension. Parsing individual response is at least as complex a challenge as the task of increasing or decreasing the amount of a specific protein, fatty acid, or other component of the plant itself (Brigelius-Flohe (2006). There is also evidence that early food regimes can effect later-life health; for example, some children who survived famine conditions in certain regions of Africa grew into adults battling obesity and related problems, presumably due to the selective advantage of the thrifty gene in their early food-stressed environment becoming a hazard during more abundant times especially if later diets are calorie dense. Functional food components are of increasing interest in the prevention and/or treatment of a number of the leading causes of death: cancer, diabetes, cardiovascular disease, and hypertension. Many food components are known to influence the expression of both structural genes and transcription factors in humans (Go et al. 2005; Mazzatti et al. 2008, Bidlack and Rodriguez 2011). The large diversity of phytochemicals suggests that the potential impact of phytochemicals and functional foods on human and animal health is worth examining as targets of biotechnology efforts.

9.2 Methodologies

It is well recognized that absolute safety is not an achievable goal in any human endeavor, and this is relevant with respect to food and feed. The safe use of food or feed has typically been established either through experience based on its common use or by application of generally recognized scientific assessment procedures. Starting in the 1990s, novel (especially GM) food and feed crops have been held to the standard that they should be as safe as an appropriate counterpart with a history of safe use. Based on this with respect to potential risks to human health, the consensus of scientific opinion and evidence is that GM foods and feeds present no new or unusual dangers to human health (Food and Agriculture Organization FAO/WHO 1997; OECD 2003; Seven Academies Report; Académie Des Sciences Française 2003; Royal Society of London 2002; US National Research Council 2000; Society of Toxicology 2003). The main principles of the international consensus approach are listed below. They serve to illustrate the variety of principles that have been at the center of the discussions and that are continuously being updated:

Substantial equivalence This is the guiding principle for the safety assessment. In short, substantial equivalence involves the process of comparing the GM product to a conventional counterpart with a history of safe use. Such a comparison commonly includes agronomic performance, phenotype, expression of transgenes, and composition (macro- and micronutrients), and identifies the similarities and differences between the GM product and the conventional counterpart. Based on the differences identified, further investigations may be carried out to assess the safety of these differences. These assessments include any protein(s) which are produced from the inserted DNA. Reports have demonstrated that GM crops are often more closely related to the isogenic parental strain used in their development than to other members of the same genus and species.

Potential gene transfer Where there is a possibility that selective advantage may be given to an undesirable trait from a food safety or environmental impact perspective, this should be assessed; for example, in the highly unlikely event of a gene coding for a plant made pharmaceutical is transferred to commodity to corn. Where there is a possibility that the introduced gene(s) may be transferred to other crops, the potential environmental impact of the introduced gene and any conferred trait must be assessed.

Potential allergenicity Since most food allergens are proteins, the potential allergenicity of newly expressed proteins in food must be considered. A decision-tree approach introduced by ILSI/IFBC in 1996 has become internationally acknowledged and updated by Codex (FAO/WHO 2003). Allergy safety assessments use a panel of characteristics to evaluate novel proteins for allergenic potential. The starting point for this approach is the known allergenic properties of the source organism for the genes. Other recurrent items in this approach are structural similarities between the introduced protein and allergenic proteins, digestibility of the newly introduced protein(s), and eventually if needed, sera-binding tests with either the introduced protein or the biotechnology-derived product. If a novel food protein is not similar to allergens and is not derived from allergenic protein families, it is unlikely to provoke allergy. Moreover, if a protein is degraded in simulated gastric fluid, is small in size (e.g., < 10 kDa), and is not glycosylated, the likelihood that it will be an allergen is also unlikely (Lehrer and Bannon 2005; Sánchez-Monge and

Salcedo 2005). There are many factors that may elicit food protein allergy, including extrinsic factors such as food use and processing; however, this is not well understood and is a continuing topic of research and discussion. It is important to note that many proteins possess one or more of the properties characteristic of an allergen, and yet they do not cause allergy. The likelihood that a novel biotechnology-derived protein will be an allergen is very low for proteins without sequence similarity and physical properties comparable to known allergens. From a safety perspective, the risk of allergy due to introduction of a novel protein into a crop remains low, regardless of whether proteins are introduced through conventional crop breeding or modern biotechnology.

Potential toxicity Some proteins are known to be toxic, such as enterotoxins from pathogenic bacteria and lectins from plants. Commonly employed tests for toxicity include bioinformatic comparisons of amino acid sequences of any newly expressed protein(s) with the amino acid sequences of known toxins with those of introduced proteins, as well as rodent toxicity tests with acute administration of the proteins. Currently, appropriate toxicity studies are typically needed for the introduced protein in GM crops. The assessment often includes studies such as conducting a bioinformatic comparison of the new protein's amino acid sequence with a database of all publicly available protein sequences (e.g., protease inhibitors, lectins), and testing the newly introduced protein's stability to heat or processing and to degradation in suitably representative gastric and intestinal model systems. Appropriate toxicity studies may be needed in cases where the protein present in the food is dissimilar to proteins that have previously been consumed safely in food. Such studies should also take into account the biological function of the protein in the plant, when known. In addition to purified proteins, whole grain from GM crops has been tested in animals, commonly in subchronic (90-day) rodent studies.

Unintended effects Besides the intended effects of the modification, interactions of the inserted DNA sequence with the plant genome are possible sources of unintended effects. Another source might be the introduced trait unexpectedly altering plant metabolism. Unintended effects can be both predicted and unpredicted. For example, variations in intermediates and endpoints in metabolic pathways that are the subject of modification while undesirable are predictable, while switch on of unknown endogenous genes through random insertion in control regions is both unintended and unpredictable. The process of product development that selects a single commercial product from hundreds to thousands of initial transformation events eliminates the vast majority of situations that might have resulted in unintended changes. The selected commercial product candidate event undergoes additional detailed phenotypic, agronomic, morphological, and compositional analyses to further screen for such effects.

Long-term effects It is acknowledged that premarket safety assessment should be rigorous to exclude potentially adverse effects of consumption of foods or feeds derived from GM crops. Nevertheless, some have insisted that such foods should also be monitored for long-term effects by post-market surveillance. The cost to benefit of such surveillance is severely skewed toward the cost side. This observation is predicated on a number of facts. In the first place, no international consensus exists as to whether such surveillance studies are technically possible without a testable hypothesis in order to provide meaningful information regarding safety. Secondly, any GM crop with a testable safety concern will not pass regulatory review as the deregulation process is so thorough. In fact, there is probably a greater probability of a non-GM crop slipping through as they are not subject to any premarket surveillance. Finally, the practical feasibility of implementing such a process is also not trivial. One potential method would be the use of measurable biomarkers, however, these would need to be determined for all foods and feeds, whatever the source, and the question of reasonable economic burden arises. Most probably, the marketplace would not support such a requirement as the value gained would not be considered a counter to the enormous cost of enforcement.

An example of using GM to improve the level of a macronutrient is work being done to increase the level of protein in plants. Protein energy malnutrition is the most lethal form of malnutrition (Food and Agriculture Organization 2006) and affects every fourth child worldwide, according to the World Health Organization (2006). The Food and Agriculture Organization estimates that 850 million people worldwide suffer from undernutrition, to which insufficient protein in the diet is a significant contributing factor. While they forecast that this would drop to 582 million by 2015, market circumstances have pushed back that estimate. The Millennium Development Goal of halving the number of those facing "extreme poverty and hunger" (United Nations 2000) is currently receding. At the end of the first decade of the twenty-first century, an estimated 1.2 billion people lacked food security (Nature 2010).

Most plants have a poor balance of essential amino acids relative to the needs of animals and humans. The cereals (maize, wheat (*Triticum aestivum*), rice, etc.) tend to be low in Lys, whereas legumes (soybean, pea (*Pisum sativum*), etc.) are often low in the sulfur-rich amino acids Met and Cys. Poultry, swine, and other nonruminant animals have specific requirements for each of the essential amino acids. The primary requirements for maize and soybean meal-based diets are Lys in mammals and Met in avian species. High-Lys and high-Met maize and soybeans could allow diet formulations that reduce animal nitrogen excretion by providing an improved balance of essential amino acids. When out of balance, the amino acid in excess results in increased nitrogen excretion. That balance can be accomplished now, but only by adding costly synthetic Lys and Met to the diet. Successful examples of improving amino acid balance include high-Lys maize (Eggeling et al. 1998; O'Quinn et al. 2000) canola (*Brassica napus*), and soybean (Falco et al. 1995). Free Lys is significantly increased in high-Lys maize by the introduction of the dapA gene from *Corynebacterium glutamicum*, which encodes a form of dihydrodipicolinate synthase that is insensitive to Lys feedback inhibition. Some novel indirect approaches have also been taken to improve protein content. Uauy et al. (2006) "rescued" an ancestral wheat allele that encodes a transcription factor (NAM-B1) which accelerates senescence and increases nutrient remobilization from leaves to developing grains (modern wheat varieties carry a nonfunctional allele). Reduction in RNA levels of the multiple NAM homologs by RNA interference delayed senescence by more than 3 weeks and reduced wheat grain protein, zinc, and iron content by more than 30%. Consumption of foods made from these crops potentially can help prevent malnutrition in developing countries, especially among children.

Conversely, genetic modification is used to limit the level of toxins in plant foods by downregulating or even eliminating the genes involved in the metabolic pathways for the production, accumulation, and/or activation of these toxins in plants. For example, the solanine content of potato has been reduced substantially using an antisense approach, and efforts are under way to reduce the level of the other major potato glycoalkaloid, chaconine (McCue et al. 2003). Work has also been done to reduce cyanogenic glycosides in cassava through the expression of the cassava enzyme hydroxynitrile lyase in roots (Siritunga and Sayre 2003). When "disarming" natural defenses of plants in this way, we need to be cognizant of potentially increased susceptibility to pests and diseases, so the base germplasm should have input traits to counter this.

In the larger context, the ability to modify the nutritional content of food crops can have a significant part to play in addressing issues related to the disability-adjusted life year (DALY), which is a measure of overall disease burden, expressed as the number of years lost due to ill health, disability, or early death and the quality-adjusted life year (QALY), which not only measures disease burden but also the quality and quantity of life lived. The DALY extends the concept of potential years of life lost due to premature death to include equivalent years of "healthy" life lost by virtue of being in states of poor health or disability. In so doing, mortality and morbidity are combined into a single, common metric. Traditionally, health liabilities were expressed using one measure: (expected or average number of) "Years of Life Lost" (YLL). This measure does not take the impact of disability into account, which can be expressed by: "Years Lived with Disability" (YLD). DALYs are calculated by taking the sum of these two components. DALY = YLL + YLD.

The DALY relies on an acceptance that the most appropriate measure of the effects of chronic illness is time; both the time lost due to premature death and the time spent disabled by disease. One DALY, therefore, is equal to one year of healthy life lost. Both DALY and QALY measurements clearly have the potential to be affected by nutrition and food functionality intervention since knowledge about, and delivery of, those traits can have major effects on health and longevity.

9.3 Critical Assessment

While translation of biotech research into value added products for producers and consumers is a challenge in the EU and even the US, it is exponentially more difficult in least developed countries. A problem facing Africa in particular is the lack of a dynamic private sector to take technologies to the farmer. It has also been estimated that regulatory costs exceed the costs of research and experimentation needed to develop a given GM crop, which is a major problem in releasing such crops to the market. A way to reduce the costs of generating food and environmental safety data is to develop regional "centers of excellence" with complementary facilities for biosecurity compliance. This can be done reliably and could help with reduction of regulatory costs. The economic gains from using GM crop technology in Sub-Saharan Africa (SSA) are potentially large according to the World Bank Group (Anderson (2005). The results suggest the welfare gains are potentially very large, especially from Golden Rice (beta-carotene-enhanced rice) and nutritionally enhanced GM wheat, and that those benefits are diminished only slightly by the presence of the EU's current position GM foods. Using the global economy-wide computable general equilibrium model known as Global Trade Analysis Project GTAP, Anderson et al. (2005) specifically noted that if SSA countries impose bans on GM crop imports in deference to EU market demand for non-GM products, the domestic consumer's net loss from that protectionism would be more than the small gain derived from greater market access to the EU.

Given the current regulatory climate, it is difficult to imagine many future traits ever reaching the marketplace. This discourages research on anything but the most mundane of crops and traits and is a real disincentive to creative research. A case in point is the BASF decision to end its development of the late blight-resistant "Fortuna" potato in the EU. Two genes from a wild variety of potato confer robust resistance against *Phytophthora infestans* and are a much preferred and more sustainable system than the use of topical fungicides, natural or otherwise. In October, 2011, the company requested cultivation and marketing approval as a feed and food from the EU Food Safety Authority (EFSA) for the potato with a notion of commercializing by 2015 (BASF 2011). However, in March 2012, BASF announced the discontinuation of its breeding efforts for all GM crops adapted to European conditions (Dixelius 2012). It will close its agricultural branch in Europe and is choosing instead to focus on the American and Asian markets. Now instead of adopting the GM Fortuna cultivar and the subsequent reduction of the use of harmful chemicals, European farmers must rely on the continued use of fungicides, some of the least friendly biocontrol chemicals. Ironically, this choice obstructs further expansion of organically produced potatoes and tomatoes because adopting the GM Fortuna cultivar in "conventional" agriculture could have led to reduced disease pressures benefitting alternative farming systems (Dixelius 2012). In addition, as a major consumer of potatoes, the EU will become increasingly dependent on imports from other regions, as they inevitably lose the battle against *P. infestans*. Over time, these imported potatoes are likely to be GM Fortuna so Europe is still left with the problem of tackling political resistance against it or any other GM crop.

In India, the further development of nutritionally enhanced potato varieties with balanced amino acid compositions is also awaiting approval for commercial release. However, despite the success of Bt cotton in India, the experience with Bt brinjal (eggplant) creates optimism about the prospects for commercialization of food crops difficult. Bt brinjal is effective against fruit and shoot borer (FSB), with 98 % insect mortality in shoots and 100 % in fruits compared to less than 30 % mortality in non-Bt counterparts. Multilocation research trials confirmed that Bt brinjal required, on average, 77 % less insecticides than non-Bt counterparts for control of FSB, and 42% less for the control of all insect pests of brinjal. The benefits of Bt brinjal translate to an average increase of 116% in marketable fruits over conventional hybrids, and 166% increase over popular open-pollinated varieties. Furthermore, the significant decrease in insecticide usage reduced the farmers' exposure to insecticides and results in a substantial decline in pesticide residues on brinjal fruits (Bricknel 2010). It is estimated that Bt brinjal will deliver farmers a net economic benefit ranging from US$ 330 to 397 per acre with national benefits to India exceeding US$ 400 million per year. However, in February 2010, the Indian Environmental Minister announced a 6-month moratorium citing that "There is no overriding food security argument for Bt brinjal. Our objective is to restore public confidence and trust in Bt brinjal," clearly articulating the fact that the decision was not based on scientific analysis or risk assessment. Reinforcing this notion, in 2012, an interim report of the technical expert committee (TEC) appointed by the Supreme Court of India recommended a ban on open field trials, including any ongoing trials, for 10 years. However, in October 2012, the Supreme Court decided that before considering this interim recommendation they would seek the views of all stakeholders, including the agriculture ministry and the GM crop industry, on the issue (Times of India 2012).

The above case studies illustrate that there is a need to address the issue that barriers exist to commercialization. The actual commercialization of biotech products may have little to do with technical limitations and more to do with external constraints primarily the process of regulatory approval. Most of the crops approved to date support that notion that the deregulation process is prohibitive for any but well-financed companies whose focus is primarily the large commodity crops as just discussed. Worldwide there is clear asymmetry and lack of consensus in regulatory systems. Non-hypothesis-based evaluations have become the standard around the world, and these are being enshrined in the Cartagena Protocol's roadmap, with the result that cost of safety assessment has skyrocketed without any discernible gain in safety. Given the current regulatory climate, it is difficult to imagine many of the traits described ever reaching the marketplace. As noted, this discourages research on anything but the large commodity crops, marginalizing some of the most crucially important crops and traits, and is especially a disincentive for translatable research in public institutions. For all intent and purposes there is just one trait from a public institution that has successfully traversed the regulatory minefields and been translated into a commercially viable commodity and that is the viral coat protein protection system initially developed for the papaya ringspot virus (PRSV) pandemic in Hawaii. Papaya is an important tropical fruit crop in the Asian region. However, production in many countries is set back by the prevalence of the PRSV disease as well as post-harvest losses. The PRSV-resistant papaya, based on RNAi suppression of the coat protein expression, literally saved the economy US$ 17 million in Hawaii, and while the disease is of significant importance in Taiwan and other Southeast Asian countries, it has yet to be approved there. Rather interestingly, it has been reported anecdotally that organic papaya growers now surround their plots with the transgenic rainbow variety as the post transcriptional gene silencing (PTGS) system proves to be a most effective method to reduce the viral reservoir thus protecting susceptible varieties through a mechanism that is similar to herd immunity in mammalian systems.

The clear importance of improved nutrition goes beyond the mere delivery of crops with improved nutritional content, the secondary implications for improved quality, and access to improved nutrition has an additive effect on the socioeconomic impact. When determining regulatory oversight the focus is solely on risk of introduction especially within the EU where the precautionary principal prevails. The "cost" of precautionary inaction needs to be addressed from an ethical viewpoint relative to the "cost" of failing to adopt a new product that could significantly benefit sizable world populations. As noted in Chassy et al. (2008), similar to risk assessment, the methodology for assessing other impacts, particularly positive nutritional effects, has to be sufficiently detailed and science based. In addition, it is conceivable that some positive impacts may be dependent upon certain variable conditions, such as regional differences in processing, consumption, and health affections. The DALY and QALY assessment methods discussed above, which measure disability and QALYs respectively, may provide suitable metrics to carry out this analysis. The assessment of impacts most likely is carried out by different scientists than the risk assessment, and can be instigated, for example, by a need to verify claims of positive health effects on a given food product. It is also important to examine the use and the intake of nutritionally enhanced foods and feeds, once they are commercially available, to determine whether the desired nutritional benefit has been achieved in the target populations. As noted non-scientific issues may also be linked with positive effects, which therefore are within the domain of risk managers, rather than that of scientific assessors.

The case studies of novel nutritionally improved crops in this chapter focus on recommended scientific assessments of possible risks associated with the new nutritionally improved food or feed. However, science-based premarket assessments, as currently performed, do not balance the assessment of potential risks against the intended benefits that accrue from use (Boobis 2006; Chassy et al. 2008). For nutritionally enhanced crops, it is particularly important to balance the intended substantial benefits (e.g., significant improvements in health, significant decreases in disease, suffering, and/or death) against the outcome of the risk characterization. The perceived hazards often represent relatively small risks, whereas the potential nutritional benefits are relatively large. An obvious case in point is Golden rice as discussed above. An additional example is the development of iron- and zinc-dense varieties of rice and wheat for India and Bangladesh which, it is estimated, could prevent 44 million cases of anemia over 10 years.

In December 2012, FAO's Director-General, José Graziano da Silva, noted that food insecurity in Africa's Sahel region is closely linked to peace and stability, and he stressed that short-term humanitarian efforts in the Sahel needed to be replaced with longer-term development (Da Silva 2012). Apart from the suggested implication of food and agricultural markets as being one of the trigger factors in catalyzing the "Arab Spring," the most recent global food crisis was in 2008, and 2013 may bring an even greater one. During the 2008 crisis, the Gates Foundation announced US$ 306 million in grants to boost agricultural yields in the developing world, with nearly US$ 165 million to replenish depleted soils in Africa. As noted by US News and World reports, these efforts are not without controversy as critics consider that western philanthropists are violating African "food sovereignty" and promoting American interests at the expense of peasant farmers knowledgeable about local practices (Lavelle and Garber 2008). Local practices, however, have yielded scarcity. A farmer in India grows three to four times as much food on the same amount of land as a farmer in Africa; a farmer in China, roughly seven times as much.

9.4 International Arena

Besides the international organizations such as FAO/WHO, OECD, ILSI, and IFBC, other organizations have also formulated their views and recommendations on oversight of GM crops. In 2003, the Organization for Economic Co-operation and Development (OECD) developed international consensus documents on the particular components that could be analyzed for specific crops (OECD (2003). For the analyses of these components, a number of validated methods are available that are well understood through their long history of use. Such approaches work well to assess the concentrations of specific predetermined compounds that, because of their selective nature, may miss other unintended changes. In the Codex Alimentarius model, based on consensus across the above organizations, risk analysis is composed of three elements (i.e., risk assessment, risk management, and risk communication). During risk analysis, the risks are to be weighed against other issues, such as the benefits, with the aim to ensure the highest appropriate level of public health protection and to strive for risk management transparency and continuous communication between assessors and managers during the process. Implementation should be examined and reviewed for its effectiveness in protecting public health.

In all countries with biosafety protocols in place, a full risk assessment has to be carried out, irrespective of the positive effects that may accrue from a particular food item.

It is thus recommended that the risk assessment of food follow elaborate, internationally harmonized procedures.

9.5 Administrative Consequences

Commercialization of the products of recombinant DNA technology was another facet in a long history of human intervention in nature for agricultural and food production purposes. As such, the same parameters of risk-based assessment should apply. Commercialization of products must be undertaken within a regulatory framework that ensures adequate protection of the consumer, the environment, and alternate production systems while not stymieing innovation.

There is almost universal agreement that innovation is essential for sustaining and enhancing agricultural quality and productivity. There also would be general concurrence that this involves at some level, new, science-based products and processes that contribute reliable methods for improving quality, productivity and environmental sustainability. Most of the innovative technologies of that have been applied to production agriculture have come into common usage without much controversy or even knowledge by the average consumer. However, thanks to the globalization and democratization of knowledge afforded by the internet, the cult of the amateur as noted by Trewavas (2008) is in danger of giving equal weight to uninformed opinions and blatant fear mongering when it comes to applying the tools of modern technology to improving agricultural efficiencies and effectiveness and enhancing food security.

In the past, we have not regulated changes in agriculture based on unpredictable SECs or public opinion. However, recombinant DNA technology has inspired debate in a manner unlike any other previous technological development where demand for safety assurance is held to unreasonable standards. There are now worldwide data from more than 15 years of commercial use of GM crops and more than two decades of research experience and no verified adverse consequences have been reported. Some scientists have begun asking if the premarket safety assessment used in many countries is attempting to achieve a standard of absolute safety by continually adding data requirements as newer analytical methods come available and indeed, they recommend that the extent and type of data that is part of a current safety assessment be updated to reflect this long-term safe experience with GM crops coupled with new information about plant genome plasticity (Bradford 2005a, b; Kalaitzandonakes 2007; McHughen 2007; McHughen and Smyth 2008). These scientists suggest that, from a scientific perspective, the cumulative experience over a couple of decades of assessing GM crop safety should allow us to determine which tests need to be applied to new GM varieties to determine if they are as safe as their traditional counterparts with a history of safe consumption. They suggest that refinements to the process could include incorporation of factors such as "familiarity" (e.g., for commonly used proteins such as CP4 EPSPS, Cry1Ab, and PAT) and the source of the gene (e.g., when the gene is from the same crop species or is one with a history of safe use) into the overall safety assessment, influencing the extent to which event-specific data are needed. This process is relatively simple for some crop species, while others may require more extensive analysis—yet today all crops species undergo the same assessment process, regardless of risk. The fundamental concept to the comparative risk assessment process is that it enables a reasonable certainty that a new GM variety of a crop is as safe as the conventional varieties currently being safely consumed by humans and animals. We consider this to be a sensible and pragmatic approach to regulatory oversight.

In 2001 and 2011, the European Commission released two reports that cover 25 years of research on GM crops or food on human health or the environment: "A decade of EU-funded GMO research (2001–2010)" (EU Commission Report 2010) and "EC-Sponsored research on the safety of genetically modified organisms (1985–2000)" (EU Commission Report 2001). The more recent was a compendium of 50 research projects on the safety of GMOs over the last decade. The Commission funded research from 130 research projects involving 500 independent research groups over 25 years, concluding that "There is, as of today, no scientific evidence associating GMOs with higher risks for the environment or for food and feed safety than conventional plants and organisms" (Europa Press Release 2010). Indeed, they concluded, the use of more precise technology and the greater regulatory scrutiny probably make them even safer than conventional plants and foods. A declaration signed by more than 3,500 scientists including 25 Nobel Laureates reiterates this position.

Right now, for essential macronutrients and micronutrients that are limited in various regional diets, the strategies for improvement are clear and the concerns, such as pleiotropic (unintended) effects and safe upper limits, are easily addressed by traditional safety risk assessments. However, for many other health-promoting phytochemicals, clear links with health benefits remain to be demonstrated. Such links, if established, will make it possible to identify the precise compound or compounds to target and which crops to modify to achieve the greatest nutritional impact and health benefits. The achievement of this aim will be a truly interdisciplinary effort, requiring expertise and input from many disparate fields, ranging from the obvious human physiology and plant research to the less obvious "omics" and analytic fields. With rapidly emerging technologies, the increase in our understanding of and ability to manipulate plant metabolism during the coming decades should place plant researchers in the position of being able to modify the nutritional content of major and minor crops to improve many aspects of human and animal health and well-being. Methodologies to measure the positive and negative impacts on public health will be needed for these more long-term health impacts of GMOs to be properly included in an SEC assessment.

9.6 Summary/Synthesis

- Technical and scientific challenges to modifying intricate metabolic pathways and networks as opposed to single genes are not trivial and will need to be met.
- Metabolic engineers must not only understand the fundamental physiology of the process to be affected but also the level, timing, subcellular location, and tissue or organ specificity that will be required to ensure successful trait modification.
- Acute safety concerns raised by nutritional changes to GM plants and livestock are addressed by current risk assessments.
- Although there is a growing knowledge base indicating that elevated intakes of specific phytochemicals may reduce the risk of disease, such as certain cancers, cardiovascular diseases, and chronic degenerative diseases associated with aging, further research and epidemiological studies are still required to prove definitive relationships. New methodologies will be needed to measure potential benefits to human and animal health and well-being from improved nutritional and functionality qualities of food.

References

Académie Des Sciences Française (2002) Les plantes génétiquement modifiées Rapport sur la science et la technologie n 13. http://www.academie-sciences.fr/publications/rapports/rapports_html/RST13.htm

Anderson K, Jackson LA, Nielsen CP (2005) GM rice adoption: implications for welfare and poverty alleviation, J Econ Integr 20(4):771–788

BASF (2011) Business BASF applies for approval for another biotech potato. http://www.research-in-germany.de/84190/2011-11-17-business-basf-applies-for-approval-for-another-biotech-potato.html. Accessed 25 Feb 2013

Beckman M, Enot DP, Overy DP et al (2007) Representation, comparison and interpretation of metabolome fingerprint data for total composition analysis and quality trait investigation in potato cultivars. J Agric Food Chem 55: 3444–3451

Bidlack W, Rodriguez RL (2011) Nutritional genomics: dietary regulation of gene function and human disease. Taylor and Francis/CRC Press December 2011

Bino RJ, de Vos CHR, Lieberman M et al (2005) The light-hyperresponsive high pigment-2dg mutation of tomato: alterations in the fruit metabolome. New Phytol 166:427–438

Block G, Patterson B, Subar A (1992) Fruit, vegetables, and cancer prevention: a review of the epidemiological evidence. Nutr Cancer 18:1–29

Botella-Pavı´a P, Rodriguez-Conceptı´on M (2006) Carotenoid biotechnology in plants for nutritionally improved foods. Plant Physiol 126:369–381

Bradford KJ, Gutterson N, Parrott W et al (2005a) Reply to "Regulatory regimes for transgenic crops." Nat Biotechnol 23:787–789

Bradford KJ, Van Deynze A, Gutterson N et al (2005b) Regulating transgenic crops sensibly: lessons from plant breeding, biotechnology and genomics. Nat Biotechnol 23:439–444

Brigelius-Flohe R, Joost HG (2006) Nutritional Genomics: impact on health and disease. Wiley-VCH Verlag, Weinheim, pp 3–17

Bruce W, Folkerts O, Garnaat C et al (2000) Expression profiling of the maize flavonoid pathway genes controlled by estradiol-inducible transcription factors CRC and P. Plant Cell 12:65–80

Carrari F, Baxter C, Usadel B et al (2006) Integrated analysis of metabolite and transcript levels reveals the metabolic shifts that underlie tomato fruit development and highlight regulatory aspects of metabolic network behavior. Plant Physiol 142:1380–1396

Chassy, B.; Egnin, M; Gao, Y.; Glenn, K.; Kleter, G.A.; Nestel, P.; Newell-McGloughlin, M.; Phipps, R.H.; Shillito, R. 2008 Nutritional and Safety Assessments of Foods and Feeds Nutritionally Improved through Biotechnology: Case Studies Comprehensive reviews in food science and food safety 7 . - p. 50–99.

Codex Alimentarius (2003) Proposed draft principles and guidelines for the conduct of microbiological risk management (at step 3 of the procedure). Joint FAO/WHO Food Standards Programme, Codex Committee on Food Hygiene, Food and Agriculture Organization, Rome, Italy. ftp://ftp.fao.org/codex/ccfh35/fh03_07e.pdf

Corthe´sy-Theulaz I, den Dunnen JT, Ferre´ P et al (2005) Nutrigenomics: the impact of biomics technology on nutrition research. Ann Nutr Metab 49:355–365

da Silva José Graziano (2012) Remarks by FAO's Director-General, Rome, UN officials stress link between food security and peace in Sahel. UN News Centre http://www.un.org/apps/news/story.asp?NewsID=43713&Cr=sahel&Cr1=#.UMl1mXfsLF0. Accessed 25 Feb 2013

Davidovich-Rikanati R, Sitrit Y, Tadmor Y et al (2007) Enrichment of tomato flavor by diversion of the early plastidial terpenoid pathway. Nat Biotechnol 25:899–901

Davies KM (2007) Genetic modification of plant metabolism for human health benefits. Mutat Res 622:122–137

DellaPenna D, Pogson BJ (2006) Vitamin synthesis in plants: tocopherols and carotenoids. Annu Rev Plant Biol 57:711–738

Dixelius C, Fagerström T, Sundström JF (2012) European agricultural policy goes down the tubers. Nat Biotechnol 30:492–493. doi:10.1038/nbt.2255

Eggeling L, Oberle S, Sahm H (1998) Improved L-lysine yield with Corynebacterium glutamicum: use of dapA resulting in increased flux combined with growth limitation. Appl Microbiol Biotechnol 49:24–30

EU Commission Report (2001) EC-sponsored research into the safety of genetically modified organisms. Fifth Framework Programme—External Advisory Groups "GMO research in perspective". Report of a workshop held by External Advisory Groups of the Quality of Life and Management of Living Resources Programme. http://ec.europa.eu/research/fp5/eag-gmo.html. Accessed 25 Feb 2013

EU Commission Report (2011) A decade of EU-funded GMO research (2001–2010) Reference: IP/10/1688 Event Date: 09/12/2010 http://europa.eu/rapid/press-release_IP-10-1688_en.htm. Accessed 25 Feb 2013

Europa Press Release (2010) Commission publishes compendium of results of EU-funded research on genetically modified crops http://europa.eu/rapid/press-release_IP-10-1688_en.htm. Accessed 17 Mar 2013

Falco SC, Guida T, Locke M et al (1995) Transgenic canola and soybean seeds with increased lysine. Biotechnology 13:577–582

Food and Agriculture Organization (2006) State of food insecurity in the world. Food and Agricultural Organisation, Rome

FAO/WHO (1997) Risk management and food safety. Report of a joint FAO/WHO consultation, January 27–31, 1997. Food and Agriculture Organization/World Health Organization, Rome, Italy. ftp://ftp.fao.org/docrep/fao/w4982e/w4982e00.pdf

Fraser PD, Enfissi EMA, Goodfellow M et al (2007) Metabolite profiling of plant carotenoids using the matrix-assisted laser desorption ionization time-of-flight mass spectrometry. Plant J 49:552–564

Giliberto L, Perrotta G, Pallara P et al (2005) Manipulation of the blue light photoreceptorcryptochrome 2 in tomato affects vegetative development, flowering time and fruit antioxidant content. Plant Physiol 137:199–208

Go VLW, Nguyen CTH, Harris DM et al (2005) Nutrient-gene interaction: metabolic genotype-phenotype relationship. J Nutr 135:3016S–3020S

Goldberg I (1994) Functional foods, designer foods, pharmafoods, nutraceuticals. Chapman and Hall, New York

Haake V, Zrenner R, Sonnewald U et al (1998) A moderate decrease of plastid aldolase activity inhibits photosynthesis, alters the levels of sugars and starch, and inhibits growth of potato plants. Plant J 14:147–157

Hall RD, Brouwer ID, Fitzgerald MA (2008) Plant metabolomics and its potential application for human nutrition. Plant Physiol 132:162–175

Hajirezaei M, Sonnewald U, Viola R et al (1994) Transgenic potato plants with strongly decreased expression of pyrophosphate:fructose-6-phosphate phosphotransferase show no visible phenotype and only minor changes in metabolic fluxes in their tubers. Planta 192:16–30

Herman EM, Helm RM, Jung R et al (2003) Genetic modification removes an immunodominant allergen from soybean. Plant Physiol 132:36–43

International Life Sciences Institute (2004a) Nutritional and safety assessments of foods and feeds nutritionally improved through bio- technology. Compr Rev Food Sci Food Saf 3:35–104

International Life Sciences Institute (2004b) Nutritional and safety assessments of foods and feeds nutritionally improved through biotechnology: an executive summary. J Food Sci 69: CRH62–CRH68

International Life Sciences Institute (2008) Nutritional and safety assessments of foods and feeds nutritionally improved through biotechnology: case studies. Compr Rev Food Sci Food Saf 7:50–99

Kalaitzandonakes N, Alston J, Bradford K (2007) Compliance costs for regulatory approval of new biotech crops. Nat Biotechnol 25(5):509–510

Kinney AJ, Knowlton S (1998) Designer oils: the high oleic acid soybean. In: Roller S, Harlander S (eds) Genetic modification in the food industry. Blackie Academic and Professional, London, pp 193–213

Lavelle M, Garber K (2008) Eight ways to fix the global food crisis. US News&World Report posted May 9, 2008. http://www.climos.com/news/articles/8waysto.htm.
Liu Q, Singh S, Green A (2002) High-oleic and high-stearic cottonseed oils: nutritionally improved cooking oils developed using gene silencing. J Am Coll Nutr 21:205S–211S
Long M, Millar DJ, Kimura Y et al (2006) Metabolite profiling of carotenoid and phenolic pathways in mutant and transgenic lines of tomato: identification of a high antioxidant fruit line. Phytochemistry 67:1750–1757
McCue KF, Allen PV, Rockhold DR et al (2003) Reduction of total steroidal glycoalkaloids in potato tubers using antisense constructs of a gene encoding a solanidine glucosyl transferase. Acta Hortic 619:77–86
McHughen A (2007) Fatal flaws in agbiotech regulatory policies. Nat Biotechnol 25:725–727
McHughen A, Smyth S (2008) US regulatory system for genetically modified organism(GMO), rDNA or transgenic crop cultivars. Plant Biotechnol J 6(1):2–12
Mathers JC (2006) Plant foods for human health: research challenges. Proc Nutr Soc 65:198–203
Marti A, Goyenechea E, Martínez JA (2010) Nutrigenetics: a tool to provide personalized nutritional therapy to the obese. J Nutrigenet Nutrigenomics 3(4–6):157–169
Mazzatti DJ, van der Ouderaa F, Brown L (2007) The future of food: nutrigenomics and nutrigenetics. Agro Food Ind Hi Tech 18:63–66
Memelink J (2004) Tailoring the plant metabolome without a loose stitch. Trends Plant Sci7:305–307
Mutch DM, Wahli W, Williamson G (2005) Nutrigenomics and nutrigenetics: the emerging faces of nutrition. FASEB J 19:1602–1616
Nature (2010) How to feed a hungry world. Nature 466(7306): 531–532
Newell-McGloughlin M (2008) Nutritionally improved agricultural crops. Plant Physiol 147:939–953
OECD (Organization for Economic Cooperation and Development) (2003) Considerations for the safety assessment of animal feed stuffs derived from genetically modified plants. ENV/JM/MONO. 2003.11. Organisation for Economic Co-operation and Development, Paris, France
O'Quinn PR, Nelssen JL, Goodband RD et al (2000) Nutritional value of a genetically improved high-lysine, high-oil corn for young pigs. J Anim Sci 78:2144–2149
Oksman-Caldenty KM, Inze´ D (2004) Plant cell factories in the post genomic era: new ways to produce designer secondary metabolites. Trends Plant Sci 9:433–440
Paul MJ, Knight JS, Habash D et al (1995) Reduction in phosphoribulokinase activity by antisense RNA in transgenic tobacco: effect on CO2 assimilation and growth in low irradiance. Plant J 7:535–542
Potter JD, Steinmetz K (1996) Vegetables, fruit and phytoestrogens as preventive agents. IARC Sci Publ 139:61–90
Potrykus I (2010) Regulation must be revolutionized. Nature Jul 29, 466(7306):561–561
Royal Society Report (February (2002) Genetically modified plants for food use and human health—an update. http://www.royalsoc.ac.uk/policy/index.html http://www.royalsoc.ac.uk/files/statfiles/document-165.pdf
Society of Toxicology (SOT) (2003) The safety of genetically modified foods produced through biotechnology. Toxicol Sci 71(1):2–8. http://www.asa-europe.org/pdf/gmfoods.pdf
Schauer N, Fernie AR (2006) Plant metabolomics: towards biological function and mechanism. Trends Plant Sci 11:508–516
Siritunga D, Sayre RT (2003) Generation of cyanogen-free transgenic cassava. Planta 217:367–373
Times of India (2012) No moratorium on GM crop trials till stakeholders heard: SC. http://timesofindia.indiatimes.com/india/No-moratorium-on-GM-crop-trials-till-stakeholders-heard-SC/articleshow/17011950.cms. Accessed 25 Feb 2013
Trewavas A (2008) The cult of the amateur in agriculture threatens food security. Trends Biotechnol 26(9):475–478. doi://000259324200002 AND http://www.ask-force.org/web/Discourse/Trewavas-Cult-Amateur-Agriculture-2008.pdf
Uauy C, Distelfeld A, Fahima T, Blechl A, Dubcovsky J (2006) A NAC gene regulating senescence improves grain protein, zinc, and iron content in wheat. Science 314:1298–1301

United Nations (2000) United Nations Millennium Declaration. http://www.un.org/millennium/declaration/ares552e.htm
Wambugu F (1999) Why Africa needs agricultural biotech. Nature 400:15–16
World Health Organization (2006) Nutrition for Health and Development. World Health Organization. http://www.who.int/nutrition/challenges/en/index.html. Accessed 15 April 2008

Chapter 10
Impact on Biodiversity

José Falck-Zepeda, Patricia Zambrano and Melinda Smale

10.1 Introduction

The impact of genetically modified organisms (GMOs) on "the conservation and sustainable use of biological diversity," as stated in Article 26 of the Cartagena Protocol on Biosafety (Secretariat of the Convention of Biological Diversity 2000, pg. 19), has generated a heated and polarized debate. On one side is the position of those who believe that not only economic but ethical, religious, and cultural considerations related to biodiversity should be taken into account. On the other side, stand those who believe that the impact of GMOs on the environment should be circumscribed to environmental assessments. Whatever the position, generating a useful analysis of the impacts of GMOs on biological diversity requires thoughtful definition of concepts and selection of appropriate valuation methods.

The Convention on Biological Diversity (CBD) defines biodiversity as the variability among living organisms from all sources including intra- and interspecific, and ecosystem diversity (Convention on Biological Diversity 1992; Hawsworth 1995). Biodiversity has and will continue to fluctuate significantly over time going through both periods of growth and mass extinction (Cockell 2006). Mass extinction periods, and to some degree, overall composition of biodiversity, have resulted from rapid environmental changes mainly associated with changing and degrading habitats.

Surveys of the literature have demonstrated that the impacts of biodiversity loss on proposed ecosystem functions vary significantly from negligible to great in magnitude, depending on the function, location, and scale of analysis (Cardinale et al. 2012). Although scholars disagree about whether overall biodiversity is declining (e.g., Rabosky 2009; Millennium Ecosystem Assessment 2005), there is a knowledge-based consensus that humans have negatively influenced biodiversity through accelerated impacts on habitats and ecosystems (Tilman et al. 2001; Foley

J. Falck-Zepeda (✉) · P. Zambrano
International Food Policy Research Institute (IFPRI), 2033 K Street NW,
20006-1002, Washington, DC, USA
e-mail: j.falck-zepeda@cgiar.org

M. Smale
207 Agriculture Hall, Michigan State University, 48824, East Lansing, MI, USA

K. Ludlow et al. (eds.), *Socio-Economic Considerations in Biotechnology Regulation*,
Natural Resource Management and Policy 37, DOI 10.1007/978-1-4614-9440-9_10,

et al. 2005; Phalan et al. 2011). Tracking ecosystem degradation may now be more complex due to multiple, complex, and interactive causes and linkages. Damage to some ecosystems may be so great that it poses a barrier for future development (United Nations 2005, p. 43).

In fact, one of the United Nations' development goals is to ensure environmental sustainability and reduce biodiversity loss as a requisite to ensuring food security and for the eradication of diseases. Ecosystem degradation can be reversed in many instances by taking corrective actions to address those human activities that have negative impacts on ecosystems but also by providing appropriate measures for conservation within existing biodiversity reserves and hotspots.

The convergence of biodiversity losses, increasing demand for ecosystem services, and the need to eradicate poverty and hunger demands appropriate policies. This chapter examines the nexus where these issues converge and argues for the need for a systems approach to address complexity. We focus on two distinct scales of analysis: (1) the broad ecosystem, which incorporates all biological species; and (2) the agricultural system, viewed from the perspective of the farmer. By definition and practice, the second scale focuses on domesticated species and/or those that are in use by farmers and others.

10.1.1 Ecosystem Services and Assessment

Biodiversity benefits human society in multiple ways, including the provision of goods and services such as renewable resources, mitigation of environmental and other biophysical challenges, and cultural and aesthetic enjoyment. Biodiversity is thus important for sustaining agricultural systems, human health, scientific innovation, business, and industry. Many of the benefits of biodiversity are tangible, but some experts also consider that biodiversity has an intrinsic value whichmay not be captured by conventional economic measures of progress (Gomez-Baggethun and Martin-Lopez 2012). Valuing the multifaceted benefits of biodiversity is difficult in part due to conceptual challenges, but also because most are not valued in markets.

Box 10.1 summarizes some CBD concepts and definitions that are relevant to biodiversity assessment. Ecosystems refer to all biological communities and the non-living environment, which can be considered a functional unit (Gomez-Baggethun and Martin-Lopez 2012). Natural capital is the concept of an ecosystem as a stock or in terms of providing flows of goods and services that are valuable to humans. Goods and services can be tangible or intangible. Ecosystem functions are thus the components and processes capable of generating value to human society. Functions may include provisioning, supporting, and regulating aspects of biodiversity.

Ecosystem services are the direct and indirect benefits that contribute to humanity's welfare. In this chapter, we will use the term ecosystem services interchangeably with the term "ecosystem goods and services" as we consider all tangible and intangible goods and services of value to humanity. Changes in ecosystem services

Box 10.1 Terms and Definitions (Source: Article 2 of the Convention on Biological Diversity (Convention on Biological Diversity 1992, p. 3)

Biological Diversity

Means the variability among living organisms from all sources including, inter alia, terrestrial, marine and other aquatic ecosystems and the ecological complexes of which they are part; this includes diversity within species, between species and of ecosystems.

Biological Resources

Includes genetic resources, organisms or parts thereof, populations, or any other biotic component of ecosystems with actual or potential use or value for humanity.

Sustainable Use

Means the use of components of biological diversity in a way and at a rate that does not lead to the long-term decline of biological diversity, thereby maintaining its potential to meet the needs and aspirations of present and future generations.

Natural Capital

Economic conceptualization of ecosystems as stocks able to generate flows of ecosystem services on a sustained basis over time. The concept has precedents in the way land is conceived of as a factor of production in classical economics.

Ecosystems

A dynamic complex of plant, animal and micro-organism communities and their non-living environment interacting as a functional unit.

Ecosystem Functions

From an anthropocentric perspective, these are all the ecosystem components and processes capable of generating ecosystem services benefiting human welfare.

Ecosystem Services

Direct or indirect contributions of ecosystems to the welfare of society. The concept of "ecosystem goods and services" is equivalent to that of "ecosystem services," but includes both tangible and intangible contributions.

have thus direct and indirect impacts on human welfare. In many instances, we depend on the provision of ecosystem services for survival such as access to clean water, use of good quality soil, crop pollination, and access to genetic resources for use in food production, among others.

Figure 10.1 delineates the different steps involved in evaluating ecosystem services. Valuation is achieved by examining the biophysical and socio-cultural

Fig. 10.1 Steps in ecosystem services evaluation. (Source: Figure based on Haines-Young and Potschin (2010) and Gomez-Baggethun and Martin-Lopez (2012))

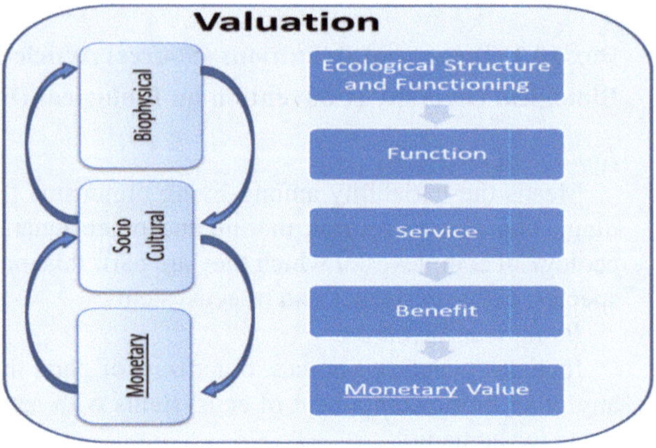

contexts, and assessing monetary value. Valuation metrics depend on the context and disciplinary perspective of the analyst.

One of the most widely used frameworks is Total Economic Value (TEV), a fundamental concept in environmental and resource economics. TEV takes into account both the utilitarian or anthropocentric value and the non-utilitarian, intrinsic, or existence value that can complement or compete with the utilitarian value (Fig. 10.2).

In the specific case of GM crops, it is critical to establish a line of causality and to define the mechanism by which impact occurs. Introduction of a GM crop, as that of any other crop, can potentially have positive or negative, direct or indirect impacts on the stock and flow of ecosystem services of value to humanity.

An alternate line of causality is that adoption of GM crops may increase the agricultural frontier and encroach on protected habitats. This impact on biodiversity would be considered to be negative. But at the same time GM crops could contribute to averting the expansion of the agricultural frontier, with benefits for the environment. As discussed in other chapters of this book, this is because introduction of a GM crop may have other impacts that are unrelated to biodiversity.

10.2 Methodologies

Several method categories can be applied to assess the impact of GMOs on biodiversity. This chapter focuses on three of these. The first type assigns biodiversity indicators to specific components of the value of ecosystem services, as depicted in Figs. 10.1 and 10.2. The second type includes methods that measure taxonomic diversity, usually over space and/or time. Given that the focus is on GMOs, the discussion will be limited to those that are used in the context of agricultural biodiversity. The third method type provides an economic value to the taxonomic diversity. Value measurement may be in monetary or non-monetary terms or include tangible

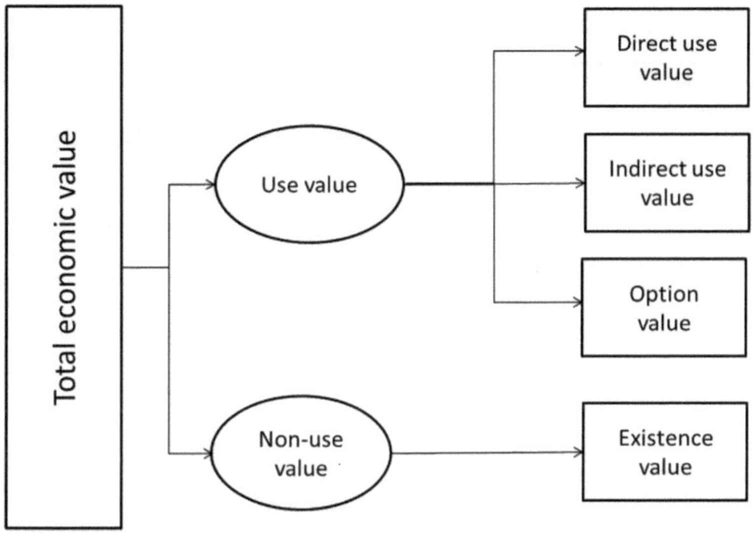

Fig. 10.2 Total economic value. (Source: Figure based on Fig. 6.1, Millennium Ecosystem Assessment (2005))

and intangible values. Methods and approaches used to assess socio-cultural aspects are described in other chapters in this book.

10.2.1 Using Biodiversity Indicators

Biodiversity indicators use quantitative data to measure different policy issues related to biodiversity and its status over time and space, ecosystem services, and policy consequences. Indicators are estimated at the macro, regional, and national levels, and their purpose is to assist in policy formulation and decision-making. For example, they can be utilized to provide a quantifiable baseline and to permit monitoring and tracking of changes over time.

A major driver behind the development of biodiversity indicators is the Biodiversity Indicators Partnership (BIP).[1] The BIP is a coalition of international organizations that strive to develop and improve existing information and help monitor biodiversity trends. Table 10.1 introduces the focal areas and the headline indicators used by the BIP as guidance for development of biodiversity indicators at the global, regional, national, and ecosystem levels.

[1] http://www.bipindicators.net/.

Table 10.1 Biodiversity indicators. (Source: 2010 Biodiversity Indicators Partnership (2010))

Focal areas	Headline indicators
Status and trends of the components of biodiversity	Trends in extent of selected biomes, ecosystems, and habitat
	Trends in abundance and distribution of selected species
	Coverage of protected areas
	Change in status of threatened species
	Trends in genetic diversity
Sustainable use	Area of forest and agricultural and aquaculture ecosystems under sustainable management
	Proportion of products derived from sustainable sources
	Ecological footprint and related concepts
Threats to biodiversity	Nitrogen deposition
	Invasive alien species
Ecosystem integrity and ecosystem goods and services	Marine trophic index
	Water quality of freshwater ecosystems
	Trophic integrity of other ecosystems
	Connectivity/fragmentation of ecosystems
	Incidence of human-induced ecosystem failure
	Health and well-being of communities who depend on local ecosystem goods and services
	Biodiversity for food and medicine
Status of traditional knowledge, innovations, and practices	Status and trends of linguistic diversity and numbers of speakers of indigenous languages
	Other indicators of the status of traditional and indigenous knowledge
Status of access and benefit sharing	Indicator of access and benefit sharing
Status of resource transfers	Official development assistance provided in support of the Convention
	Indicator of technology transfer

10.2.2 Measuring Taxonomic Diversity

Measurements of taxonomic diversity are a biophysical evaluation of ecological structure as shown in Fig. 10.1. The metrics that have been proposed to measure species-level biodiversity include quantitative indexes that capture the important concepts of species richness and species abundance. Examples include richness indices (such as the Margalef), and indices of proportional abundance (such as the Simpson, Shannon-Wiener, Mean Species Abundance, and the Berger-Parker indices). A crucial parameter in these indices is the scale of the unit analyzed, or the size of the geographic area in which biodiversity is measured over time.

Taxonomic diversity indexes can be calculated at the global, regional, national, ecosystem and production system or farm scales. For example, Smale (2006) describes the application of these indices to crop production by farmers in developing countries. The emphasis in this approach is thus the apparent diversity reflected in those attributes visible by farmers and those that may be managed by farmers. Thus, the emphasis among the studies in that collection is on biodiversity among "domesticated" rather than "wild" species.

10.2.3 Performing Economic and Non-pecuniary Valuations

When performing economic and non-pecuniary valuations of biodiversity and ecosystem services, the first step is to define value. There are different definitions and types of value that have been used. Direct value in use are those products arising from ecosystem services that can be consumed or used directly including water, timber, lumber, biomass, and others. In turn, indirect value in use refers to those ecosystems services whose indirect effect draws value. This includes nutrient production and retention, climate regulation, and others.

Option values refer to the premium difference between using a resource now and that drawn from maintaining ecosystem goods and services for future value in use. Option value may include an estimate of those ecosystems services thatare not known at the time of the valuation. Intrinsic, existence, or non-use values refer to those that exist irrespective of their use and are thus associated with cultural, aesthetic, and traditional customs, which may be valuated using economic and non-pecuniary valuation methods (Fig. 10.2).

There are three major types of economic and non-pecuniary valuation methods. The three major approaches include revealed preferences, stated preferences, and cost-based methods. A fourth category, systems approaches, referring to approaches that are integrated or which use a broader set of methods has also been included (See TEEB 2012 for as discussion on these approaches).

10.2.3.1 Revealed Preferences

Revealed preference methods examine the observed behavior of actors on a market. For example, hedonic pricing method relates the price of goods in markets to attributes that are implicitly traded in markets. This approach uses econometric methods to determine the relative contribution of each attribute or characteristic to the value of the good or service in question. This requires that markets function well, and that market prices reflect the values of each one of the attributes that contribute to the overall value of a good or service.

There are three major revealed preference methods that are used in the valuation of ecosystem services. The market price method provides an estimate of ecosystems services that are traded in established markets. The productivity method examines the economic values for ecosystems services that are used to produce market goods.

The third is the surrogate market method, which may include the benefit transfer, hedonic pricing, and travel cost methods. The benefit transfer method examines ecosystem services values by extrapolating from studies in similar locations or issues. Hedonic pricing estimates values by examining ecosystem services that directly have an impact on market prices of a good. In turn, the travel cost method estimates value associated with an ecosystem or ecosystem services by estimating the cost of traveling to the specific site in question. This method assumes that the value that humans are willing to pay to travel to a site or for ecosystem service reflects the value of the site or ecosystem services.

Revealed preference approaches have two major limitations. First for some issues, real or proxy markets may not exist for biodiversity attributes or characteristics. Even if there is such market, prices may not reflect the value of biodiversity attributes due to incomplete markets. Furthermore, markets may be thin as they may lack standards by which consumers can judge quality and/or governments may intervene through price control mechanisms. Second, market values may not be a good approximation of biodiversity value as by definition they are value in use and thus would not necessarily reflect intrinsic values.

10.2.3.2 Stated Preferences

Stated preferences are used to address limitations of revealed preferences as they may help identify the complete value especially of non-market goods. These methods explicitly attempt to measure the value of a change in the supply of a non-market good in carefully designed and implemented studies. There are two major methods used in stated preference: contingent valuation and contingent choice.

Contingent valuation methods estimate ecosystem services value by attempting to estimate non-use values. These methods carefully ask respondents their willingness to pay for a specific ecosystem service. The procedure usually involves questions and hypothetical scenarios that are used to estimate value. Contingent choice methods estimate ecosystems services value by asking respondents to make trade-offs between different sets of ecosystem services and/or their attributes or characteristics. This method does not directly elicit willingness to pay from respondents; rather it is inferred from trade-off responses while considering the costs of such trade-offs.

10.2.3.3 Cost-based Models

Cost-based models are those which estimate ecosystems services values based on estimates of the cost of avoiding damage to a specific ecosystem, cost of replacing ecosystem services, and/or the cost of providing alternate or substitute services. These models are largely based on direct and indirect cost estimates and opportunity costs.

10.2.3.4 Systems Approaches

These are a set of approaches used in the literature to examine different aspects of biodiversity and ecosystem services (Daniel et al. 2012; UNEP-WCMC 2011) These tend to integrate different methods in order to address value of such resources. The methods have been used to examine agricultural production choices in terms of variety used and ex-situ and in-situ conservation of agricultural biodiversity (see Smale 2006 for several examples). These have also been used to examine household- and village-level choices for biodiversity conservation, and the institutional economics involved in examining the relationship between policy, diversity, and productivity, among others.

10.3 Critical Assessment[2]

The methods outlined above have both advantages and disadvantages. Revealed preference methods suffer from collinearity among attributes, as well as other modeling issues. Stated preference methods have been criticized due to their hypothetical nature and due to the fact that actual behavior is not being observed. The most promising research direction would be perhaps those efforts attempting to merge revealed and stated preferences. Merging both approaches may increase statistical efficiency of estimation and enhance validity (Table 10.2).

Estimating the value of ecosystem services and biodiversity is a valuable tool for policy formulation and decision-making. Providing such value estimates may help provide incentives for biodiversity and ecosystems services conservation, rational use, and improvement for future use. Thus, ecosystem services valuation help make better informed policy choices. Based on the state of nature in the ecosystem services valuation literature, we recognize that much needs to be done in terms of methods development.

Any study will be by nature incomplete as we have neither comprehensive data (economic and biophysical) nor sufficiently mature tools to examine the full value of biodiversity and ecosystem services. In fact, most valuation studies will tend to underestimate the value of ecosystem services, especially when conducted across a larger scale of analysis as compared to an intensive case study. Some ecosystems may be too difficult or impossible to replace, so that attempting to estimate a replacement cost will be futile. In other cases, the complexity of value determination will be such that attempting to estimate impacts may not be feasible.

Addressing non-use value such as those related to cultural, religious, ethical, and/or aesthetic value may be difficult. Furthermore, focusing only on an economic or pecuniary valuation may underestimate the true value of biodiversity and ecosystems services. This may be especially problematic in those situations where there may be difficulties in estimating non-use values.

[2] This section draws significantly from Smale (2006).

Table 10.2 Advantages and disadvantages of different valuation methods

Approach	Advantages	Disadvantages
Revealed preference	Data parsimonious	Only works where there is a market for the resource
	Relatively easy implementation	Does not estimate intrinsic values
	Ample history of use	
Stated preference (contingent valuation)	Very flexible	Results sensitive to survey design and choice
	Estimates of option and existential value	Not clear if Contingent Valuation measures willingness to pay
	Ample history of use	
Cost-based approaches	Useful to measure ecosystem services	Problematic to implement when there is no market for the resource
	Relatively easy implementation and interpretation of results	Provides rough estimates of value as it is quite difficult to implement by finding a substitute
System approaches	Helps address issues in a holistic manner	Relatively harder to implement as they use indicators and other approaches to estimate value

Method triangulation that includes cost methods, institutional analysis, and other qualitative methods should help elicit better valuations in the future. Developing extended applications of stated preference methods that may be used in developing countries with less literate populations is needed to bridge a large gap in the literature (see, for example, Bennett and Birol 2010). Developing better models to examine the dynamics and inter-temporal impacts of biodiversity and ecosystem services is urgently needed for future use. Furthermore, other models that consider not only values and impacts of biodiversity and ecosystems services, but which also relate these to other measures of ecosystem, ecosystem services, and the agrobiodiversity system such as resiliency are also needed. Such models should also incorporate different parts of biodiversity and ecosystem services especially those related to agriculture, livestock, and forestry systems.

Results from ecosystem services valuation may be ecosystem and even site specific. Results from such valuations may not be extrapolated or used to estimate value in other ecosystems or sites. These estimates may also be specific for a particular group being studied, and may reflect a time dimension representing opinions in a particular point in time. As with any other valuation approach there is no guarantee that results from such studies will be used in policy making or that that they may improve such decision-making.

10.4 International Arena

The CBD and in its subsidiary agreement the CPB as well as other international treaties have introduced in their texts the potential inclusion of socioeconomics as related to biodiversity. The CBD is the main international treaty relevant to biodiversity valuation and socio-economics, particularly its Article 8j:

> (j) Subject to its national legislation, respect, preserve and maintain knowledge, innovations and practices of indigenous and local communities embodying traditional lifestyles relevant for the conservation and sustainable use of biological diversity and promote their wider application with the approval and involvement of the holders of such knowledge, innovations and practices and encourage the equitable sharing of the benefits arising from the utilization of such knowledge, innovations and practices.

Article 8j of the CBD has a relationship with the CPB and Socio-Economic Considerations. In the CPB, Article 26 describes the potential inclusion of socio-economics in decision-making. This article reaffirms the sovereign right of all countries to potentially implement such procedure pending their international obligations. Article 26 suggests a line of causality originating from the potential release or presence of a GMO that may have an impact on the value of biodiversity to local and indigenous communities, although it is not limited to such impact. There are several ongoing discussions on whether and how to implement this article and what its potential impact could be for different countries, particularly developing economies.

Article 27 of the CPB considers liability and redress issues related to the potential release and use of a GMO. Parties to the CPB negotiated the Nagoya–Kuala Lumpur Supplementary Protocol on Liability and Redress, the ratification of which is pending by the parties to the CPB. This supplementary protocol to the CPB requires the parties to evaluate a biodiversity baseline, provide definitions of scope, adverse effect (damage), significant effect, who is the operator and the competent authority, response measures, administrative approaches, among other issues. This is not a compensatory mechanism, rather it provides for restitution to the previous state of nature of an affected ecosystem and presumably ecosystems services. Thus, the inclusion of SECs derived from biodiversity and ecosystem services may have a relevant role in valuating such components.

There are other international treaties relevant to the valuation of biodiversity and ecosystem services, and their relationship with SECs. The WTO agreements (SPS, TBT, and GATT) may limit the potential inclusion of socio-economics. In turn, the International Treaty on Plant Genetic Resources for Food and Agriculture (ITPGRFA), particularly Article 5 on the "Conservation, exploration, collection, characterization, evaluation and documentation of Plant Genetic Resources for Food and Agriculture," Article 6 on "Sustainable use of plant genetic resources," and Article 9 on "Farmers' rights" may have an influence on policy formulation and decision-making at the national and international levels.

10.5 Administrative Consequences

Any nation considering the inclusion of any particular SEC factor in a domestic regulatory framework will have to consider the legal and administrative requirements that would need to be in place to be able to assess impacts on biodiversity, particularly in relation to Articles 26 and 27 of the CPB.

From the standpoint of legal requirements, the national biosafety framework will need to include rules and regulations that define the competent authority, the operator, and what constitutes adverse effects or damage to biodiversity. Furthermore, these will have to determine what can be accepted as a significant effect as compared to a baseline and the inherent dynamics over time of ecosystems, as well as the measures that will be used for compensation or restitution of ecosystem and ecosystem services. Finally, countries will have to ensure that these measures are consistent with their national legislation and legal frameworks as well as with their international obligations.

From an administrative perspective, countries will have to ensure they have appropriate human and financial resources, as well as the technical capability to implement such regulatory procedures. The administrative burden will include identification of the operator, damage evaluation, identification and implementation of response measures, and maintaining appropriate knowledge sharing and communication flow to inform the public and/or potentially affected communities.

10.6 Summary

- Robust estimates of biodiversity value are quite valuable for decision-making but are currently hampered by the available research methods that require further refinement.
- Several research method approaches are available to researchers focused on the evaluation of biodiversity and the value of ecosystem services. There is no research method that can be considered better than others. All approaches have advantages and disadvantages that need to be carefully used as it may need to be implemented for specific ecosystems.
- Additional efforts will need to be invested to help advance the current state of methods especially with regard to their application to policy questions and its integration with other socio-economic approaches and methods in order to support policy and decision-making.
- Due to the relatively immature state of the research method approaches, application of such methods will require extensive expertise and capacity for implementation especially if its intent is to help drive policy and/or become part of a regulatory process. This may limit their application in developing agricultural economies.

- Applied researchers are advised to follow those elements of best research practice. The need exists to identify implementation challenges especially in developing countries and explore innovative approaches to improve the robustness of current research methods and approaches.

References

2010 Biodiversity Indicators Partnership (2010) Biodiversity indicators and the 2010 target: experiences and lessons learnt from the 2010 Biodiversity Indicators Partnership (Technical Series No. 53). Secretariat of the Convention on Biological Diversity, Montréal

Bennett J, Birol E (eds) (2010) Chapter I. Using choice experiments to investigate environmental conservation and economic development trade-offs. In: Choice experiments in developing countries implementation, challenges and policy implications. Edward Elgar Publishing, Northampton

Cardinale B, Duffy E, Gonzalez A et al (2012) Biodiversity loss and its impact on humanity. Nature 486:59–67. doi:10.1038/nature11148

Cockell C (2006) Biological processes associated with impact events ESF IMPACT, 1st edn. Springer, Berlin, pp 197–219 (ISBN 978-3-540-25735–6)

Convention on Biological Diversity (1992) United Nations. http://www.cbd.int/doc/legal/cbd-en.pdf. Accessed 20 June 2013

Daniel TC, Muhar A, Arnberger A et al (2012) Contributions of cultural services to the ecosystem services agenda. Proc Natl Acad Sci 109(23):8812–8819. doi:10.1073/pnas.1114773109

Foley JA, DeFries R, Asner GP et al (2005) Global consequences of land use. Science 309:570–574

Gomez-Baggethun E, Martin-Lopez B (2012) The socio-economic costs of biodiversity loss. Lychnos 3:68–73

Haines-Young R, Potschin M (2010) The links between biodiversity, ecosystem services and human well-being. In Raffaelli DG, Frid LJ (eds) Ecosystem ecology—a new synthesis. Cambridge University Press, Cambridge

Hawksworth. D. L. 1995. Biodiversity: Measurement and Estimation. The Royal Society and Chapman and Hall Publishers, London, UK. ISBN 0 412 75220 4

Millennium Ecosystem Assessment (2005) Concepts of ecosystem value and valuation approaches. http://www.millenniumassessment.org/documents/document.304.aspx.pdf. Accessed 10 May 2013

Phalan B, Balmford A, Green RE et al (2011) Minimising the harm to biodiversity of producing more food globally. Food Policy 36(Suppl 1):S62–S71

Rabosky DL (2009) Ecological limits and diversification rate: alternative paradigms to explain the variation in species richness among clades and regions. Ecology Letters 12(8):735–743. doi:10.1111/j.1461-0248.2009.01333.x

Smale M (ed) (2006) Valuing crop biodiversity: on-farm genetic resources and economic change. CABI Publishing, Wallingford

Secretariat of the Convention on Biological Diversity (2000). Cartagena Protocol on Biosafety to the Convention on Biological Diversity: text and annexes. Montreal: Secretariat of the Convention on Biological Diversity. Extracted from the Internet May 31, 2013 at http://bch.cbd.int/protocol/text/

TEEB (2010) Mainstreaming the economics of nature: a synthesis of the approach, conclusions and recommendations of TEEB. http://www.teebweb.org/wp-content/uploads/Study%20and%20Reports/Reports/Synthesis%20report/TEEB%20Synthesis%20Report%202010.pdf. Accessed 20 June 2013

Tilman D, Fargione J, Wolff B et al (2001) Forecasting agriculturally driven global environmental change. Science 292:281–284

UNEP-WCMC (2011) Developing ecosystem service indicators: experiences and lessons learned from sub-global assessments and other initiatives (Technical Series No 58). Secretariat of the Convention on Biological Diversity, Montréal

United Nations (2005) Millennium Development Goals Report 2005. United Nations Department of Public Information, New York. http://unstats.un.org/unsd/mi/pdf/MDG%20Book.pdf Accessed 31 May 2013.

Chapter 11
Traditional Knowledge

Peter WB Phillips

11.1 Introduction

There is an increasing interest in the nature, value, use, preservation and ownership of a wide range of genetic resources that are embodied in populations of microbes, plants, animals, and humans. These resources can be found in situ in organisms in all climates and cultures on land, in the sea, and in the air or ex situ in botanical gardens, gene banks, and public and private research collections. Genetic resources are inextricably intertwined with the environment (including human populations as hosts and users), complicating an already difficult discussion about how to manage them and how to arrange appropriate access and benefits sharing to both the primary genetic resources and any complementary or resulting inventions and innovations.

As interest rises, conflicts emerge. While national claims can be problematic, for the most part they can be adjudicated within the individual countries where they emerge. In contrast, products that may use or be related to claims of traditional knowledge (TK) that enter the international marketplace can trigger a wide range of challenges, which may be adjudicated by national courts, international tribunals, and various dispute settlement systems. While the World Trade Organization (WTO) and World Intellectual Property Office (WIPO) both have experience of adjudicating other socioeconomic considerations (SECs), the advent of the Cartagena Protocol on Biosafety (CPB), especially Art 26, has raised expectations that the issue of the "value of biological diversity to indigenous and local communities" will be realized (Secretariat of the CBD 2000, p 19). The difficulty is that the expectations are high, but the models, methods, and metrics that might be introduced as evidence in support or in defense of a challenge are not well established.

The purpose of this chapter is to outline the scale and scope of the problem, the array of definitions and conceptualizations of TK, the institutional structures that govern any claims and disputes, and the array of approaches different disciplines have used to frame the scale and scope of claims and to align rights with obligations.

P. W. Phillips (✉)
University of Saskatchewan, Saskatoon, Canada
e-mail: peter.phillips@usask.ca

K. Ludlow et al. (eds.), *Socio-Economic Considerations in Biotechnology Regulation*,
Natural Resource Management and Policy 37, DOI 10.1007/978-1-4614-9440-9_11,

11.2 Methodologies

There is a large body of scholarly work on this topic (e.g., see Phillips and Onwuekwe 2007; Bubela and Gold 2012). Interestingly, much of the work to date has a strong normative framing, as the scholars involved are both committed to and motivated by a desire to find a more equitable regime for TK, in an effort to support the advancement of the rights of indigenous peoples. While this work is valuable, it does tend to be more aspirational and exhortative than analytical and descriptive. This poses a significant challenge to those tasked with managing and adjudicating systems related to access and benefits sharing (ABS) of TK, as the prescriptive perspective has not offered much in the way of confirmed and validated models, methods, and metrics that can be relied upon to deal with claims and disputes about TK and related ABS.

Legal and common law studies tend to advocate for new institutions or examine the roles of patents and copyright and the potential for using *ordre public* in patent systems to address conflicting claims. Meanwhile, economists have examined the potential incentive or disincentive effects of the emerging property system and assessed the scale of benefits that might be shared while sociologists and some economists have examined the potential impacts of different rights regimes on the socio-economic systems of indigenous communities. Crosscutting all of these efforts, a range of legal and social scholars and NGOs has interjected concerns about the ethical underpinnings of the current or any prospective ABS system.

Although these issues are all valid and important, they largely ignore how ABS systems currently operate. Much of the debate about ABS has been narrowly constructed to examine the role of patents and treaties in delimiting rights and facilitating exchange of codified inventions. Although this is clearly a relevant issue, a full accounting of the ABS issue requires examination of a full slate of types of knowledge—including TK (which is usually tacit, context laden, and seldom codified in any formal sense and frequently known only by a limited group in a community)—and a complexity of different types of institutions.

Dutfield (2004) asserts there are three compelling reasons to protect TK: to fulfill moral obligations toward indigenous and local communities; to comply with legal requirements embodied in international treaties and emerging norms (e.g., the CPB); and for more utilitarian goals such as local, national, and global economic and welfare benefits and for improved sustainable management of biodiversity and conservation.

11.2.1 Definitions and Moral Obligations

Stone (1989) asserts that policy problems are not found, but rather they are constructed through the use of causal stories. The policy issues around TK remain somewhat loosely framed, as there is no universally accepted definition or causal story that offers hard boundaries for what it is and how one might work with it. Phillips and Onwuekwe (2007) note that neither the Convention on Biological Diversity (CBD 2013), the source of the international agenda related to TK and ABS, nor the Rio Declaration or Agenda 21 defines the term, even though they use it in various forms. Article 8j of the CBD goes the furthest, identifying a range of definitions, rights, and obligations related to "TK, innovations, and practices". Article 8j asserts that the parties concur that this involves "knowledge, innovations and practices of indigenous and local communities embodying traditional lifestyles relevant for the conservation and sustainable use of biological diversity". As with any definition, the devil is in the details. The provision, in and of itself, does not define "indigenous and local communities", " traditional lifestyles," or "conservation and sustainable use". To fill in the gaps, the Secretariat of the CBD offers this elaboration:

> Traditional knowledge refers to the knowledge, innovations and practices of indigenous and local communities around the world. Developed from experience gained over the centuries and adapted to the local culture and environment, traditional knowledge is transmitted orally from generation to generation. It tends to be collectively owned and takes the form of stories, songs, folklore, proverbs, cultural values, beliefs, rituals, community laws, local language, and agricultural practices, including the development of plant species and animal breeds. Traditional knowledge is mainly of a practical nature, particularly in such fields as agriculture, fisheries, health, horticulture, and forestry (Convention on Biological Diversity 2006).

Dutfield (2001) contends that TK exists notwithstanding its definitional dilemma. He adopts Martha Johnson's definition of traditional ecological knowledge, as a starting point: "A body of knowledge built by a group of people through generations living in close contact with nature. It includes a system of classification, a set of empirical observations about the local environment, and a system of self-management that governs resource use."

One reason TK is difficult to define is that it exhibits an array of dichotomies: it can be both explicit and implicit; it can be local or global; and it can be individual or collective (Jensen et al. 2007). This array of attributes means the functional space encompassed by the term is wide and variable.

There have been various attempts in academia to parse the concept of knowledge into tractable and measureable components by considering their degree of codification (tacit or codified) and the degree of dispersion (held by individuals or extending to the collective). Crookshanks and Phillips (2012) extended Malecki's typology to include six discernible types of knowledge: know-why, know-what, know-how, know-who, know-where, and know-when (Table 11.1).

Each type of knowledge has special features. Know-why refers to explanatory knowledge of the principles and laws of nature, which in the scientific domain is generally undertaken in publicly-funded universities and not-for-profit research institutes and is subsequently codified and published globally in academic or professional journals. Traditional communities also have their causal stories that are preserved and transmitted, usually orally or in cultural practices. Know-what refers to knowledge about techniques and recipes, which in the scientific domain can be codified and transferred through the commercial marketplace for ideas; traditional communities also have some capacity to develop instructions. Know-how refers

Table 11.1 Classification of types of scientific and traditional knowledge. (Source: Adapted from Crookshanks and Phillips 2012)

	Degree of codification	Scientific knowledge	Traditional knowledge
Know-why	Completely codified, either in journals or oral tradition	Developed by universities and public labs and fully published and disbursed in scientific journals or books	Developed, preserved, and transmitted by shaman or community leaders
Know-what	Completely codified in patents, business practices, or tradition	Universities, public labs, and private companies often advance basic knowledge to application and use patents (which disclose methods) or trade secrets to exploit	The recipes for successful exploitation are often collectively normalized, conserved, and transmitted (by leaders and community members)
Know-how	Tacit, not codified	Developed hands-on in labs or teams (i.e., learned by doing) and often has only limited dispersion as it becomes an effective trade secret	Often widely held in hands of community members
Know-who	Tacit, not codified	Developed and sustained within firms or research communities	Often limited to single communities
Know-where	Traditionally tacit; can be codified on maps	Often not formally part of system; most important for users of scientific knowledge	Developed and preserved in communities; recently codified on maps
Know-when	Traditionally tacit; can be codified	Seldom part of formal system; most important for users of scientific knowledge	Developed and preserved in traditional communities; often managed by temporal or civil leadership; can be codified

to the combination of intellectual, educational, and physical dexterity, skills, and analytical capacity to effectively combine know-why and know-what knowledge. This capacity is often learned through education, technical training, or other forms of tutelage and perfected by doing, which in part generates a barrier for the uninitiated and makes it difficult to transfer to others and, hence, more difficult to codify. Know-who, which involves information about "who knows what and who knows how to do what", is especially important in traditional communities and is becoming increasingly important in the scientific enterprise as the breadth of knowledge required to innovate expands and collaboration become essential. Know-who knowledge is seldom codified but accumulates often within an organization or, at times, in communities. To these four "modern" types of knowledge, one can add know-where and know-when, which are the foundation of much of the TK claimed by indigenous groups. Essentially, entire communities, their leaders, or special members of traditional communities frequently have knowledge of where and when one might find and use specific plants, animals, or microbes that are important for their food or culture. These two types of knowledge often are indistinguishable, as the time and place can be distinct for each species or occurrence. While these features can be important in advanced agricultural and resource development, the underlying knowledge is increasingly disconnected from the scientific enterprise—instead it becomes the unique asset of the users of the technologies, such as farmers who know which fields have the best returns with specific crops or practices.

Another complementary way to classify knowledge is through the use of modalities. According to Michael Gibbons (1994) and a group of colleagues at Science Policy Research Unit (SPRU) at University of Sussex, focus should be put on the institutions governing knowledge generation rather than the types of knowledge. The Gibbons model posits two modes of knowledge generation. Mode-1, labelled modern or scientific knowledge (following on the lines of the classics of the sociology of science by Robert Merton, Karl Popper, and Thomas Kuhn), is generated within disciplinary and cognitive contexts. Mode-2, in contrast, is created in a broader, transdisciplinary social and economic context and tends to lead to a more socially distributed knowledge. Crookshanks and Phillips (2012) assert neither of the modes put forward by Gibbons adequately describes TK. They argue TK emerges from a Mode-0 system: it is generated only within a stable collaborative social context; it is inductive in nature; it is fundamentally inseparable from its social and environmental context; and it is seldom developed or manipulated with the aim of attaining commercial goals. Mode-0 knowledge is distributed throughout society, similar to Mode-2 knowledge, except that it is adapted to changing environments and community pressures.

It is clear from the foregoing that the policy area is ill defined and probably expansive and that expectations are high. Given the rather fuzzy and nebulous definition of TK, the debate about who owns and controls it and how ABS should occur is highly complex and fully engages a wide range of groups in global society. Twarog (2004) concludes that this highlights the need for a common and clearer definition that narrows the range of issues under discussion, sharpens the focus on the economic dimension, and reduces the potential for confusion and failed coordination among forums and agencies.

11.2.2 Legal and Institutional Approaches

Much of the legal scholarship and most of the international negotiations are founded on a normative set of principles that vest indigenous peoples, TK and related plants and animals with intrinsic value, above and beyond what might be measureable in any explicit applied use. These principles are derived from a variety of sources, including philosophy, common law, human rights, and sociology. For the most part, this style of scholarship has generated a range of valid and informative arguments in favor of extending and affirming collective rights to TK and to manage an ABS regime. While much of this is focused on the formal IP rights (IPRs) system—

embodied in national patent and plant breeders' laws and in international treaties such as the WIPO and the TRIPS Agreement of the WTO—a complementary approach has been to examine the institutions that give effect to policies that relate to TK and ABS. Along the way, the literature has helped to frame how one might view ownership and what principles and approaches might be used to examine contested claims of ownership or to resolve disputes over unauthorized uses of TK.

There is quite an extensive literature that examines the definition of rights for TK and the framing of claims to benefits from those accessing those rights. The root of this debate was the 1983 International Undertaking on Plant Genetic Resources (IUPGR), which strove "to ensure that plant genetic resources of present or potential economic and/or social importance, particularly for agriculture, will be explored, preserved, evaluated, and made available for plant breeding and other research purposes." Its fundamental assumption was "that plant genetic resources are a common heritage of mankind and consequently should be available without restriction." Many developing nations and quite a few scholars were concerned about that blanket appropriation of the TK embodied in seeds (Sullivan 2004). In 1996 the UN Environmental Program in the context of the CBD developed a set of guidelines on appropriate roles and structures for ABS. In addition to 8j of the CBD, a number of other articles offer direction and advice, including: 16 (provisions on access to and transfer of technology), 17 (exchange of information), 18 (technical and scientific cooperation), 19 (the handling and distribution of the benefits of biotechnology), and 20 and 21 (financial resources and mechanisms). These provisions suggest benefits that could involve: monetary compensation in the forms of fees, research support, royalties and salaries; or nonmonetary benefits, such as in-kind support for institutions and communities, information, transfer of equipment, software and knowhow, training, joint R&D, capacity building, and local employment.

Against this institutional backdrop, scholars have attempted to refine how TK might be managed internationally and what rights, if any, claims of TK might have on benefits sharing. At one extreme, Craig (2007) examines the international law and policy relating to human rights to determine whether it could provide the basis for a *sui generis* system for protecting the integrated rights of indigenous people. She asserts it is becoming increasingly clear that this type of system accords closely with a growing body of international law and policy specifically relating to indigenous rights and the aspirations of indigenous peoples for self-determination. She posits that much work remains to be done to understand human rights, environment rights, and specific indigenous rights—the sometimes strained relationship between the environmental and indigenous rights movements raises fundamental ethical, legal, and moral issues, which could come to play in any dispute. At the other extreme, Mgbeoji (2007) argues that appropriation of indigenous peoples' knowledge is rooted in the "colonial assault" on indigenous and TK systems. In effect, the appropriation of indigenous knowledge systems cannot be divorced from their cultural and economic underpinnings. Accordingly, he argues modern scholarship seeking to devise regimes for the protection and promotion of indigenous knowledge of the uses of plants would necessarily need to re-examine the historical roots. He is especially concerned about the mischaracterization of indigenous knowledge as

Table 11.2 Typology of ABS regimes. (Source: Adapted from Crookshanks and Phillips 2012)

		Benefits sharing	
		Commercial	*Noncommercial*
Access	*Legal*	1: Brazil, Panama, Mexico, Malawi	2: Uganda, The Philippines, Peru, Costa Rica, Indonesia
	Nonlegal	3: Samoa, Bangladesh Ghana, Nigeria, Vietnam, Laos, South Africa	4: Pacific Islands Framework Agreement, OAU Model Law

superstition, which he views as an unfair generalization of a legitimate alternative worldview. He sees the emergence of indigenous knowledge systems as legitimate systems of knowledge that can and should be recognized in modern international law, but he acknowledges that in the final analysis effective protection requires significant work at the domestic level. He asserts the first step will need to be to explore the juridical resources already recognized by indigenous peoples in their daily production, use, sharing, and propagation of knowledge. Somewhere in the middle, Castle and Gold (2007) use a set of legal and philosophical arguments to assess claims for compensatory benefits, concluding that "justifications for benefit sharing cannot be derived from claims to property rights in traditional knowledge, if not because natural property rights are themselves problematic, then because property is normally considered free unless there is a normative justification for restricting access, particularly in the case of knowledge assets".

Dutfield (2004) bridges between the normative and positive approaches, asserting that the moral and legal obligations intertwine in an array of regimes and instruments, including customary law, IPR vehicles (such as patents, copyrights, and plant variety rights), concepts in civil and common law related to unfair competition, privacy, breach of confidence and passing off, and contracts law (including provisions related to trade secrets, licenses, and material transfer agreements).

Crookshanks and Phillips (2012) adopted a more positivist approach in a survey of the various institutional approaches used in a range of countries, seeking to find common or best practices of instruments. Using a variant of the framing of TK discussed above, they developed a typology for differentiating different access and benefit sharing regimes, based on the type of access afforded (legal or nonlegal) and the primary type of benefits (commercial and noncommercial) targeted. Table 11.2 shows the 2×2 matrix of possibilities this generates.

Quadrant one, commercial legal systems, involves a strict system of rules that bio-prospectors would have to comply with in order to gain access. The primary aim of these rules is to ensure that financial benefits are gained through each bio-prospecting agreement. The preservation and protection of TK is only of secondary importance. Brazil, Panama, Mexico, and Malawi all have adopted this approach. Quadrant two, noncommercial legal systems, also have strict systems of rules that interested parties must follow to gain access to genetic resources and TK, but in this case there is rarely a financial incentive behind these rules; the desire to preserve and protect local communities drives these restrictions, which results in a largely conservation-based, community development approach. Uganda, The Philippines,

Peru, Costa Rica, and Indonesia all have systems like this. In quadrant three, commercial, nonlegal systems offer open access, with the host country seeking financial gains without formal legislation. Agreements between the host country and the interested party vary with each transaction; no consistent ABS rules exist. National governments will sometimes offer negotiation guidelines, but allow for a case-by-case determination of the details. Bio-prospectors often prefer this type of system as negotiating power is often in their favour, allowing them to create an agreement that maximizes their interests. Samoa, Bangladesh, Ghana, Nigeria, Vietnam, Laos, and South Africa generally operate under such rules. Quadrant four encompasses noncommercial, nonlegal systems, which generally apply in the context of regions rather than at the nation-state level. These structures offer model frameworks to guide countries toward adopting their own precise goals and enforceable rules. Regional initiatives have no enforcement power and remain nonlegal in nature. Moreover, regional organizations rarely have full support from member states, as national governments do not want to relinquish sovereignty over decision-making. As a result, participating countries seldom support regional interests, preferring that any benefits derived within the country's borders remain in the control of the national government. Both the Pacific Islands Framework Agreement and the Organization of African Unity Model Law are examples of this approach.

In practical terms, both the legal and institutional perspectives offer guidelines that might be useful in the development and management of ABS systems for TK and for adjudicating disputes about various uses. In addition to the discussion by Crookshanks and Phillips (2012) about the scale and scope of the legal and commercial relationships (which would go some way to defining "standing" in any dispute), a number of other studies have gone into some detail about appropriate steps or best practices for collecting data (Thornstrom 2009), deal making (Costanza et al. 2009), and the role of public and private registries (Hansen and van Fleet 2009). Given that trade and property disputes are frequently decided on procedural grounds rather than substantive evidence, this body of literature helps to frame the nature of various arguments and processes for gathering evidence that might be used to determine whether indigenous communities, governments, or researchers have engaged in transparent and fair dealing, including advanced informed consent and best efforts in concluding and managing equitable ABS. Ultimately, commercial best practices in terms of marketplace transactions can provide a benchmark of what constitutes "fair and equitable" (Tully 2003).

11.2.3 Economic Evidence

Most of the economic argumentation and analysis related to TK and ABS has focused on two primary issues. First, economists are vitally concerned with innovation. They usually look at the role of incentives and institutional factors on the rate of investment in R&D and subsequent improvements in productive capacity. In that sense, economists are most concerned about gaining efficient and effective access to knowledge from all available sources and not explicitly about protecting and preserving special types of property. Second, a number of economists, either directly or by inference, have attempted to measure the economic value of TK, which could be an important input into policy debate, specific commercial ventures, and adjudicating disputes.

Overall, economists (and many lawyers), while acknowledging the inherent biases of any intellectual property (IP) system, tend to justify the current property rights regime—where exclusionary rights are granted for fixed terms, in exchange for disclosure of the protected method—as a second-best approach to accelerating investment in research, development, and commercialization. The argument is that while any resulting monopoly power may generate short-term profits, intellectual property rights (IPRs) ultimately generate incentives that raise the level of productivity and social welfare of a wide range of people over the long term—the generally accepted conclusion is that the long-term gain justifies the short-term pain. In this context, then, economists are most concerned about the incentives for investment, which could include the costs of research (which can be influenced directly by public subsidies and public provision of specific research services or infrastructure or indirectly by the spillovers of localized innovation systems), the length and cost of the commercialization pathway (which may vary depending on government policy and the forward links into the market) and the durability of market power (which is conventionally modelled as a question of scope of patent or IP claims and duration of those rights). Scotchmer (2004) and Alston et al. (1995), among others, have fully delimited the economic model, formulated a range of methodological approaches to analysis, and contributed to a rapidly growing body of empirical evidence that demonstrates the aggregate and differential effect of each of the specific parameters of the system.

One important empirical result particularly relevant for TK and ABS policy is that the distribution of outputs and benefits is almost always skewed, which worries many governments and policy advisors. Most of the analyzes tend to assign minimal or no economic value to the TK and accumulated productive capacity in landraces and traditional cultivars, instead focusing on the value added by the application of proactive, modern scientific enterprise. Critics often quite convincingly point to the important contributing role of genetic resources or TK from developing countries and express serious concerns that unless some resolution is found, these resources may not (and some argue maybe should not) be available as inputs to the global agri-food research system.

Mandelsohn (1999) offers a second economic interpretation, suggesting that while the value in TK may be undervalued, there is some market potential. He points out that there are significant problems associated with joint ownership, which undercut the ability to extract whatever value may be embodied in TK, and suggests that some form of monopoly may be required to realize those benefits. The challenge is to determine the type of monopoly—he posits that one that acts in the best interests of all landowners would be able to sell both germplasm and related TK, the proceeds of which would be allocated to individual owners.

Meanwhile, a group of economists and practitioners have attempted to measure the economic significance of TK, offering estimates ranging from immense to insignificant. Ultimately, the diversity of life on earth is based on the protein-generating capacity of plants—the main question is what value one assigns to the TK that underpins that system. Richards (2008) argues it is very difficult to accurately estimate the value of TK because: (i) it is often an essential component in developing other products; (ii) most TK-derived products never enter modern markets; and (iii) most TK has cultural or spiritual value that cannot be quantified in monetary terms. Nevertheless, a number of groups and individuals have made attempts to estimate the gross value. The World Bank reports that agriculture comprises 31 % of the GDP of low-income economies and the combined annual market of plant life forms (in pharmaceuticals, crop production, botanicals, and natural care) was estimated at up to US$ 800 billion in 2007 (Mgbeoji 2006; Wynberg and Laird 2007). A 1992 UNCTAD-ICTSD Project on IPRs and sustainable development put the value of plant-based medicines in the pharmaceutical industry at US$ 61 billion annually, or about 10 % of the annual value of production. Farnsworth (1988) asserts the link to TK was obvious in that they found 119 plant-based compounds used in medicine worldwide, 74 % which had the same or related uses as the medicinal plants from which they were derived. The World Health Organization (WHO) also estimated that the global market for traditional therapies, including but going beyond medicinal compounds, at more than US$ 70 billion annually. As just one illustration of the scale of the issue, the Indian Government has estimated that worldwide more than 2,000 patents are issued annually based on traditional Indian medicines. More recently, a United Nations Development Program (UNDP) study reported that developing countries are losing as much as US$ 300 million a year in unpaid royalties from farmers' seeds and over US$ 5 billion a year in unpaid royalties for medicinal plants (based on a 2 % royalty for material and knowledge transfers) (Shiva 2001).

Meanwhile, the flow of plant genetic material has major economic effects both on the donor and recipient communities, and the flows are not uniquely one way. Evenson (1996) estimated that land races acquired from India and overseas contributed about 5.6 % or US$ 75 million annually to India's rice yields. Dutfield (2004) asserts that assuming land races contribute equally to other countries where rice is cultivated, the global value added of land races used globally could be as high as US$ 400 million per year. Pardey et al. (1996) support that notion, concluding that an overall US government investment in CGIAR wheat and rice research of US$ 134 million generated a gross return up to 1996 of US$ 14.7 billion. While they did not explicitly attempt to assign value to TK and related genetic resources, their methodology clearly showed the role specific landraces played in the generation of value for US producers. Some would assert that at least some of that incremental value was due to tapping into the knowledge and genetic resources from TK and landraces.

If one uses a simple model to calculate the net present value of even the most conservative transfers of TK embodied in medicinal plants and recently accessed landraces (at a discount rate of 5 %, for example), the base value of these accessions is in the range of US$ 65 billion. Some advocates use these kinds of numbers to justify compensation claims for misappropriated value. The alternate view is that few, if any, of the TK transfers were useful without further invention and adaptation, and that most of the value being assigned to TK is actually more appropriately assigned to the subsequent investments in making this genetic material function in a new setting or new use. Posey (1999) also notes that economists at one level miss the core issue, in that they hesitate to assign any estimates to the intrinsic cultural or spiritual value of TK and related genetic resources.

11.2.4 Case Studies

United Nations Environment Program (UNEP) (1996) asserts that "since there has been little examination of benefit-sharing to date, case studies are needed on commercial and non-commercial arrangements, between different stakeholders," which can then provide a basis for a "detailed dialogue" within countries on appropriate mechanisms.

While the literature examined in Sect. 12.2.2 suggests that legislation and regulation are the primary vehicles for obtaining access to genetic resources, in practice negotiated access arrangements are more important for understanding what TK is accessed and how benefits are calculated and distributed (Tully 2003). Tully (2003, p 92) asserts at root, "mutually agreed terms are closely related to PIC [prior informed consent] and fair and equitable benefit sharing." PIC, which is laid out as a part of the Bonn Guidelines, is the principal procedural basis for regulating trade in a wide range of other areas, including hazardous waste, pesticides, and chemicals. This usually involves information exchange, export notifications, global databases, and designation of competent national authorities. While PIC involves using best available science and practices in support of fair and equitable trade, it is not as easy to assess as one might imagine, as weak behavioral standards in the various international instruments make it hard to adjudicate disputes. Concepts of respect and dignity, which underlay full equity, are vague and aspirational rather than operational.

There has been an array of case studies. Twarog and Kapoor (2004) present a range of studies on medicinal plants in Vietnam, traditional medicine in Burkina Faso and Vietnam, and food crops in Ethiopia and Brazil. Other more general geopolitical case studies include Costa Rica (Medaglia 2007) and Kenya (Nnadozie 2012), First Nations in Canada (Phillips et al. 2012), Brazil (Rodriques 2012), and South Africa (ten Kate and Laird 1999; Vermeylen 2007). The main message from this scan of the literature is that much of the work is focused on a subset of countries (often centers of genetic diversity), that most of the cases examine medicinal plants and traditional medicines and that while many of the systems examined are operational, they offer somewhat limited evidence that can be used to develop and adjudicate other ABS systems.

11.3 Critical Assessment

The main message from this analysis of the literature is that the scope of what is envisaged by TK and ABS is ill defined, the institutional landscape is still evolving and diversifying, and the methods for assessing economic value and related evidence can support a wide range of interests and opinions. Moreover, the existing body of practical cases are for the most part idiosyncratic and offer limited insight into how effective and efficient ABS systems might operate more widely.

While each of the main methods—definitional efforts through knowledge framing, institutional analysis, economic assessment, and case studies—offers some understanding of the scale and scope of the issue, they are inextricably interconnected. What one defines as TK and how one defines the rights and rules related to ownership will define what value can or should be attributed to TK and what practical, on-the-ground structures might be appropriate to realizing that value. None of the methods can stand alone.

A number of factors complicate the application of Art 8j. First, TK and the related genetic resources are not documented or managed in any way that would meet the usual regulatory and legal standards. Hansen and Van Fleet (2003) assert that the lack of formal documentation of knowledge in local and indigenous communities is a key impediment. The challenge is that as information accumulates, we are finding that there are overlapping and interlocking claims to TK and related genetic resources that frequently transcend specific indigenous communities and often span national boundaries and sometimes put different regions in conflict. While we have some sense of the original centers of origin, in many cases indigenous peoples have added significantly to those initial endowments. Although TK may be unique (or at least differentiable), the underlying genetic resources may be indistinguishable at the molecular level.

The link between TK and genetic resources in modern scientific knowledge varies widely. Many *ex situ* genetic resources have become disconnected from their roots. The recent "omics" revolution is advancing the effort to identify genes of interest in existing organisms and make use of that genetic material in the development of new and useful formulations, quite often involving sequencing the species and then dry science of data mining the genetic code. The recently launched International Barcoding of Life Project (iBOL) project may amplify concerns as this international scientific community seeks to collect, identify, and genetically barcode all of the species on earth (www.ibol.org). Meanwhile, the emergence of new bibliometric and biometric tools and methods are revealing the roots of many key genetic lines and establishing a chain of custody through ethnographic and scientific explorers' reports. Some argue that this is important in the governance of modern research, as the qualitative nature of TK has the potential to provide missing context surrounding quantitative studies (Economic Commission for Africa 2002).

Given the incomplete package of models, methods, and metrics, it is difficult to say when and where in the research pipeline that TK should be managed. While optimally one would hope that appropriately-structured ABS agreements could be constructed before research advances significantly, the cost and energy required to develop ABS arrangements would undoubtedly discourage most research teams from either accessing the related TK or negotiating an *ex ante* contract. The international effort to enact disclosure rules in national patent systems has gained some adherents, with strong disclosure rules enacted in a range of biodiverse countries (e.g., Costa Rica and the Andean countries), successively less stringent rules in a number of small, rich nations (e.g., Norway, Denmark, Switzerland, and New Zealand), and weak rules in the EU but no formal rules in the US and many other countries intensively engaged in biotechnology research (Queen Mary Intellectual Property Research Institute 2004). These rules could offer a signal to researchers about the need to consider the provenance of their genetic materials, but the staging of these rules in the IP system is probably too late in the innovation pipeline to overcome the "hold-up" problem that results from sunk costs and asymmetric bargaining (Just and Hueth 1993). In effect, these rules could simply have a chilling effect on private investment. While some granting agencies have encouraged scholars and researchers to be proactive about disclosure at earlier stages in the process, the rules are a bit hit and miss and have yet to lead to major changes in practice.

If negotiations are for the time being likely to be too little and too late to address the concerns of indigenous communities, then disputes are likely to arise. The challenge is that the adjudication of those disputes will be complicated and delayed, if adjudicated at all. The absence of unambiguous ownership, clear guidelines for standing, and a plethora of forums, none of which can adjudicate all the claims, and a blancmange of models, methods, and metrics that can provide some evidence for almost any point of view, will likely lead to complicated and contested solutions, which would complicate an already over-subscribed policy landscape.

11.4 International Arena

A large number of governments, institutions, and organizations are involved in addressing concerns about plant genetic resources ABS. In some respects, there are both too many and not enough efforts. Although individual countries have attempted to control access to their genetic resources for hundreds of years, recently there has been a diffusion of effort so that now we have an alphabet soup of international agencies, treaties, and committees that have produced a wide range of proposals, guidelines, and commitments, but little in the way of operational or judiciable structure. Meanwhile, the commitment of some key nation-states and much of the industrial research system is variable, at best. The USA, for example, is not a party to the CBD and hence is not bound by the provisions of Article 8j; while key countries such as Canada are signatories to the CBD, they have declined to fully engage or commit to hard measures that would entrench rights to TK and force more extensive ABS programming; the EU supports the effort to strengthen the system but many of its member states are less committed; and even countries representing centers of origin or diversity have differing approaches and perspectives and seldom are willing to combine their efforts to advance the international agenda.

Recently, there has been significant debate and effort invested in negotiating a range of international conventions or treaties to delimit and protect indigenous rights to genetic resources, involving the International Labour Organization, the UN and Inter-American Draft Declarations on Rights of Indigenous Peoples, the UNDP/UNCTAD, and the European, Asian, and African Development Banks. In the context of plant genetic resources, in particular, there are a number of special institutions involved in delimiting rights and facilitating ABS. These include the CBD, the Bonn Guidelines on Access to Genetic Resources and Fair and Equitable Sharing of the Benefits Arising out of the Their Utilization, Agenda 21 and the CPB, the IUPGR and ITPGR, the CGIAR centers and related genebanks, and various national programs (e.g., CIDA and SEDA). The 2010 Nagoya Protocol on ABS created a clearing house mechanism to facilitate the collection and sharing of information on the national implementation of ABS, including legislative and other measures, permits, relevant authorities, and institutions as well as codes of conduct and best practices.

In effect, the world has created a relatively dense system of national and international rights, claims, and processes for handling disputes, which all rely on a body of evidence to adjudicate (Phillips and Ryan 2010). Tully (2003) argues that while these efforts may yield some value, ultimately what is protected and how it is protected and exploited is a matter for nation-states to decide domestically, which renders international efforts to benchmark norms and standardize benefit-sharing requirements difficult. Ultimately, these measures in a worst-case scenario could simply add to negotiation costs and generate more disputes.

In the absence of formal, governmental, or intergovernmental structures, civil litigation seems to be the only way to advance the agenda. A wide range of private concerns have claimed IPRs (mostly patents in the US and EU) on a range of genetic resources (ranging from plant cultivars to human genes), which has precipitated a range of legal disputes in domestic courts (e.g., Moore v. Regents of the University of California 793 P.2d 479, Cal. 1990) and with domestic patent offices (e.g., EU Patent Office decision to revoke a patent on a fungicide derived from the neem tree).

11.5 Administrative Consequences

While national institutions in many of the case studies have demonstrated that they can strike ABS agreements and gain some direct or indirect benefits, there is little evidence that the international institutions that are being vested with authority for administering policies and adjudicating disputes have the appropriate standing, membership, staff, models, methods, or metrics to accomplish those tasks. While negotiation of treaties and conventions and drafting of law may be important, it is just as important to build the human capacity to structure appropriate ABS systems that will stand the test of adjudication and use. That capacity building effort has yet to be fully engaged.

11.6 Summary/Synthesis

- TK is incompletely defined, creating challenges for assigning rights and adjudicating disputes.
- The value of TK embedded in plant genetic resources is hotly contested as knowledge claims are both ill defined and overlapping and the net incremental value to modern breeding programs and the drugs business of using TK is estimated to range from the trivial to billions of dollars.
- ABS systems are working, but are highly idiosyncratic; best practices are yet to emerge.
- An array of international undertakings, treaties, and guidelines have created significant expectations that past and current use of TK in plant genetic resources will be used to empower and enrich indigenous communities around the world but the institutional structures are incomplete.
- There is inadequate human or institutional capacity to manage effective and resilient ABS systems.

Acknowledgment This work was undertaken as part of the VALGEN Project, administered by Genome Canada through Genome Prairie.

References

Alston J, Norton G, Pardey P (1995) Science under scarcity: principles and practice of agricultural research evaluation and priority setting. Cornell University Press, Ithaca

Bubela T, Gold ER (2012) Genetic resources and traditional knowledge: case studies and conflicting interests. Edward Elgar, Cheltenham

Castle D, Gold ER (2007) Traditional knowledge and benefit sharing: from compensation to transaction. In: Phillips P, Onweukwe C (eds) Accessing and sharing the benefits of the genomics revolution. Kluwer/Springer, Dordrecht

Convention on Biological Diversity (2013) What is traditional knowledge? http://www.biodiv.org/programmes/socio-eco/traditional/default.asp. Accessed 3 January 2013

Costanza C, Christoffersen L, Anderson C and Short JM. 2007. Deal Making in Bioprospecting. In Intellectual Property Management in Health and Agricultural Innovation: A Handbook of Best Practices (eds. A Krattiger, RT Mahoney, L Nelsen, et al.). MIHR: Oxford, U.K., and PIPRA: Davis, U.S.A. Available online at www.ipHandbook.org.

Craig D (2007) Biological resources, Intellectual Property Rights and International Human Rights: impacts on indigenous and local communities. In: Phillips P, Onwuekwe C (eds) Accessing and sharing the benefits of the genomics revolution. Kluwer/Springer, Dordrecht

Crookshanks R, Phillips P (2012) A comparative analysis of access and benefits-sharing systems. In: Bubela T, Gold ER (eds) Genetic resources and traditional knowledge: case studies and conflicting interests. Edward Elgar, Cheltenham

Dutfield G (2001) TRIPS-related aspects of traditional knowledge. Case W. Res. J. Int'l L 33:233–274

Dutfield G (2004) Developing and implementing national systems for protective traditional knowledge: Experiences in selected developing countries. In: Twarog S, Kapoor P (eds) UNCTAD summary publication protecting and promoting traditional knowledge: systems, national experiences and international dimensions. United Nations Press, Geneva, pp 141–153 www.unctad.org/en/docs/ditcted10_en.pdf. Accessed 3 January 2013

Economic Commission for Africa (2002) Why industrial revolution missed Africa: a 'traditional knowledge' perspective. http://www.uneca.org/docs/Conference_Reports_and_Other_Documents/espd/2002/TKB.pdf. Accessed 3 January 2013

Evenson R (1996) Economic valuation of biodiversity for agriculture. In: Pan American Health Organization (ed) Biodiversity, biotechnology, and sustainable development in health and agriculture: emerging connections. PAHO, Washington DC

Farnsworth NR (1988) Screening plants for new medicines. In: Wilson EO, Peter FM (eds) Biodiversity. National Academy Press, Washington DC, pp 83–97

Gibbons M, Limoges C, Nowotny H, Schwartzman S, Scott P, Trow M (1994) The new production of knowledge: the dynamics of science and research in contemporary societies. Sage, London

Hansen S, van Fleet J (2003) Traditional knowledge and Intellectual property: a handbook on issues and options for traditional knowledge holders in protecting their intellectual property and maintaining biological diversity. American Academy for the Advancement of Science, Washington DC, http://shr.aaas.org/tek/handbook/. Accessed 3 January 2013

Hansen S, van Fleet J (2009) Issues and options for traditional knowledge holders in protecting their intellectual property. In: Krattinger A et al (eds) ipHandbook of best practices. MIHR (Oxford, UK), PIPRA (Davis, US) and Oswaldo Cruz Foundation (Rio de Janeiro, Brazil) and bioDevelopments-International Institute (Ithaca, US). http://www.iphandbook.org/handbook/ch16/p06/. Accessed 15 October 2012

Jensen M, Johnson B, Lorenz E et al (2007) Forms of knowledge and modes of innovation. Resear Pol 36:680–693

Just R, Hueth D (1993) Multimarket Exploitation: The Case of Biotechnology and Chemicals. Am J Agri Eco 75:936–945

Mandelsohn R (1999) The market value of farmers' rights. In: Santaniello V et al (eds) Agriculture and Intellectual Property Rights: economic institutional and implementation issues in biotechnology. CAB International, Oxford

Medaglia J (2007) Bioprocessing partnerships in practice: a decade of experiences at INBio in Costa Rica. In: Phillips P, Onweukwe C (eds) Accessing and sharing the benefits of the genomics revolution. Kluwer/Springer, Dordrecht

Mgbeoji I (2007) Lost in translation? The rhetoric of protecting indigenous peoples' knowledge in International law and the omnipresent reality of biopiracy. In: Phillips P, Onweukwe C (eds) Accessing and sharing the benefits of the genomics revolution. Kluwer/Springer, Dordrecht

Nnadozie K (2012) Old wine in new skin: traditional knowledge and customary law under the evolving normative environment in Kenya. In: Bubela T, Gold ER (eds) Genetic resources and traditional knowledge: case studies and conflicting interests. Edward Elgar, Cheltenham

Pardey P, Alston J, Christian J, Fan S (1996) Hidden harvest: US benefits from International research aid. IFPRI, Washington DC, USA http://www.ifpri.org/sites/default/files/publications/pr6.pdf. Accessed 15 October 2012

Phillips P, Onwuekwe C (eds) (2007) Accessing and sharing the benefits of the genomics revolution. Kluwer/Springer, Dordrecht

Phillips P, Ryan C (2010) Governance of International networks: a social network analysis of International institutions related to plant genetic resources. Political Studies Association, Edinburgh

Phillips P, Zhang S, Williams T DeBusschere L (2012) Canada's first Nations' policies and practices related to managing traditional knowledge. In: Bubela T, Gold ER (eds) Genetic resources and traditional knowledge: case studies and conflicting interests. Edward Elgar, Cheltenham

Posey DA (ed) (1999) Cultural and spiritual values of biodiversity. United Nations Environment Programme, Nairobi, Kenya

Queen Mary Intellectual Property Research Institute (2004) Report on disclosure of origin in patent applications. European Commission, DG-Trade, Brussels, Belgium http://trade.ec.europa.eu/doclib/docs/2005/june/tradoc_123533.pdf. Accessed 15 October 2012
Richards M-A (2008) Cultural and economic importance of traditional knowledge, traditional cultural expressions and genetic resources for Caribbean countries. Presentation at WIPO/GRTK/KIN/08 Meeting, Kingston, Jamaica, March 18. http://www.wipo.int/meetings/en/doc_details.jsp?doc_id = 114456. Accessed 15 October 2012
Rodriques E Jr (2012) Property rights, biocultural resources and two tragedies: some lessons from Brazil. In: Bubela T, Gold ER (eds) Genetic resources and traditional knowledge: case studies and conflicting interests. Edward Elgar, Cheltenham
Scotchmer S (2004) Innovation and incentives. MIT Press, Cambridge
Secretariat of the Convention on Biological Diversity (2000) Cartagena protocol on biosafety to the convention on biological diversity. Secretariat of the Convention on Biological Diversity, Montreal
Shiva V (2001) Protect or plunder? Understanding intellectual property rights. Zed Books, London
Stone D (1989) Causal stories and the formation of policy agendas. Polit Sci Quart 104(2):281–300
Sullivan S (2004) Plant genetic resources and the law: past, present, and future. Plant Physiol 135(1):10–15
ten Kate K, Laird SA (1999) The commercial use of biodiversity: access to genetic resources and benefit sharing. Earthscan, London
Thornstrom C-G (2009) Access and benefit sharing: understanding the rules for collection and use of biological materials. In: Krattinger A et al (eds) ipHandbook of Best Practices, MIHR (Oxford, UK), PIPRA (Davis, US) and Oswaldo Cruz Foundation (Rio de Janeiro, Brazil) and bioDevelopments-International Institute (Ithaca, US). http://www.iphandbook.org/handbook/ch16/p02/. Accessed 15 October 2012
Tully S (2003) The Bonn guidelines on access to genetic resources and benefit sharing. Rev Eur Commun & Int Environ Law 12(1):84–98
Twarog S (2004) Preserving, protecting and promoting traditional knowledge: national actions and international dimensions. In: Twarog S, Kapoor P (eds) UNCTAD summary publication protecting and promoting traditional knowledge: systems, national experiences and international dimensions. United Nations Press, Geneva. www.unctad.org/en/docs/ditcted10_en.pdf. Accessed 3 January 2013
UN Environmental Program (UNEP), Convention on biological diversity (1996) Fair and equitable sharing of benefits arising from the use of genetic resources. Conference of the Parties to the Convention of Biological Diversity UNEP/CBD/COP/3/Inf53. http://www.cbd.int/doc/meetings/abs/absep-01/other/absep-01-equitable-en.pdf. Accessed 5 March 2013
Vermeylen S (2007) Contextualizing 'fair' and 'equitable': the San's reflections on the Hoodia benefit-sharing agreement. Local Environ 12:423–436
Wynberg R, Laird S (2007) Bioprospecting. Environ 49(10):20–32

Chapter 12
Intellectual Property

Charles Lawson

12.1 Introduction

Intellectual property (IP), broadly defined, is a series of privileges accorded to inventors and creators. These privileges are recognized through a series of international agreements that establish minimum standards. The most significant of these agreements is probably the World Trade Organization (WTO) *Agreement on Trade Related Aspects of Intellectual Property Rights* (TRIPS)[1] that binds WTO Member States to comply with a range of existing international IP agreements,[2] *and then* imposes minimum standards for copyright and related rights, trademarks, geographical indications, industrial designs, patents, layout designs (topographies) of integrated circuits, and undisclosed information (TRIPS Articles 9–39). These IP privileges are generally recognized and enforced through national laws that are consistent with these international norms. The result is a patchwork of national laws, each attempting to articulate at least the minimum standards (albeit many are more generous—the so-called "TRIPS-plus") in the context of national and regional choices. The TRIPS standards adopted and applied by WTO Member States are then subject to WTO dispute resolution and penalties that include retaliatory trade sanctions where states have not implemented and applied their obligations to maintain TRIPS' minimum IP standards (TRIPS Articles 64). The WTO dispute resolution and sanction mechanism makes TRIPS one the few enforceable international laws, and hence its gravity in assessing the impacts of IP.

[1] Albeit there are others, such as, *International Union for the Protection of New Varieties of Plants* (plant variety rights), *Nagoya Protocol on Access to Genetic Resources and the Fair and Equitable Sharing of Benefits Arising from their Utilization* (traditional knowledge), *International Treaty on Plant Genetic Resources for Food and Agriculture* (farmers' rights), and so on.

[2] These are the *Paris Convention for the Protection of Industrial Property*, the *Stockholm Act of the Paris Convention for the Protection of Industrial Property (1967)*, the *Berne Convention for the Protection of Literary and Artistic Works*, the *Paris Act of the Berne Convention for the Protection of Literary and Artistic Works (1971)*, the *International Convention for the Protection of Performers, Producers of Phonograms and Broadcasting Organizations*, and the *Treaty on Intellectual Property in Respect of Integrated Circuits (1989)*: TRIPS Article 2.

C. Lawson (✉)
Australian Centre for Intellectual Property in Agriculture, Griffith Law School, Griffith University, Gold Coast, QLD, Australia
e-mail: c.lawson@griffith.edu.au

K. Ludlow et al. (eds.), *Socio-Economic Considerations in Biotechnology Regulation,* Natural Resource Management and Policy 37, DOI 10.1007/978-1-4614-9440-9_12,

IP is important to living modified organisms (LMOs) under the Cartagena Protocol on Biosafety (CPB) as many of these organisms are, or embody, inventions and creations that are protected by IP. For example, a genetically modified (GM) LMO might contain patent-protected gene constructs and sequences, be the product of a patent-protected process, be marketed using a trademark, and be associated with confidential information provided to satisfy regulatory standards in an importing country.

The consequence of IP is to grant exclusivity to the IP holder and exclude all others, preventing them from using or applying the product or process protected by IP. Thus, patent protected LMOs can only be used by the patent owner, or those licensed or authorized by the patent owner (e.g., TRIPS Article 28). All others will be subject to sanction (an account of profits and damages that can be up to treble damages in some jurisdictions) for using, applying, copying (making), importing, or otherwise infringing the IP owner's exclusive rights (e.g., TRIPS Article 28). The effect of the IP can be either to prevent others accessing and using the IP-protected product or process (such as using the patented LMO), or doing something as a consequence of the IP-protected product or process (such as importing the patented LMO). The result is that for the trade in LMOs under the CPB and TRIPS, there are potentially internationally contested inherent conflicts about IP protection, such as:

a. *Equality of access to genetic resources*—IP-protected genetic resources may not be available at a reasonable price. The result is that IP might limit access to a particular genetic resource or impose barriers to the uses of a particular genetic resource.
b. *Sharing the benefits from the accessed genetic resources*—IP may not adequately reward genetic-resource providers through royalties or preferred access to the IP-protected products or processes. In essence, the IP appropriates ownership of the genetic resource and inhibits the adequate sharing of the benefits that might flow from the exploitation of that genetic resource.
c. *Freedom to conduct research*—Key products and processes necessary for research may be IP protected. The result is that others wanting to conduct research might be limited in the research they might conduct and the commercialization of any outcomes of their research because of IP protections. This is likely to be particularly problematic where the IP protects and limits the uses of key research inputs and materials. The problem can be either limited access to IP-protected products or processes, or having to negotiate too many licenses to access and use a material (such as a "patent thicket").
d. *Freedom to adopt and apply technology*—Key products and processes necessary for exploiting the LMOs may be IP protected, including both the LMO itself and other products and processes, such as pesticides and herbicides. This includes the problems associated with transferring technology, and the know-how to enable the technology to be exploited.

e. *The increasing consolidation and monopolization of agriculture and food production (especially among seeds and chemical companies)*—The positioning of IP-protected products or processes into vertically integrated production and supply chains (especially where corporations have built their businesses around the exclusivity delivered by IP-protected products or processes) limiting supply and increasing price. This can be particularly problematic where there is not adequate competition on the market (and there is a lack of alternative substitutes).
f. *The terms and conditions by which IP-protected products and processes can be exploited*—The restrictive terms and conditions applying to users of IP-protected products or processes, especially onerous license agreements (such as maintaining records for 20 years) and the avoidance of exhaustion so that the purchaser never actually owns the purchased item. This raises particularly difficult problems for saving, reusing, selling, and exchanging seeds.
g. *Traditional knowledge and folklore*—Either the failure to take into account or address traditional knowledge and folklore in the discovery, research, and development and commercialization of IP-protected products and processes, or the deleterious effects LMOs might have on traditional knowledge and folklore, such as displacing traditional farming and cultural practices, and so on.

While the CPB does not specifically address IP, other than "confidential information" (Article 21), there is express recognition that "trade and environment agreements should be mutually supportive with a view to achieving sustainable development" and "this [CPB] shall not be interpreted as implying a change in the rights and obligations of a Party under any existing international agreements" (Preamble). Further, any "socio-economic considerations" taken into account in reaching decisions about importing or regulating LMOs under the CPB must be "consistent with their international obligations" (Article 26(1).[3] Thus, the CPB must be applied and implemented consistent with the Parties' other commitments to international IP agreements, and this specifically includes among others the WTO's TRIPS.

Thus, IP does have the potential to limit the CPB framework's reconciling of environmental and trade needs *and* any measures adopted under the aegis of the CPB that must be "consistent with their international obligations." Importantly though, the impact of IP is likely to be considerably reduced in practice as many countries, including those adopting and applying domestic IP arrangements, are not required to enforce IP *unless* the IP is expressly protected within that country. While some developed countries register and recognize large numbers of IP-protected products, processes, and creations, significant amounts of IP (particularly patents and plant variety rights) are not registered and protected in most developed and developing countries. IP remains, however, a significant limitation on the CPB.

[3] Noting also the Preamble: "Recognizing that trade and environment agreements should be mutually supportive with a view to achieving sustainable development."

12.2 Methodologies

The IP issue in the context of the CPB is best framed as whether there are exemptions, exceptions, and flexibilities in the international IP agreements that might accommodate decisions involving socioeconomic considerations (SECs) made under the CPB. This contemplates IP as a grant recognized in national laws that is either extinguished or restricted where a SEC in a decision made under the CPB imposes on the IP by limiting the exclusive rights of the owner. Framed this way there are a range of exceptions and flexibilities that allow SECs to limit IP, and this in turn frames the possible content of those SECs. Those affecting patents and plant breeders' rights under TRIPS are considered here as they are likely to be the main IPs affecting LMOs.

A further critical distinction is also necessary. TRIPS generally provides for the grant of negative rights so that the privilege awarded is not to exploit or use a particular protected subject matter *but rather* to prevent certain acts with that subject matter, such as copying (making), using, offering for sale, selling, or importing—"This fundamental feature of [IP] protection inherently grants Members freedom to pursue legitimate public policy objectives since many measures to attain those public policy objectives lie outside the scope of [IP] and do not require an exception under the TRIPS Agreement" (European Communities 2005, (7.210)). This perspective justifies a range of laws applied to subject matters generally, independent of the IP, such as health and safety laws for medicines and devices. For example, the Australian *Therapeutic Goods Act 1989* (Cth) ss 41B-41MR regulates the safety and performance characteristics of medical devices preventing any devices, whether IP protected or otherwise, from being imported, exported, used, or supplied unless they satisfy various standards.

The WTO context of TRIPS is also important. Most WTO agreements (except TRIPS) are directed toward removing trade barriers setting limits on national regulation based on national treatment and non-discrimination. This allows regulation that does not go *above* the threshold set by the WTO agreements. Meanwhile TRIPS establishes and imposes minimum standards requiring regulation that does not go *below* the threshold set by TRIPS. The effect of this structural restraint in TRIPS is that there is little room for national differences about how to satisfy a minimum standard, contrary to the alternative that allows a range of measures up to the threshold. The norms for the minimum IP thresholds are generally well known and so provide little room for significant regulatory differences.[4]

Further, the kinds of privileges exercised by an IP owner are private interests albeit they have a broader public benefit (promoting innovation, creation, investment, and dissemination). This means, however, that governmental decisions about LMOs made under the CPB are affecting the private interests of an IP owner, and this is likely to be a third party to the governmental LMOs' decision under the CPB.

[4] Albeit TRIPS provides: "Members shall be free to determine the appropriate method of implementing the provisions of this Agreement within their own legal system and practice": TRIPS, Article 1(1).

In essence, the CPB decisions (and regulation) are imposed by governmental bodies setting a legal environment while the IP owners are private entities (persons and corporations) whose property interests are directly impacted by the CPB decisions (and regulation). This is significant because it means that the concessions being sought from IP owners under the CPB's SECs are generally being made by non-governmental IP owners, and the concession being made will be either a financial detriment (such as the loss of a royalty payment) or a commercial disadvantage (such as allowing competition).

The basic methodology is, therefore, to:

(1) Identify what IP is embodied in the LMO
(2) Identify who owns the IP embodied in the LMO
(3) Check whether the IP embodied in the LMO is enforceable in the jurisdiction
(4) Check whether the CPB's SECs will affect the IP embodied in the LMO

Only if the answer to the last question is "Yes" is there an issue about the conflict between the CPB and IP. This assessment of the methodologies suggests three avenues by which the CPB's SECs, as they relate to IP, might be addressed: (1) within the exemptions and exceptions in TRIPS; (2) within the flexibilities in TRIPS; and (3) the financial and technology transfer mechanisms in TRIPS and the *Convention on Biological Diversity* (CBD) to offset the effects and consequences of IP. Similar avenues might also be considered for other international IP agreement commitments. These avenues under TRIPS and the CBD are now considered in turn.

12.2.1 Exemptions and Exceptions in TRIPS: Articles 27, 30, and 31

The distinction between exemptions and exceptions is that the former does not require a patent be available for a subject matter (such as a plant or animal) at all, while the latter requires that patents for a particular subject matter might not be available according to the particular circumstances (such as particular microorganisms in a public health emergency). The scope of these exemptions and exceptions are available within the bounds of the TRIPS text. Unfortunately the text is not entirely clear and there are contested understandings about what may or may not be allowable. So while these exemptions and exceptions are theoretically possible their actual scope remains uncertain.

The express exemptions for patents in TRIPS Article 27 are available for *ordre public* or morality (Article 27(2)), diagnostic, therapeutic, and surgical methods for the treatment of humans or animals (Article 27(3)(a)), plants, and animals other than microorganisms and certain essentially biological processes (Article 27(3)(b)). Where patents are not available for plants then "an effective *sui generis* system or by any combination thereof" must be available. A generally accepted *sui generis* system is the *International Convention for the Protection of New Varieties of Plants* that provides a broad exemption for the "public interest" (Article 17(1)).

This Convention also provides exceptions for acts done privately and for non-commercial purposes, acts done for experimental purposes, and certain acts done for the purpose of breeding other varieties (Article 15).

A general and more limited exception is available in TRIPS Article 30 for patented inventions that satisfy the cumulative thresholds of not unreasonably conflicting with a normal exploitation of the patent and not unreasonably prejudicing the legitimate interests of the patent owner, taking account of the legitimate interests of third parties. This is definitely an available, and not an illusory exception, although it is not a general basis for avoiding patents. This exception has been relied on, for example, to allow the making and using of the patent protected invention in gaining regulatory approvals for manufacturing and sale after the patent expired (the so-called "Bolar exception") (Canada—Patent Protection of Pharmaceutical Products 2000, (7.38)), but not for the manufacturing and stockpiling of a patent protected product in anticipation of a patent expiry (Canada—Patent Protection of Pharmaceutical Products 2000, (7.84)).

Another general exception is available in TRIPS Art 31 for laws allowing "other use of the subject matter of a patent without the authorization of the right holder," subject to respecting conditions and procedures aimed at protecting the "legitimate interests" of the rights holder. This is essentially a provision that allows for compulsory licensing for government and third-party uses. In other words, Members may determine the *grounds* for an award of a compulsory licence,[5] but must accord with the *conditions and procedures* required by TRIPS.[6] These are that each authorization is to be "considered on its…merits" and subject to review; that "efforts to obtain authorization on reasonable commercial terms and conditions" have been unsuccessful within a "reasonable…time"; the authorization has a limited "scope and duration"; the authorized use is not exclusive; the authorized use is not assignable; the authorized use is "predominantly for the supply of the domestic market"; and the authorized use may be terminated when the circumstances requiring authorization cease and there is adequate remuneration and this decision is reviewable (Article 31(a)–(j)). The issuing of authorizations for anticompetitive conduct is treated separately (Article 31(k)), and additional requirements are imposed for the proper working of another patent (dependent patents) (Article 31(l)).

12.2.2 Flexibilities in TRIPS: Articles 7 and 8

Within TRIPS, the main flexibility in dealing with socioeconomic questions is Art 8 that provides:

1. Members may, in formulating or amending their laws and regulations, adopt measures necessary…to promote the public interest in sectors of vital import-

---

[5] This was confirmed in the Doha Declaration: WTO Ministerial Conference (2001), *Declaration on the TRIPS Agreement and Public Health*, WT/MIN(01)/DEC/2, (5(b)).

[6] These conditions and procedures are generally addressed in Article 31, although there may be other relevant conditions imposed by TRIPS such as national treatment (Article 3), most-favored nation treatment (Article 4), and so on.

ance to their socioeconomic and technological development, provided that such measures *are consistent with the provisions of this [TRIPS] Agreement.*
2. Appropriate measures, provided that they *are consistent with the provisions of this [TRIPS] Agreement*, may be needed to prevent … the resort to practices which … adversely affect the international transfer of technology (emphasis added).

The key limitation in these TRIPS provisions is the phrase "are consistent with the provisions of this [TRIPS] Agreement." This might be interpreted to mean that any measures adopted that limits TRIPS addressing socioeconomic factors must first satisfy the TRIPS minimum standards, in effect nullifying the apparent exceptions and flexibility provided by Article 8.[7] Alternatively, a narrower meaning might have been intended, and presumably this is the case as these words were intended to have some effect otherwise they would not have been included. The issue then is what does the term "consistent" mean? Or in the alternative, how inconsistent with the substantive parts of TRIPS (that is, the non-Art 8 parts) can a provision be before it passes the "not consistent" threshold, and is a breach of TRIPS? Unfortunately there is presently no definitive interpretation by a WTO dispute settlement panel or Appellate Body, albeit the existing decisions suggest Article 8 does allow "certain adjustments" but not "a renegotiation of the basic balance of [TRIPS]" (Canada—Patent Protection of Pharmaceutical Products 2000, (7.26)). Both the subsequent *Doha Ministerial Declaration* and the *Declaration on the TRIPS Agreement and Public Health* reinforce the importance of Article 8 (see WTO Ministerial Declaration 2001, (19); WTO Declaration 2001, (5)). The effect of these declarations on future interpretation is uncertain, albeit they will certainly justify some public health actions that are contrary to the non-Art 8 TRIPS provisions. Whether other clearly inconsistent measures are allowable seems unlikely given the panel and Appellate Body preference for the perspectives and expectations of IP holders (Okediji 2003, pp. 914–915). This analysis demonstrates, however, that there are credible avenues to assert a role for Art 8 in justifying flexibility in applying TRIPS.

Closely linked to the TRIPS main exception and flexibility in dealing with SECs are the objectives set out in Art 7 that provides:

> The protection and enforcement of intellectual property rights should contribute to the promotion of technological innovation and to the transfer and dissemination of technology, to the mutual advantage of producers and users of technological knowledge and in a manner conducive to social and economic welfare, and to a balance of rights and obligations.

There appear to be six objectives: technological innovation; transfer and dissemination of technology; use of technological knowledge; the mutual advantage of producers and users; social and economic welfare; and a balance of rights and obligations. Together with the text of TRIPS' Preamble these objectives arguably apply to

[7] Some commentators consider this provision "is thus essentially a policy statement that explains the rationale for measures taken under Articles 30 (exceptions to rights conferred), 31 (other use without the authorization of the right holder) and 40 (control of anti-competitive practices]": Gervais 2008, p. 209.

a global interpretation of TRIPS and the measures implementing TRIPS in domestic laws. (For advocates of this proposition, albeit not universally accepted, see, Yusuf 2008, p. 14; Correa 2007, p. 104. Other proponents assert that these objectives are necessary to prevent the abuse of exemptions and exceptions: see, for example, UNCTAD-ICTSD 2005, p. 133). This means that in considering any particular apparent inconsistency between a domestic law or practice and TRIPS there needs to be an assessment taking into account the totality of the domestic arrangements. This perhaps suggests considerable potential to adopt measures that can be justified by a demonstrable social and economic welfare or balance of rights and obligations in a particular domestic regulatory setting. Importantly, both the *Doha Ministerial Declaration* and the *Declaration on the TRIPS Agreement and Public Health* reinforce the importance of Art 7 (WTO Ministerial Declaration 2001, (19); WTO Declaration, (5)). Again, the effect of these declarations on future interpretation is uncertain, albeit they will certainly justify some public health actions that are contrary to the non-Article 8 TRIPS provisions.

In short, there are avenues under TRIPS for SECs to be taken into account according to the flexibilities in TRIPS. This is particularly so where the regulation addresses how the IP will be exercised or exploited. More problematic and uncertain are regulations that might limit or conflict with the IP itself. In these circumstances, TRIPS may not be very flexible and will generally need to be complied with.

12.2.3 Offset the Effects and Consequences of IP using the Financial and Technology Transfer Mechanisms in TRIPS and the CBD

Where IP does pose a problem for the CPB then, there are potential mechanisms in the CPB, CBD, and TRIPS to ameliorate these problems. At this stage, these are postulated, and undoubtedly they will remain hotly contested, especially by those advocating the primacy of IP. Perhaps the most significant problem for the CPB will be where IP must be respected, but the IP is owned by a third party rather than a governmental entity, and requires compensation for the derogation of the IP as a consequence of the decision under the CPB. In these circumstances the CPB, CBD, and TRIPS do contemplate financial and technology transfer mechanisms, albeit the opportunity and scope will be contestable.

The CPB expressly provides that in addressing the financial resources necessary for the implementation of the CPB the provisions of CBD about financial resources are applicable (Article 28), and this includes the CBD Article 20(4):

> The extent to which developing country Parties will effectively implement their commitments under this [CBD] will depend on the effective implementation by developed country Parties of their commitments under this [CBD] related to financial resources and transfer of technology and will take fully into account the fact that economic and social development and eradication of poverty are the first and overriding priorities of the developing country Parties.

And Article 20(5):

> The Parties shall take full account of the specific needs and special situation of least developed countries in their actions with regard to funding and transfer of technology.

These CBD provisions suggest that for implementing the CPB there is an express obligation to consider financial and technology transfer issues and that this is the developed country Parties making concessions to the developing country Parties.[8] Further, the CBD's financial mechanisms expressly contemplate economic and social development as relevant objects.

Along similar lines, in addition to TRIPS Article 8 (set out above), Article 66(2) provides:

> Developed country Members shall provide incentives to enterprises and institutions in their territories for the purpose of promoting and encouraging technology transfer to least-developed country Members in order to enable them to create a sound and viable technological base.

Following the Doha Ministerial Conference commitment in 2001[9] to "put in place a mechanism for ensuring the monitoring and full implementation of the (Art 66(2)) obligations" (WTO Ministerial Conference 2001, (11(2))), the TRIPS Council has implemented a scheme of annual reporting "on actions taken or planned in pursuance of their commitments under [Art 66(2)]" (Council for Trade-Related Aspects of Intellectual Property Rights 2003, (1)). This reporting addresses the legislative, policy, and regulatory framework for incentives, types of incentives, the entities providing the incentives and information about the functioning (efficiency and effectiveness) of the incentives (Council for Trade-Related Aspects of Intellectual Property Rights 2003, (3)). Within this broad remit the incentives to enterprises and institutions might well include financial and technology transfer mechanisms that address any conflicts between IP and the CPB.

12.3 Critical Assessment

The methodologies considered above essentially accept that in implementing SECs under the CPB there must be respect for IP. This assessment accepts that IP will almost certainly take a higher priority in the hierarchy of rights and obligations because of the CPB's express statement that it be applied and implemented "consistent with their international obligations" (Article 26(1)).[10] There are various exemptions and exceptions available under the IP agreements, albeit they are presently contest-

[8] There is recent interest under the CBD about effective access to and transfer of relevant technology: see Conference of the Parties to the CBD pp. 42 and 166–168.

[9] This commitment has been reiterated in the context of TRIPS and public health: see WTO General Council 2005, (6); WTO General Council 2003, (7).

[10] Noting also the Preamble: "Recognizing that trade and environment agreements should be mutually supportive with a view to achieving sustainable development."

ed and have not been clearly articulated. While in practice these apparently onerous IP obligations are likely to be diminished because the international IP obligations often depend on domestic registration, and the like, there still remains a potential problem. Where there is a problem this is likely to be magnified because IP owners are generally third parties and not governmental entities that are the parties to the CPB decisions. This might be ameliorated, however, through a creative consideration of the financial and technology transfer mechanisms available under the CBP and other agreements.

12.4 International Arena

While the analysis here has focused on the WTO's TRIPS agreement there are numerous other international IP agreements. These include the *Paris Convention for the Protection of Industrial Property*, the *Stockholm Act of the Paris Convention for the Protection of Industrial Property*, the *Berne Convention for the Protection of Literary and Artistic Works*, the *Paris Act of the Berne Convention for the Protection of Literary and Artistic Works*, the *International Convention for the Protection of Performers, Producers of Phonograms and Broadcasting Organizations*, the *Treaty on Intellectual Property in Respect of Integrated Circuits*, and so on. There are also agreements that extend, modify, or articulate further IP obligations such as the *International Union for the Protection of New Varieties of Plants* (plant variety rights), *Nagoya Protocol on Access to Genetic Resources and the Fair and Equitable Sharing of Benefits Arising from their Utilization* (traditional knowledge), *International Treaty on Plant Genetic Resources for Food and Agriculture* (farmers' rights), and so on. Within this multiparty international IP domain, however, there are also further entrenched obligations under bilateral and regional trade agreements. These bilateral agreements are proliferating and often impose significantly higher IP obligations that the existing multiparty agreements. The effect in the international arena is that the particular IP landscape of each nation-state needs to be determined according to their particular commitments.

12.5 Administrative Consequences

Where decisions about SECs are being made the particular IP landscape needs to be considered. Presumably the IP content of the LMO and any related materials, processes, and so on, will need to be identified and the relevant owners determined. As IP is often owned by non-governmental third parties to the LMO decisions, the IP owners will need to be clearly identified and probably involved in the decision-making process. And where decisions adversely affect an IP owner's IP then mechanisms of review and compensation will be required.[11] Presumably, IP could be identified through the importation application process under the CPB by asking questions on the application form about the kinds of IP embodied in the LMO.

12.6 Summary/Synthesis

- The CPB must be applied and implemented consistent with the Parties' other commitments to international IP agreements, and this specifically includes the WTO's TRIPS agreement.
- International IP agreements do not expressly exclude SECs.
- Where there are conflicts between the CPB and IP, there are various exemptions, exceptions, and flexibilities available under the IP agreements, albeit they are presently contested and have not been clearly articulated.
- Where IP is a problem for implementing SECs under the CPB this is likely to be magnified because IP owners are generally third parties and not governmental entities that are the parties to the CPB decisions.
- Any conflicts between the CPB and IP agreements might be ameliorated through a creative consideration of the financial and technology transfer mechanisms available under the CBP and other agreements.

References

Canada—Patent Protection of Pharmaceutical Products (2000) WT/DS114/R

Conference of the Parties to the CBD (Convention on Biological Diversity) (2011) Report of the tenth meeting of the conference of the parties to the convention on biological diversity UNEP/CBD/COP/10/27 (Decision X/16)

Correa C (2007) Trade related aspects of intellectual property rights: a commentary on the TRIPS agreement. Oxford University Press, Oxford

Council for Trade-Related Aspects of Intellectual Property Rights (2003) Implementation of Art 66.2 of the TRIPS Agreement IP/C/28

European Communities (2005) Protection of trademarks and geographical indications for agricultural products and foodstuffs. WT/DS174/R

Gervais D (2008) The TRIPS agreement: drafting history and analysis, 3rd edn. Sweet and Maxwell, London

Okediji R (2003) Public welfare and the role of the WTO: reconsidering the TRIPS agreement. Emory Int Law Rev 17:819–918

UNCTAD-ICTSD (International Centre for Trade and Sustainable Development) (2005) UNCTAD-ICTSD Resource Book on TRIPS and Development. Cambridge University Press, Cambridge

World Trade Organization (2001) Declaration on the TRIPS agreement and public health. WT/MIN(01)/DEC/2

World Trade Organization (2001) Ministerial declaration. WT/MIN(01)/DEC/1

[11] See, for example, TRIPS, Article 41.

WTO General Council (2003) Implementation of para 6 of the Doha Declaration on the TRIPS Agreement and Public Health WT/L/540
WTO General Council (2005) Amendment of the TRIPS Agreement WT/L/641
WTO Ministerial Conference (2001) Implementation-related issues and concerns WT/MIN(01)/17
Yusuf A (2008) TRIPS: background, principles and general provisions. In: Correa C, Yusuf A (eds) Intellectual Property and International Trade: The TRIPS Agreement, 2nd ed. Kluwer Law International, London

Chapter 13
Labor Impacts

Marnus Gouse

13.1 Introduction

The Agricultural Revolution in Britain and the subsequent Industrial Revolution stretched between the fifteenth and the end of the nineteenth centuries and rapidly spread through most of Western Europe, North America, and parts of Asia. Improvements in seed quality, soil fertility management, and a shift to mechanized agriculture resulted in increased agricultural productivity and ensured an acceptable level of food security in the then, developing economies. Mechanization released rural labor, which was readily absorbed in the growing industrial and ensuing services sectors as the developing economies underwent a lengthy process of de-agrarianization. This same agricultural-productivity-driven economy transformation process has been observed more recently in countries in Asia and Latin America but has to a large degree been absent in Africa and especially sub-Saharan Africa (SSA). The reasons why most SSA countries have failed or have been slow to follow the 'usual' development pathway are numerous and complex and fall outside the scope of this chapter. The main point of interest for this chapter is that agricultural production in most SSA countries is still characterized by labor-intensive but low-input, low-output production systems compared to capital-intensive, profit-driven mechanized production systems in developed countries and actively developing countries in Latin America and Asia (Table 13.1).

Pingali (2007, pp 2779–2805), following work by Pingali et al. (1987), explains the persisting low levels of mechanization (Table 13.2) in relatively land-abundant SSA according to the driving forces of agricultural intensification and the incentives for productivity growth. "Agricultural areas facing relatively inelastic demand conditions, due to low population densities and/or poor market infrastructure, tend to persist in low intensity, low yield subsistence production systems. The move to mechanical technologies for land preparation is not cost-effective in such societies." In agricultural systems where there is no incentive to expand production, due to limited profitable marketing opportunities, there is thus a low or no demand for substituting out of existing power sources, which in Africa is mainly family labor.

M. Gouse (✉)
Department Agricultural Economics, Extension and Rural Development, University of Pretoria, Pretoria, South Africa
e-mail: marnus.gouse@up.ac.za

K. Ludlow et al. (eds.), *Socio-Economic Considerations in Biotechnology Regulation*, Natural Resource Management and Policy 37, DOI 10.1007/978-1-4614-9440-9_13,

Table 13.1 African agriculture compared to other developing countries (Source: FAO/UNIDO 2008)

Region	Cereal yield kg/ha	Fertilizer use kg/ha	Irrigation % of arable land	Tractors per 1,000 ha
Average for African countries	1040	13	5	28
Average of nine selected non-African developing countries*	3348	208	38	241

*Brazil, Bangladesh, China, India, Pakistan, Philippines, Republic of Korea, Thailand, and Vietnam

Table 13.2 Farm power sources. (Source: FAO/UNIDO 2008)

Region	Manual	Animal	Engine
Sub-Saharan Africa	65	25	10
Average of Asia, Near East and North Africa, Latin America and Caribbean	25	25	50

Analyzing 40 years of data, De Janvry and Sadoulet (2010) showed that though still high, the share of the labor force in agriculture has dropped substantially in SSA countries. This rural-urban migration has occurred, not because of rising employment opportunities in the urban economy triggered by agricultural growth, but by the opposite—a lack of sufficient agricultural growth to overcome rapid population growth, resulting in stagnant rural incomes.

While developed-country economies are less dependent on agriculture for economic growth or employment, agriculture remains the main contributor to gross domestic production for most African countries (varying between 20–60 %) and supports up to 75 % of the African population by providing a livelihood and a means of trade and subsistence (UNEP GEO Data 2006; UNIFEM 2010). Though crop type plays a substantial role in farm-worker-per-hectare country comparisons (Asian rice requires more labor than African maize or sorghum production), Fig. 13.1 below illustrates the labor-intensive nature of mainly grain production in African countries.

Farming in the developed world is generally large scale, technology and capital (machinery) intensive, and increasingly more corporate with profit as the main objective, while agriculture in SSA is generally small scale, diversified (to address production risk and nutritional needs), subsistence based, and strongly linked to household and rural community nutrition, vitality, and survival. Due to the substantial difference in production systems and the role agriculture plays in the economy and people's livelihoods, it can be expected that concerns regarding the introduction of a new agricultural technology would vary between developed and developing countries.

Due to the nature of the novel characteristics of first-generation genetically modified (GM) crops, namely enhanced input traits, the direct technology-associated impacts are on farm-level input factors and the farm-level production system. Second-generation crops with value-added traits for consumers are yet to reach the commercial food, feed, and fiber market and current international GM crop plantings are dominated by first-generation insect resistant (Bt) cotton and maize and herbicide tolerant (HT) soybeans, cotton, maize, and canola.

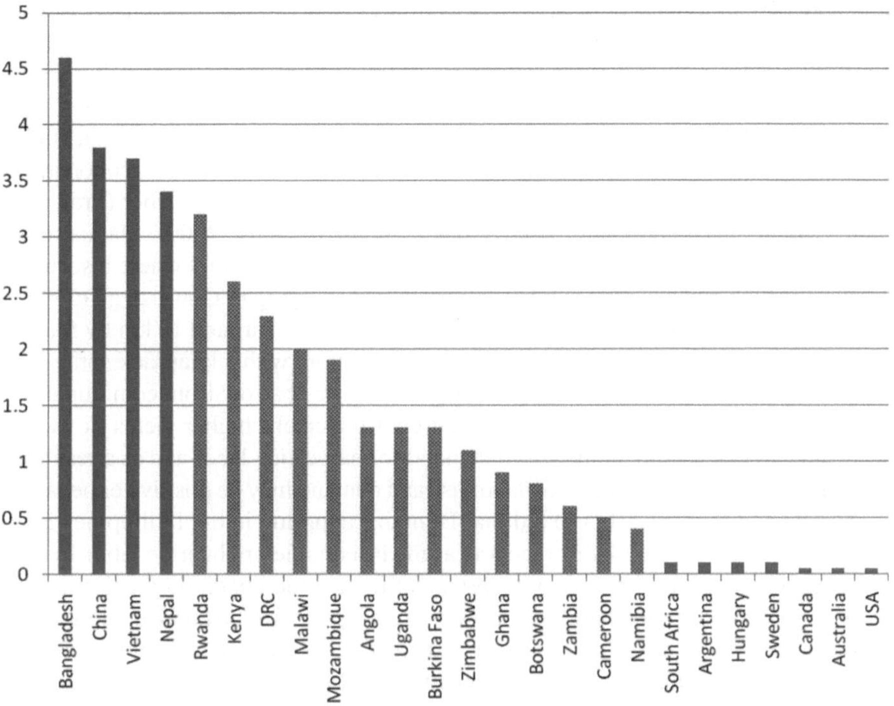

Fig. 13.1 Farm workers per hectare of arable and permanent crop land. (Source: World Resource Institute 2000)

In most GM-technology-adopting countries, Bt cotton and maize resulted in an increase in yield due to more effective insect control (less insect-related damage), a decrease in insecticide use and an increase in seed cost due to the additional Bt technology fee (for a summary of peer-reviewed publications and findings, see Gomez-Barbero and Rodriguez-Cerezo 2006; Smale et al. 2009; Brookes and Barfoot 2011; Finger et al. 2011). Publications on the use of herbicide-tolerant crops are dominated by research on HT soybeans, but based on these studies and a few on HT maize, it is clear that farmers who adopt HT crops benefit due to lower expenditure on herbicides, labor, machinery, and fuel (Carpenter and Gianessi 1999; 2000; 2002; Fernandez-Cornejo et al. 2002; Marra et al. 2004; Qaim and Traxler 2005; Gouse 2012). In Argentina, increased profitability and ease of management with HT soybeans have resulted in a substantial expansion in the soybean area planted (Trigo 2011).

It thus follows that when considering the labor impact of Bt and HT crops it is important to distinguish between farmers in developed countries using mechanized production systems and farmers in developing countries where agriculture is in many cases the sole source of livelihood and employment. In countries where insect and weed control is mechanized, Bt adoption has led to less tractor-sprayer insecticide applications and HT to fewer tillage operations. Even though tractor-operator laboursaving has been reported, this physical labor component is minimal and readily absorbed in other farming activities. In mechanized production systems, the laboursaving effect of Bt and HT technology tends to manifests itself as ease of management, saving on management time, and increased off-farm income (Fernandez-Cornejo et al. 2005; Gardner et al. 2009; Hurley et al. 2009) rather than a significant decrease in demand for physical manual family and or hired labor.

In more labor-intensive agricultural production systems like China, India, Burkina Faso, and the smallholder production sector of South Africa, Bt cotton adoption has led to substantial savings on insecticide-application-related labor (Pray et al. 2002; Subramanian and Qaim 2009; Vitale et al. 2010; Gouse 2009). Manual labor associated with insecticide application can be substantial especially where insecticide pressure is high and clean spraying water limited. According to Gouse et al. (2005) a South African smallholder farmer has to cover a distance of around 10 km by foot for a single insecticide application on one hectare of cotton, with a knapsack sprayer on his/her back and water has to be collected by hand/wheelbarrow from communal water sources, a job usually reserved for children. Conversely, higher yields (lower insect damage) can lead to an increased demand for harvesting labor and as a result the total or net effect of Bt adoption on labor use and demand may be positive or negative.

Yorobe and Smale (2012) found that Bt maize adoption in the Philippines is associated with increased off-farm income as the insecticide-application labor freed is significant enough for farmers to undertake other income-generating activities like carpentry or trade in merchandise. In contrast, evidence from Bt maize adoption by subsistence farmers in South Africa suggests that due to minimal use of insecticides to control borers (linked to low borer pressure) and related minimal insecticide application labor, the laboursaving benefit of Bt maize is limited. It can however, be expected that in seasons and areas with substantial borer pressure, insecticide use on conventional maize will be more substantial and the laboursaving impact of Bt more considerable. Generally, the number of insecticide applications necessary to control stalk borers on maize is lower than that required in the control of bollworms on cotton. Higher yields with Bt maize has led to a higher harvest labor demand in some seasons, but the effect has been less obvious than with Bt cotton due to smaller yield effects (Gouse 2012).

Though Bt adoption has had a significant impact on labor use in labor-intensive production systems, the effect of HT can be expected to be more substantial. Unlike *Lepidoptera* infestations that vary over seasons and regions, weeds can be described as the universal agricultural problem. This is true especially for Africa where weeds compete for limited soil moisture and expensive nutrients. According to Kent et al. (2001) the principal limiting factor to farm sizes in Africa is the number of necessary weedings during the period following planting. Kibata et al. (2002) showed that in Kenya, delayed weeding caused by a shortage of labor early in the season, resulted in maize yield losses of 15–90%, and Marais (1992) argues that poor weed control is the single biggest contributor to low maize yields for smallholder farmers in Africa. Subsistence crops like maize and sorghum compete with cash crops like groundnuts and cotton for labor, resulting in delayed weeding and crop losses. According to Chikoye et al. (2007), smallholder farmers spend between 50 and 70% of their total labor time hand pulling, hand slashing, and hoeing. Citing publications by Akobundu (1987), Ishaya et al. (2007a), Mavudzi et al. (2001) and

Ishaya et al. (2007b), Gianessi (2009) reports that to prevent weed damage, maize requires 276 hours of hand weeding per hectare, groundnuts 378 h/ha, sorghum 150 h/ha, cotton 200–400 h/ha, and rice 200–418 h/ha. Gouse (2012) found over a three season period that smallholder maize farmers in South Africa spent between 152 and 267 h/ha on manual weeding. Gender-based labor division for agricultural activities varies between countries (White et al. 1981; Tsikata 2009; FAO 2011) but women are involved in most activities, especially in the production of subsistence crops. The share of women in the agricultural labor force ranges from 36 % in Côte d'Ivoire to over 60 % in Lesotho, Mozambique, and Sierra Leone (FAO 2011). According to the World Bank (2009), 65 % of the African agricultural labor force is between 15 and 24 years of age and 25 % of children between 5 and 14 have to work to earn a living.

There are currently no peer-reviewed findings on the labor impact of HT cotton in labor-intensive developing country agriculture, and soybeans are generally not a smallholder crop. Findings on HT maize are limited to the South African smallholder experience where Gouse (2012) followed GM-maize-planting smallholders in KwaZulu-Natal for eight seasons. Farmers planting HT maize seed and using a no-tillage practice locally referred to as "planting without ploughing", where planting station soil is opened by hand-held hoe, a seed planted and the soil closed by hoe and weeds controlled with a post-emergent glyphosate application, enjoyed a yield increase due to more effective weed control but importantly saved approximately 50 % on labor. The largest laboursaving is on manual weeding, which is predominantly performed by household members. Female household members indicated that the laboursaving in the maize field enabled them to spend more time at home, caring for children or in their vegetable plots.

In countries or production systems where labor is a limiting factor, the use of a HT-based production system might enable farmers or households with limited labor to expand their production area and/or spend more time on other crops or off-farm activities. Over the longer term and where markets are functioning effectively, the use of a labor-saving technology thus holds the potential to increase production and result in labor opportunities for the rural poor. However, in countries where the landless rural unemployed depend on seasonal weeding for their livelihood, and where market or geographical limitations hinder expansion, introduction of a weed control laboursaving technology could have a substantial negative impact on employment and related sub-groups of the rural community. The magnitude and duration of the impact of the potential drop in employment will depend on the growth potential of the sector (expansion of specific crop area, other crops, processing, and the rest of supply chain) and employment opportunities in other sectors, but it is possible that the impact of a labor-saving technology over the short term (one or two seasons) could impact labor sellers (seasonal laborers) negatively. It is clear that a laboursaving technology can be beneficial to some role players but detrimental to others. Herein lies the importance of considering labor implications during regulation of GM crops.

13.2 Methodologies

As the objective of conducting a 'non-biodiversity-linked' potential labor impact assessment would be to mitigate against potential adverse impacts through the development and implementation of policies or activities or at the very least take notice of the technology's potential labor impacts, the assessment has to be ex ante. An ex post assessment can be done a reasonable period after introduction to test the accuracy of the ex ante study, to determine the success of the implemented policies, or to establish whether new or additional regulatory, mitigating, or enabling action is required.

In neo-classical economic labor theory, rural households divide their time between working on the farm, working off farm, and leisure depending on agricultural income possibilities, opportunity cost of farm labor (off-farm wages) and household decision-makers' utility functions. Like other production factors, supply and demand curves or functions for labor can be estimated and changes in the equilibrium caused by policies or technologies, modeled. However, modeling labor markets can be complicated and agricultural labor markets even more so due to the integration of the production system, different enterprises, and the rural household.

There are numerous methodologies that can be used to assess potential impacts technology adoption can have on labor. Most of these approaches however can generally be divided into two groups—an econometric modeling type approach or a rapid impact assessment type approach.

13.2.1 Econometric Modeling Approach

Ideally this approach should utilize disaggregated time series rural household data that is collected, updated, and is available from a central bank or national research institution. Alternatively, data can be collected through farmer-category-specific, representative farm-level surveys focusing on household characteristics, endowments, and crop-production labor activities. With time-series data, labor demand and supply curves can be estimated and changes in the equilibrium caused by technology adoption modeled. Collected cross-section micro/farm-level data can be analyzed through statistical inference and changes in production systems can be simulated. These findings can be used as is or be extrapolated to the broader economy with economy or sector-wide computable general equilibrium (CGE) models or a social accounting matrix (SAM).

13.2.2 Rapid Impact Assessment Approach

By conducting semi-structured key-role-player interviews and farmer-category-and gender-specific focus-group-discussions to study the production systems and labor division of the potential adopters of the new technology, it should be possible to identify labor-impact levels associated with the novel traits of the proposed technology. It should be possible to determine the magnitude and likelihood of possible impacts and make inferences about the potential impact for certain community groups in the district, region, and country.

13.3 Critical Assessment

Econometric modeling analysis utilizing time-series data can be considered to be statistically (scientifically) the most rigorous technique, but this method is inherently data intensive and for most countries in the world, disaggregated time-series household labor data does not exist. Where labor data does exist, in most cases the level of disaggregation is not sufficient to simulate household-level impacts. It can be argued that in the absence of time series data, the second best approach to follow would be to collect data through household surveys in the communities where the technology potentially will be adopted. If the intention is to use survey findings in an economy-wide model or when extrapolatable inferences need to be made, sample statistical representativeness according to farmer characteristics, production systems, and regions need to be considered. These economy-wide models should however have been independently and formerly developed, updated, and in use by other national or academic institutions. In developing countries, these models are usually updated and run by the central banks as they take a long time to develop and are extremely data intensive (requires economy-wide time-series data) and as a result expensive to maintain. However, it is not essential that farm-level potential impact findings be linked to an economy or sector-wide model for the findings to be of value for decision- and policy-makers.

For a biosafety regulatory ex ante impact assessment study, it is likely that only one season of surveys will be done due to time and cost implications. However, various ex post impact assessment studies show and admonish that due to variability in climate over seasons and between regions, performance, and resulting impacts of GM crops adoption vary substantially and results based on one season's findings need to be considered in that light. It can be expected that variability between seasons and regions will also have a significant impact on an ex ante assessment study and the representativeness of the season thus also needs to be considered.

Though data collection through surveys of a statistically representative number of farmers making use of different production practices in different regions should enable a country-wide assessment, it is argued that for an ex ante potential labor impacts study where the number of uncertainties are substantial, the cost and time investment in a comprehensive and statistically generalizable study might not be justifiable. The decision on whether an in-depth study is required should depend on the crop, novel GM trait, and its impact level as well as on the specific country's national strategic objectives, but it is suggested that a rapid impact approach is utilized for an ex ante labor impact study. Depending on the findings of a rapid impact assessment, a comprehensive study (ex ante or ex post) might be warranted.

There are a number of rapid impact assessment approaches and some are more data intensive and rigorous than others. A useful tool to identify potential labor impacts in a more qualitative fashion is the sustainable livelihood methodology's seasonal activity calendar. The seasonal calendar can be completed during focus group discussions with participants divided according to gender or other interest groups. When considering all produced crops' production systems and labor requirements via a seasonal calendar, it is possible to identify periods of labor shortage and the labor force (males, female, children, and family or hired labor) involved with the different farming activities. It should thus be possible to identify and link the potential technology specific impacts to activities, specific times during the season, and individual groups in the labor force. A labor-focused seasonal calendar is particularly useful in identifying cultural constraints that control divisions of labor among family members or particular practice patterns and timing of planting, cultivating, and harvesting.

After identifying all the potential impacts on labor demand and supply, the potential positive and negative impacts might need to be compared in order to determine the potential net impact on labor. This can be done through a type of cost/benefit comparison approach comparing the technology under consideration's potential laboursaving/using advantages with the potential laborusing/saving disadvantages.

13.4 International Arena

Under the strict interpretation (possibly the only interpretation) of the Cartagena Protocol's Article 26.1, labor issues (SECs) can only be considered in biosafety decision-making if the adverse labor impact can be linked to a scientifically established likely potential impact of the proposed LMO on biodiversity. A more liberal interpretation points to the "impact of living modified organisms on the conservation and sustainable use of biological diversity" in which "conservation" and "use" include farming and farming practices and thus labor. So under the more liberal interpretation it can be argued that a direct adverse impact of the LMO on labor can be used as grounds for not allowing importation or general release of proposed LMO.

Article 26.1 also states that a country's decision has to be in line with its other international obligations. Generally this refers to the World Trade Organization (WTO) trade regulations and the sanitary and phytosanitary measures. WTO trade regulations do not make any mention of labor impacts of trade beside reference to "products of prison labor." The WTO has been criticized for its limited guidance on labor, but issues mainly surround working conditions, health and safety standards, and child labor. There are no provisions for potential negative impacts of technology adoption on labor and a decision not to allow importation of a LMO, even with proof of potential negative impacts on labor, will still be in breach of WTO regulations.

The International Labour Organisation (ILO) conventions mainly provide for standards, rights, and elimination of discrimination and exploitation. ILO convention 169, Indigenous and Tribal People's Convention (1989), refers to non-discrimination and government responsibility for developing co-ordinated and systematic action to protect the rights of indigenous and tribal peoples (Article 3) and ensuring that appropriate mechanisms and means are available (Article 33). It might be possible to argue that a specific laboursaving technology might infringe the protection and rights of indigenous tribes and people, but the validity of this argument has not been tested. It is not clear if this Convention holds any implications for decision-making on LMOs and it has been ratified by only a small number of countries, mainly in South America.

13.5 Administrative Consequences

Even though it is unlikely that a potential overall (net) negative labor impact could be presented as acceptable reason for not importing or releasing a GMO in a WTO trade dispute, it does not mean that a potential labor impact study should not be conducted. Such a study is vital to inform the development of mitigation strategies to limit potential undesirable impacts on labor and enabling strategies for farmers and rural households to benefit from the proposed technology's labor impacts. Depending on a country's biosafety regulation provisions or other national objective-driven legislation or requirements, it might be beneficial, when an application for general release of an GMO is received, for a general socio-economic impact assessment study to be conducted focusing on issues like supply chains (enabling and limiting institutional arrangements), potential farm level impacts (including impacts on labor), potential industry impacts, potential consumer impacts, and potential trade impacts. A socio-economic study is not recommended for other permit application like imports for food, feed, and processing or contained use because the GMO, under the conditions linked to the permit, should not enter the production system.

A general socio-economic impact assessment study should identify and assess both the potential positive and negative impacts of the proposed technology. Though such a study might assist with the interpretation of food and feed safety and biosafety study findings, the main objective of such a study would be to determine to what extent the country can benefit from the proposed technology and what measures need to be in place to maximize potential benefits and minimize potential negative impacts.

For an ex ante regulatory assessment where time, funding, and information are limited, a rapid impact assessment should be sufficient. How the study is funded and who the assessor will be will depend on each country's regulatory framework. It is suggested that because this study in a sense serves as motivation for acceptance of the application, as information on the crop and novel event is limited and in the context of developing countries' limited funding and capacity, the study should be funded and commissioned by the applicant. The study can then be reviewed by the regulatory authority and referred to commissioned experts for comment.

13.6 Summary/Synthesis

- Adoption of GM agricultural technologies can result in substantial laboursaving for farmers.
- Though a laboursaving technology might be beneficial for labor-deficit producers, a decrease in the demand for seasonal labor might negatively impact labor-selling rural poor.
- In production systems where labor is a limiting factor, increased labor productivity might result in production expansion and additional employment over the longer term.
- The impact of GM crops on labor demand and supply is a vital issue but in a non-biosafety related scenario will not be deemed a valid reason for GM crop rejection in a WTO dispute.
- A general socio-economic impact assessment, also focusing on labor, is necessary to determine to what extent the country can benefit from the proposed technology and what measures need to be in place to maximize potential benefits and minimize potential negative impacts.

References

Akobundu OI (1987) Weed science in the tropics: principles and practices. Wiley-Interscience, Chichester

Brookes G, Barfoot P (2011) GM crops: global socio-economic and environmental impacts 1996–2009. PG Economics Ltd, UK

Carpenter J, Gianessi L (1999) Herbicide tolerant soybeans: why growers are adopting roundup ready varieties. AgBioForum 2(2):65–72

Carpenter J, Gianessi L (2000) Herbicide use on roundup ready crops. Science 4:203

Carpenter J, Gianessi L (2002) Agricultural biotechnology: updated benefit estimates. National Centre for Food & Agriculture Policy, Washington, DC

Chikoye D, Udensi EU, Fontem Lum A et al (2007) Rimsulfuron for post emergence weed control in corn in humid tropical environments of Nigeria. Weed Technol 21:977–981

De Janvry A, Sadoulet E (2010) Agriculture for development in africa: business-as-usual or new departures? J Afr Econ 19(2):7–39

FAO (2011) The state of food and agriculture. Woman in agriculture, closing the gap for development. Rome, Italy

FAO/UNIDO (2008) Agricultural mechanization in Africa…Time for action. Planning investment for enhanced agricultural productivity, Report of an Expert Meeting, January, Vienna, Austria

Fernandez-Cornejo J, Hendricks C, Mishra A (2005) Technology adoption and off-farm household income: the case of herbicide-tolerant soybeans. J Agr Appl Econ 37:549–563

Fernandez-Cornejo J, Klotz-Ingram C, Jans S (2002) Farmlevel effects of adopting herbicide-tolerant soybeans in the USA. J Agr Appl Econ 34(1):149–163

Finger R, El Benni N, Kaphengst T et al (2011) A meta-analysis on farm-level costs and benefits of GM crops. Sustainability 3(5):743–762

Gardner JG, Nehring RF, Nelson CH (2009) Genetically modified crops and household labor savings in US crop production. AgBioForum 12(3, 4):303–312

Gianessi L (2009) Solving africa's weed problem: increasing crop production and improving the lives of women. Crop Protection Research Institute

Gomez-Barbero M, Rodriguez-Cerezo E (2006) The adoption of GM insect-resistant Bt maize in spain: an empirical approach, 10th ICABR conference on agricultural biotechnology, July, Ravello, Italy

Gouse M (2009) Ten years of Bt cotton in South Africa: putting the smallholder experience into context. In: Tripp R (ed) Biotechnology and agricultural development: transgenic cotton, Rural Institutions and Resource-Poor Farmers. Routledge, Abingdon Oxford

Gouse M (2012) Farm-level and socio-economic impacts of a genetically modified subsistence crop: the case of smallholder farmers in KwaZulu-Natal, South Africa. Submitted in partial fulfilment of the requirements for the degree PhD in Department Agricultural Economics, Extension and Rural Development, University of Pretoria

Gouse M, Pray C, Schimmelpfennig D (2005) The distribution and benefits from Bt cotton adoption in South Africa. AgBioForum 7(4):187–194

Hurley TM, Mitchell PD, Frisvold GB (2009) Effects of weed resistance concerns and resistance management practices on the value of roundup ready crops. AgBioForum 12:291–302

Ishaya DB, Dadari SA, Shebayan JAY (2007a) Evaluation of herbicides for weed control in sorghum (Sorghum bicolour) in Nigeria. Crop Prot 26:1697–1701

Ishaya DB, Dadari SA, Shebayan JAY (2007b) Evaluation of herbicides for weed control in three varieties of upland rice (oryza sativa L.) in the Nigerian Savannah. Crop Prot 26:1490–1495

Kent RJ, Johnson DE, Becker M (2001) The influences of cropping system on weed communities of rice in Côte d'Ivoire, West Africa. Agr Ecosyst Environ 87(3):299–307

Kibata GN, Maina JM, Thuranira EG et al (2002) Participatory development of weed management strategies in maize based cropping systems in Kenya. Thirteenth Australian Weeds Conference Proceedings 343–344

Marais JN (1992) FSR-E Newsletter No. 4, November 1992. Southern African Association for Farming Systems Research-Extension, Development Bank of Southern Africa, Halfway House, South Africa

Marra MC, Piggott NE, Carlson GA (2004) The net benefits, including convenience, of roundup ready® soybeans: results from a national survey (Technical Bulletin 2004–3). NSF Centre for IPM, Raleigh

Mavudzi Z, Mashingaidze AB, Chivinge OA et al (2001) Improving weed management in a cotton-maize system in the Zambezi Valley Zimbabwe. Brighton Crop Protection Conference, Weeds 169–174

Pingali PL (2007) Agricultural mechanization: adoption patterns and economic impact. In: Evenson R, Pingali P (eds) Handbook of agricultural economics, Volume 3. Elsevier, Amsterdam

Pingali PL, Bigot Y, Binswanger H (1987) Agricultural mechanization and the evolution of farming systems in Sub-Saharan Africa. The World Bank, Washington DC

Pray C, Huang J, Hu R et al (2002) Five years of Bt cotton in China—the benefits continue. Plant J 34:423–430

Qaim M, Traxler G (2005) Roundup ready soybeans in Argentina: farm level and aggregate welfare effects. Agr Econ 32:73–86

Smale M, Zambrano P, Gruère G et al (2009) Measuring the economic impacts of transgenic crops in developing agriculture during the first decade: approaches, findings, and future directions. Food Policy Review 10, International Food Policy Research Institute (IFPRI)

Subramanian A, Qaim M (2009) Village-wide effects of agricultural biotechnology: the case of Bt Cotton in India. World Dev 37(1):256–267

Trigo EJ (2011) Fifteen years of genetically modified crops in Argentinian agriculture. Argentine Council for Information and Development of Biotechnology—ArgenBio

Tsikata D (2009) Informalization, the Informal Economy and Urban Women's Livelihoods in Sub-Saharan Africa since the 1990s. In: Razavi S (ed) The gendered impacts of liberalization: towards "Embedded Liberalism? Routledge, New York
UNEP GEO Data Portal (2006). Accessed June 2012
UNIFEM (2010) Women's land and property rights. UN Woman, September 9. http://www.unifem.org/gender_issues/women_poverty_economics/land_property_rights.php. Accessed 27 Nov 2012
Vitale JD, Vognan G, Ouattarra M et al (2010) The commercial application of GMO crops in Africa: Burkina Faso's decade of experience with Bt cotton. AgBioForum 13(4):320–332
White DR, Burton ML, Dow MM (1981) Sexual division of labour in African agriculture: a network autocorrelation analysis. American Anthropological Association, Vol 83
World Bank (2009) African development indicators 2008/2009. Youth and employment in Africa. The potential the problem the promise. Washington DC
World Resource Institute (2000) via Nationmaster.com. Accessed 19 June 2012
Yorobe JM, Smale M (2012) Impacts of Bt maize on smallholder income in the Philippines. AgBioForum 15(2):152–162

Chapter 14
Market Access and Trade

Crina Viju

14.1 Introduction

The study of international trade is divided into three general areas—trade policy analysis, trade theory, and empirical studies of trade. Trade policy analysis focuses on the economic impacts that different forms of government intervention have on international trade. It is well known that any government intervention in the economy has a redistributive role, resulting in winners and losers. The role of trade policy analysis is to identify the two groups, to estimate the benefits and costs of any measure that impacts trade and to improve the overall policy making process (Kerr 2007). Several approaches can be utilized to quantify the impacts of different trade/domestic measures, the choice of which depends on the type of research question.

There are two means of analyzing the effect of a specific trade policy, ex ante and ex post. Ex ante analysis looks at "what if" research questions, hence in ex ante analysis, researchers simulate the effects of a trade/domestic policy change and project the potential economic impacts of that change. Among the economic effects simulated by different ex ante models are the effects on prices, quantities, and the welfare of the three market participants: consumers; producers; and government. Simulations are usually completed using partial and general equilibrium models. Alternatively, ex post analysis uses historical data to analyze the effects of a trade/domestic policy that has been already implemented. This type of analysis is represented by all econometric trade models, including the gravity model described in this chapter.

Another distinguishing characteristic differentiating the methods used in trade policy analysis is whether they are static or dynamic. A static model compares the initial and final equilibrium in an economy without taking into account the transitory phases. Partial equilibrium models and a number of general equilibrium models are static. A dynamic analysis not only examines the final equilibrium results, but also the transitory changes that occur as the economy moves from initial to final equilibrium, thus accounting for adjustment costs. This type of analysis would include capital accumulation and technological change for example. Although dynamic analysis is preferred due to its ability to estimate a full range of costs and

C. Viju (✉)
Institute of European, Russian and Eurasian Studies, Carleton University, Ottawa, Canada
e-mail: crina.viju@carleton.ca

K. Ludlow et al. (eds.), *Socio-Economic Considerations in Biotechnology Regulation*, Natural Resource Management and Policy 37, DOI 10.1007/978-1-4614-9440-9_14,

benefits, it is theoretically and computationally much more complex than a comparative static model. While a number of dynamic general equilibrium models have been developed, most simulation models in use are static in nature.

Trade policy methods are widely used in assessing the effects of the adoption of genetically modified organisms (GMOs) in various countries, by focusing on the international consequences for adopting and non-adopting countries.[1] The important regional differences between GMO regulatory frameworks represent a focal point for this literature. First, the trade studies can be divided by country of interest with a first group represented by major exporting countries that have adopted GMOs, a second group formed of major developed countries that import GMOs and have strict regulatory frameworks in place and tend to be less accepting of GMOs for various socio-economic motives and, ultimately, the group of developing countries that represent potential importers of GMOs, but do not currently produce them. Second, the trade studies can be categorized based on the main question asked, with a large group focusing on the domestic and international effects of having non-harmonized regulatory frameworks and comparisons of various policy responses by some countries to the adoption of GMOs by other countries, and another group explaining the reasons behind the different regulatory frameworks. Finally, the applied trade studies can be divided based on the methods used by authors with ex ante analysis mainly using partial and general equilibrium frameworks and ex post analysis employing various econometric techniques, including the gravity equation framework. This chapter will concentrate on the advantages and disadvantages of different methodological approaches used in applied trade policy studies.

14.2 Methodologies

The work of economists in trade theory has shown that trade liberalization is welfare enhancing under a wide range of economic environments (Perdikis and Kerr 1998). Despite this, the populist view of trade remains heavily mercantilist in its perspective (i.e., exports are good, imports are bad). There is a grudging acceptance by public policy makers that "beggar-thy-neighbor" trade wars are mutually destructive; policy makers therefore structure trade agreements to promote the benefits of cooperation over tit-for-tat retaliations (Gaisford and Hester 2007), but the general thrust of trade policy remains focused on promoting exports and eschewing imports. Protectionist policies are the usual means to limit imports.

A case for protectionism may be advocated for to achieve non-economic goals such as national defense, national sovereignty, and pride, to further foreign policy objectives, or achieve other political goals. There are some economic arguments that selectively justify protectionism. These include providing temporary protection as a means of fostering infant industries, enabling their establishment; providing

[1] For a comprehensive literature review on the trade impacts of GMOs adoption, see Gruère (2009).

temporary protection for high-tech firms to create a competitive advantage; or providing temporary protection to specific sectors or industries to avoid or reduce unemployment, particularly in a recessionary period, as exports would create domestic employment, while imports provide employment to foreign workers; providing temporary protection for food safety and environmental concerns or, as seen in GMO regulations, consumers' right to know.

Thus, despite clear historical evidence that free trade is welfare enhancing overall, governments still choose to protect their domestic markets from foreign competition. Trade policy analysis must then devise means to accurately assess the costs and benefits of such policy decisions, including those that implement forms of protectionism. Different methods of analysis are therefore used in trade literature to quantify the benefits of free trade and the impacts of protectionism. The three methods that are widely employed are presented in the following section.

14.2.1 Partial Equilibrium Models

Partial Equilibrium (PE) models are used to analyze "what if" scenarios starting from an initial state to which different circumstances are applied, such as a change in trade policy, through an exploration of the potential results. The main characteristic of PE models is that the effects of policy actions are examined only on those markets that are directly affected, ignoring effects on other branches of the economy or assuming that no other changes occur. PE models are based on the assumption of ceteris paribus, where all other factors that can affect the market under analysis are held constant. Thus, for example, assuming that a specific trade policy change in a market affects the price of a good, the income effect on other markets, prices in other markets and income available for the purchase of other goods remain constant. Examining a single market in isolation is advantageous when a particular good is highly prominent in a country's trade, employs a significant share of a country's workforce, or the good has political significance. Another assumption of PE models is that resources are unlimited; therefore, no reallocation of resources from one sector to another occurs. PE models are static, as they ignore paths of adjustment by comparing only the pre- and postchange equilibriums and are based on three components: demand of domestic consumers, supply of domestic producers, and the trading behavior of foreigners located in the rest of the world. By using PE models, policy makers are able to examine the welfare effects of a trade/domestic policy change in terms of change in consumer surplus, change in producer surplus, change in government revenues, and total welfare gain or loss.

PE models are extensively used to analyze the economic effects of certain trade responses to the inclusion of SECs in the domestic regulation of GMOs, such as the European Union (EU) 1998 de facto moratorium on the production and importation of GMOs and the new EU legislation from 2004, which requires strict GMO labeling

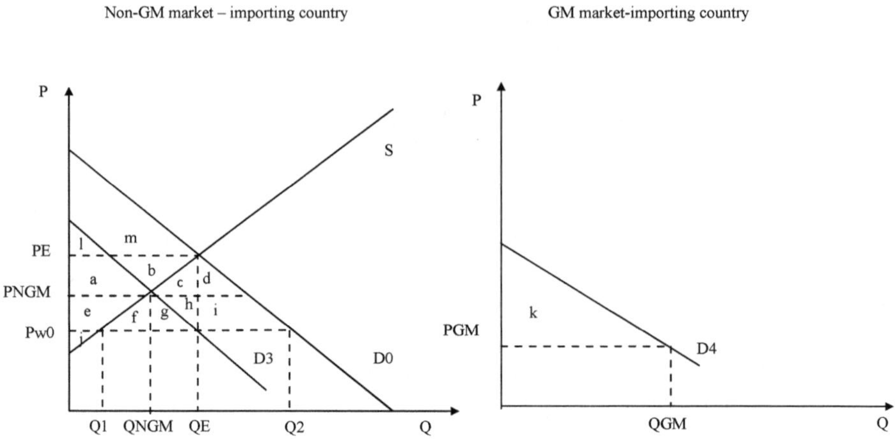

Fig. 14.1 Import ban versus labeled GMOs imports. (Source: Isaac et al. 2002, p. 144)

regulations and liability laws. Two studies are described in this section.[2] Isaac et al. (2002) compare three import policies available under World Trade Organization (WTO) and Biosafety Protocol (BSP), which importing countries are allowed to implement on environmental and food safety grounds: import of unlabeled GMOs; import of labeled GMOs; and an import ban on GMOs. The relative efficiency of the three policies is assessed by Isaac et al. (2002) by using a PE framework (see Fig. 14.1). For this chapter, the analysis is restricted to the comparison between an import ban on GMOs and labeled GMOs imports[3] based on the assumption that the GM crops are licensed only in the exporting country, but not in the importing country. In the non-GMOs market, the market relevant for this examination, D0 represents the domestic demand and Pw0 is the world price for non-GM products. Thus, the domestic production is expressed by Q1 and the imports equal the difference between Q2 and Q1. With no GM goods produced by the exporting country, the consumer surplus is l+m+a+b+c+d+e+f+g+h+i, the producer surplus is represented by j, and the total welfare of the country is given by l+m+a+b+c+d+e+f+g+h+i+j. Moving on to the situation where the exporting country starts producing GM goods and the importing country responds by imposing an import ban, the price in the non-GMOs market is increased to the equilibrium price, PE, and the new quantity produced and consumed is QE. The total welfare in this situation is reduced to l+m+a+b+e+j. Thus, the total welfare loss resulting from an import ban is represented by c+d+f+g+h+i.

In case of mandatory labeling of GM goods, the consumers are able to distinguish and choose whether to keep purchasing domestically produced non-GM goods or

[2] See also: Berwald et al. (2006); Gaisford and Lau (2001); Lapan and Moschini (2004); Sobolevsky et al. (2005); Gruère (2011).

[3] For the other comparisons see Isaac et al. (2002).

imported GM products depicted in the right-hand panel of Fig. 14.1. The price of GM products, PGM, has an effect on the consumption decision, as due to the technological cost advantage of GMOs production, it is lower than the world price for non-GM goods, Pw0. The group of consumers that do not consider the GM goods inferior will start consuming them, which will result in a decline in demand for non-GM products to D3. The decline in demand will depend on the size of the consumers' group that will choose to purchase GM goods. Thus, in the non-GMOs market, the new price for the non-GM goods is PNGM, while the quantity produced and consumed is QNGM, given that the price of GM products is PGM[4] and the quantity of GM goods consumed is QGM. The total welfare in case of mandatory labeling is represented by l+a+e+j (in the non-GMOs market)+k (in the GMOs market). In comparison to the situation where no GMOs were produced by the exporting country, the welfare change in the importer's market as a result of mandatory labeling is associated to k−(f+g+h+i). Thus, whether it is a welfare loss or gain it depends on the relative sizes of k and f+g+h+i.

By comparing the welfare effects of mandatory labeling of imported products with an import ban, the mandatory labeling is clearly superior by c+d+k. Thus, the main conclusions drawn from Isaac et al. (2002) are that an import ban will always bring welfare losses whether compared to the pre-technological or to the mandatory labeling situations. However, mandatory labeling is not necessarily welfare decreasing for the importer, the effect depending on the relative sizes of the gain in the GMOs market and the losses in the non-GMOs market. Mandatory labeling is always more economically efficient than an import ban. The authors conclude that even an import ban is always economically inefficient, vested commercial interests will gain greater rents when an import ban is implemented.

Anderson (2006) compares the effects of trade policies that would limit the imports of GMOs, such as import ban or mandatory labeling requirements with different support measures such as variable levies, export subsidies, or tariff-rate quotas (TRQs). By using a PE framework, the author shows that domestic support measures such as TRQs and export subsidies that keep domestic prices constant and prevent the price-reducing effects of biotechnology adopted abroad have a negative impact on the welfare of producers of GM products and of domestic consumers. Domestic support measures represent disincentives for investment in biotech research; however, they are less inefficient when compared to import bans.

PE models are based on economic simulation, which translates economic theory and data into mathematical equations. Changes can be quantified by using a limited number of parameters, such as demand and supply elasticities, initial quantities demanded and supplied, world price, and tariff levels. More complex PE trade models take into account multiple goods markets, by introducing cross-price and cross-quantity variables between the related markets. They do not, however, look at the entire economy. Various PE models have been also developed to simulate changes in international trade policies. These include SMART (Software for Market Analysis and Restrictions on Trade) which is part of the World Integrated Trade Solution

[4] PGM is lower than PNGM due to perceived quality differences.

(WITS) program developed by the World Bank in collaboration with the United Nations Conference on Trade and Development (UNCTAD), the Static World Policy Simulation Model (SWOPSIM) developed by the US Department of Agriculture and the Agricultural Trade Policy Simulation Model (ATPSM) of UNCTAD. PE analysis is employed when complex analysis is not necessarily appropriate and interest lies within a subsector.

14.2.2 General Equilibrium Models

Computable general equilibrium (CGE) and PE models share one similarity in that they both develop a "what if" scenario, quantifying the effects of a trade policy change under different circumstances. The major difference between the two types of models is that CGE takes into account the entire economy, including linkages between markets and focuses on the effects on the market for final goods, intermediate goods, and factors of production.

CGE offers a comprehensive assessment of trade policy change at the worldwide, regional, sectoral, or individual economy level. This is the most important feature and strength of CGE models. CGE models provide useful insights as they take into account the entire economy and thus, the possible spillovers of a policy change on sectors that are not directly targeted by the policy change. CGEs can be static or dynamic. In a static model, the economy responds only to trade/domestic policy change, keeping all other factors in the economy constant. Most CGEs are based on static models. However, modern technology allowed economists to develop dynamic models which take into account different factors that are changing over time, such as GDP, level of employment, etc. and provide a more realistic analysis.

The most widely used CGE model for analyzing trade policy changes is the Global Trade Analysis Project (GTAP), originally developed by Hertel (1997). GTAP is a static, multi-market, multi-regional model with exogenously fixed endowments and particular production and utility functions. Main assumptions of the model include perfect competition and constant returns to scale, the Armington assumption, an explicit treatment of international trade and transport margins, no link between taxes and public expenditures, and the inclusion of a global banking sector. The GTAP model also includes a database which contains SAMs (input-output data, bilateral trade flows, bilateral tariff data) for an extensive number of regions and countries.

The GTAP CGE model of world economy has been extensively used to estimate the welfare consequences of current and prospective GM crop adoption by some countries and of trade/domestic policy responses by other countries. Two studies are described in this section.[5] Anderson and Jackson (2005a) consider different scenarios such as adoption of GM crops by US, Canada, and Argentina (coarse grains

[5] See also: Anderson and Jackson (2005b); Anderson et al. (2008); Gruère et al. (2007); Huang et al. (2004); Nielsen et al. (2003).

and oilseeds) without and with an EU import ban in place and a change in the EU policy that would allow the adoption and sale of GM products. They modify the GTAP model to include productivity changes, consumers' reluctance to GMOs, and levels of substitution between GM and non-GM crops used as intermediate inputs in the production of final food products. When only US, Canada, and Argentina produce GM crops, and there is no adverse response from elsewhere, including the EU, the estimated global benefits are US$ 2.3 billion per year with 40% of benefits shared among the large importing regions, the EU and Northeast Asia, and 60% of benefits shared among the three producers of GMOs. In the scenario of EU imposing a moratorium, the gain to the three adopting countries is reduced by one-third, while the EU loses US$ 3.1 billion per year minus the value EU consumers place on having a GMOs-free market. If the EU would allow the adoption and sale of GMOs, global welfare would almost double and most of the gains would be shared among developing countries. Anderson and Jackson (2005a) continue their study by adding China and India among the countries adopting GM crops (rice and wheat). The simulations show that the global economic welfare with no EU import ban would be US$ 4.3 billion per year with two-thirds of the extra gains going to China and India, while other developing countries would benefit from lower import prices. With an EU moratorium in place, the cost for the EU will rise to US$ 5.5 billion per year, while the cost for the rest of the world would be US$ 2.9 billion per year. However, the results have to be treated with caution as the empirical model suffers from various simplifications such as no other policy response from the EU that would substitute an import ban, the imposition of moratoria by other countries, the continuous GMOs R&D or the fact that large investments have been redirected toward non-food GMOs.

Van Meijl and van Tongeren (2004) estimate using the GTAP model the effects of price insulation mechanisms such as variable import levies and export subsidies as policy responses. The domestic support mechanisms would reduce slightly the welfare of US, Canada, and Argentina. The decrease in welfare would be much lower than in the case of an import ban, results that confirm the findings of Anderson (2006) by using a PE framework.

14.2.3 Gravity Models

Gravity models are econometric methods used to estimate the impacts of various trade-related policies on international trade flows. Utilizing an ex post approach based on historical data to analyze trade flows, they are also applied to other types of analysis such as in foreign direct investment, migration, and others. The gravity model of trade expressed in Eq. 14.1 predicts that the trade volume between two countries is positively related to their economic sizes and negatively related to the distance between them.

$$X_{ij} = \beta_0 \frac{Y_i^{\beta_1} * Y_j^{\beta_2}}{D_{ij}^{\beta_3}} \tag{14.1}$$

X_{ij} = total exports or total trade (exports plus imports) between two countries *i* and *j*;
Y_i = economic size of country *i*, represented by the GDP level or population;
Y_j = economic size of country *j*, represented by the GDP level or population;
D_{ij} = distance between countries *i* and *j*, which is a proxy for trade costs.

Thus, the baseline form of the gravity model tested empirically is:

$$\ln X_{ij} = \beta_0 + \beta_1 \ln Y_i + \beta_2 \ln Y_j + \beta_3 \ln D_{ij} + \varepsilon_{ij}. \tag{14.2}$$

The basic gravity equation expressed in Eq. 14.2 is usually augmented with a number of other variables that can affect trade. Some of them are represented by dummies for islands, landlocked or neighboring countries which will affect transport costs, dummies for common language or other cultural features which will have an impact on information and search costs, and dummies for regional trade agreements (RTAs) which test whether tariff barriers play an important role.

One of their major limitations is the lack of theoretical underpinnings. In response to this shortcoming, several theories have been developed that offer a theoretical background for gravity models. First was Anderson (1979), who based on two assumptions (constant elasticity of substitution and the Armington assumption[6]), provided a theoretical foundation for the model. Bergstrand (1990) added a monopolistic competition framework with product differentiation among producing firms to Anderson's (1979) assumptions. Deardorff (1998) showed that gravity models can be explained by relative differences in factor endowments between countries following a Heckscher-Ohlin framework, while Eaton and Kortum (2002) explained gravity models based on a Ricardian trade framework, where trade takes place due to the relative differences in technology between countries. Feenstra (2004) pointed out that gravity models assume identical prices across countries, which is a strong assumption given the existence of trade barriers. Among others, the work of Anderson and van Wincoop (2003) represents an important contribution as they considered bilateral trade as being determined by relative trade costs. Helpman et al. (2007) derived the gravity equation from a heterogeneous firms' model of trade, explaining issues that previous models could not, such as zero-trade observations, asymmetric trade and the fact that more countries trade over time.

Gravity models have been widely used in the empirical literature to assess the effects of multilateral and regional trade agreements. However, ex post empirical studies and, in particular, the use of gravity models, are rare when looking at the effects of GMOs adoption on international trade. One reason behind the lack of ex post research is the scarce data on bilateral trade flows of GMOs or crops with

[6] Domestic and foreign products are not perfect substitutes and, thus, products are differentiated by the country of origin (Armington 1969).

GM content. Disdier and Fontagné (2010) employ an advanced gravity equation to estimate the trade impacts of EU's regulations on GMOs taking into account the three issues that were condemned by the WTO ruling against the EU: de facto moratorium; delays in processing product-specific applications; and inconsistency of the safeguard measures adopted by various EU Member States. Their findings confirm revenue losses for the main complainants (US, Canada, and Argentina), with the size of the losses varying across products and complainants. Their analysis extends beyond the WTO case, considering the impact of other countries' (New Zealand, Switzerland, and Norway) non-approvals of GMOs and the trade impacts of EU regulations on countries (Brazil) that did not join the WTO complaint. They conclude that market size matters when deciding to open a costly case at the WTO and, thus, the EU represented an important market for the complainants, as opposed to Norway or Switzerland. Vigani et al. (2012) concentrate on a different aspect related to international trade and GMOs. By employing a gravity equation augmented with a variable measuring the bilateral differences in GMOs regulations, they analyze whether similarity or dissimilarity in regulations between importer and exporter affects bilateral trade in GMOs. They find, firstly, that countries that have very different GMOs regulations trade significantly less and, secondly, labeling represents the most important regulatory dimension, followed by the approval process and traceability. Based on these findings, their main conclusion is that global harmonization, especially regarding labeling policies, will have a significant positive effect on global trade.

14.3 Critical Assessment

PE models are static, ex ante models, generally employed in quantifying the welfare impacts of a change in trade/domestic policy at a disaggregated level by looking at a market in isolation. PE models are particularly useful as they are transparent, less complex, and require a limited number of economic variables and data. They have the advantage of capturing changes in policy measures at a disaggregated level and are commonly used for analyzing the impacts of a policy change on the welfare of the participants in the market. They are flexible frameworks which allow the representation of a large array of institutional and market policies. The major limitations of these models are the absence of intermarket resource shifts as analysis is performed on a market in isolation and disregard for paths of adjustment. The studies employing PE frameworks to analyze the adoption of GMOs are generally based on the assumption that prices will clear through equilibrium of supply and demand and on simplistic assumptions regarding the adoption and productivity effects of the new technology. They do not attempt to assess the effects of regulations on bilateral trade flows between GMOs adopting and non-adopting countries or the trade diversion effect. One of the major problems of this literature is the limited research on developing countries. The PE simulations that assess the impacts of GMOs adoption on developing countries' welfare are based on assumptions and/or numbers (such as costs of segregations) calculated for industrialized countries. Despite these limitations, useful insights can be generated via PE models without the use of more complex and less transparent forms of analysis.

CGE models are widely used in the analysis of the welfare impacts of multilateral negotiations, formation of preferential trade agreements, and non-harmonized domestic regulatory frameworks (such as GMOs regulations) by considering the interlinkages between markets at an aggregate level. They are generally applied ex ante and can be static or dynamic. Utilizing CGE models presents a variety of important challenges that must be accounted for. First, they tend to be "black boxes" that lack transparency. Hence, even though a change in policy would result in a different set of results, the explanations for why those particular results are obtained are far from clear. They tend to be highly complex and difficult to understand or use by non-experts. Second, CGE models are sensitive to the user's choices in terms of key parameters, such as the level of substitutability between domestic and imported goods (Gaisford and Kerr 2001). Third, they suffer from a high degree of aggregation and therefore tend to be used for analyzing the linkages between a few broadly defined sectors. Fourth, the CGE model simulations of the impacts of GMOs adoption are based on simplistic ways of modeling productivity gains with no differentiation between regions, land types, or seed prices. Most of the studies do not consider the market imperfections in the input sector. Last, CGE models require an extensive amount of actual economic data to estimate the effects of different shocks to the economy, in this case, various trade/domestic policy changes.

Gruère (2009) outlines the shared limitations of the two approaches when applied in assessing the global impacts of GMOs' adoption. These are related to being ex ante studies based on uncertain assumptions, uncertainty being treated in a relatively unsophisticated manner. He points out that sensitivity analysis is limited due to models' complexity and, thus, it is performed by varying only a limited number of parameters. The high level of aggregation of both approaches hides differences between countries, sectors, and types of crops. His final criticism is based on some of the assumptions of those approaches such as perfectly competitive markets and developing countries' global market integration.

Lastly, gravity models are popular empirical tools used in analyzing the impacts of various trade policy issues on bilateral trade flows between different geographic areas. They are ex post methods of analysis and do not offer any insights on welfare re-distribution. The benefits of using gravity models are that they are simple empirical tools which utilize publicly available, historical panel data, and they allow the testing of different variables' impacts on trade. They are also highly effective in explaining bilateral trade. However, there are two major limitations of gravity models. First, it is difficult to attribute changes in trade flows to a specific trade policy change that is included in the model and not to other economic factors or policy changes that are omitted from the model. Secondly, the lack of theoretical underpinnings reduces the model's credibility. However, their use is limited when assessing the impacts of GMOs' adoption on bilateral trade flows, the main explanation for a restricted number of ex post studies being the lack of data on GMOs bilateral trade flows.

Although PE models provide only a sectoral/subsectoral analysis while CGE and gravity models provide an economy-wide picture and despite the limitations outlined in this chapter, all three methods of trade policy analysis offer information that is beneficial to policy makers and they are widely used in the policy-making process.

14.4 International Arena

From an economic perspective, regional GMOs regulatory frameworks result in global welfare losses. Mandatory labeling policies provide more efficient policy responses than trade moratoria if socio-economic factors are taken into account to accommodate consumers' right to know. However, more modeling and applied research are required to include the costs of segregation and identity preservation. As currently there are still major conflicting interests and countries have different GMOs acceptance levels, a free trading system for GMOs is far from being a goal that can be achieved in the near future. Thus, from an institutional perspective, two conclusions can be drawn from the studies outlined in this chapter. First, there is a need for a transparent, harmonized set of regulations regarding GMOs labeling and second, under the Doha round of multilateral trade negotiations, countries should seek major reductions in bound agricultural tariffs and export subsidies not to allow governments to insulate their domestic markets through the use of domestic support measures.

14.5 Administrative Consequences

One explanation for the limited ex post empirical studies analyzing the effects of GM crops adoption on international trade is the lack of data on transgenic trade flows. Bilateral trade flow data for developed countries is reported by the Organization for Economic Co-operation and Development (OECD) and international trade data for developing countries is published by FAOSTAT or the United Nations' Comtrade. These data sources frequently have large gaps in addition to being collected from various primary sources, which can make it quite difficult to consistently and confidently compare data points between the various data sources.

Data constraints do not affect only ex post studies; ex ante studies would provide more accurate results if the models' parameters are closer to reality. In terms of CGE models, the accuracy of the result has been clearly improved due to consideration of more realistic assumptions and an updated GTAP database.

The collection of data on GMOs bilateral trade flows for developing countries as well as segregation costs and adoption rates will result in more studies focused on the impacts of GMOs adoption by developing countries and more accurate effects could then be empirically estimated.

14.6 Summary/Synthesis

- Non-harmonized GMOs regulatory frameworks reduce global trade and global welfare.
- An import ban is the least efficient policy response when compared to mandatory labeling or various domestic support policies such as TRQs or export subsidies.
- There is a need for more research on the economic impacts of GMOs adoption on developing countries.
- Several methodological issues remain to be resolved in the main two models of ex ante analysis, partial and general equilibrium.
- There is a need for more ex post studies that will take into account the reality of markets.

References

Anderson JE (1979) A theoretical foundation for the gravity equation. Am Econ Rev 69(1):106–116

Anderson K (2006) Interactions between trade policies and GM food regulations. In: Just RE, Alston JM, Zilberman D (eds) Regulating agricultural biotechnology: economics and policy (Natural resource management and policy, Vol. 30). Springer, New York (Chapter 7)

Anderson K, Jackson LA (2005a) Standards, trade and protection: the case of GMOs. Paper prepared for the 41st Panel Meeting of Economic Policy in Luxembourg, April 15–16. http://ageconsearch.umn.edu/bitstream/20282/1/sp04ja06.pdf. Accessed 16 Feb 2013

Anderson K, Jackson LA (2005b) Some implications of GM food technology policies for Sub-Saharan Africa. J Afr Econ 14(3):385–410

Anderson J, van Wincoop E (2003) Gravity with gravitas: a solution to the border puzzle. Am Econ Rev 93:170–192

Anderson K, Valenzuela E, Jackson LA (2008) Recent and prospective adoption of genetically modified cotton: a global computable general equilibrium analysis of economic impacts. Econ Dev Cult Change 56(2):265–296

Armington P (1969) A theory of demand for products distinguished by place of production. Int Monet Fund Staff Pap XVI:159–178

Bergstrand JH (1990) The Heckscher-Ohlin-Samuelson model, the Linder hypothesis, and the determinants of bilateral intra-industry trade. Econ J 100(4):1216–1229

Berwald D, Carter CA, Gruère GP (2006) Rejecting new technology: the case of genetically modified wheat. Am J Agric Econ 88(2):432–447

Deardoff A (1998) Determinants of bilateral trade: does gravity work in a neoclassical framework?. In: Frankel JA (ed) The regionalization of the world economy. University of Chicago Press, Chicago

Disdier AC, Fontagné L (2010) Trade impact of European measures on GMOs condemned by the WTO panel. Rev World Econ 146:495–514

Eaton J, Kortum S (2002) Technology, geography, and trade. Econometrica 70(5):1741–1779

Feenstra R (2004) Advanced international trade: theory and evidence. Princeton University Press, Princeton

Gaisford JD, Hester A (2007) Why are there trade agreements? In: Kerr WA, Gaisford JD (eds) Handbook on international trade policy. Edward Elgar, Cheltenham, pp 57–70

Gaisford JD, Kerr WA (2001) Economic analysis for international trade negotiations—the WTO and agricultural trade. Edward Elgar, Cheltenham

Gaisford JD, Lau C (2001) The case for and against embargoes on products of biotechnology. Estey Centre J Int Law Trade Policy 1(1):83–98
Gruère GP (2009) Impacts on international trade. In: Measuring the economic impacts of transgenic crops in developing agriculture during the first decade. approaches, findings and future directions (Chapter 6), IFPRI, Food Policy Review 10. http://www.ifpri.org/publication/measuring-economic-impacts-transgenic-crops-developing-agriculture-during-first-decade. Accessed 18 Feb 2013
Gruère GP (2011) Global welfare and trade-related regulations of GM food: biosafety, markets, and politics. In: Carter CA, Moschini GC, Sheldon I (eds) Genetically modified food and global welfare (Frontiers of economics and globalization, Vol. 10). Emerald Group Publishing Limited, Bingley (Chapter 13)
Gruère GP, Bouët A, Mevel S (2007) Genetically modified food and international trade: the case of India, Bangladesh, Indonesia and the Philippines, IFPRI Discussion Paper 740. International Food Policy Research Institute, Washington DC
Helpman E, Melitz M, Rubinstein Y (2007) Estimating trade flows: trading partners and trading volumes. NBER Working Paper 12927. http://www.nber.org/papers/w12927
Hertel T (1997) Global trade analysis: modeling and applications. Cambridge University Press, New York
Huang J, Hu R, van Meijl H, van Tongeren F (2004) Biotechnology boosts to crop productivity in China: trade and welfare implications. J Dev Econ 75(1):27–54
Isaac GE, Phillipson M, Kerr WA (2002) International regulation of trade in the products of biotechnology. Estey Centre for Law and Economics in International Trade, Saskatoon
Kerr WA (2007) Introduction to trade policy. In: Kerr WA, Gaisford JD (eds) Handbook on international trade policy. Edward Elgar, Cheltenham
Lapan HE, Moschini GC (2004) Innovation and trade with endogenous market failure: the case of genetically modified products. Am J Agric Econ 86(3):634–648
Nielsen C, Thierfelder K, Robinson S (2003) Consumer preferences and trade in genetically modified foods. J Policy Model 25(8):777–794
Perdikis N, Kerr WA (1998) Trade theories and empirical evidence. Manchester University Press, Manchester
Sobolevsky A, Moschini GC, Lapan H (2005) Genetically modified crops and product differentiation: trade and welfare effects in the soybean complex. Am J Agric Econ 87(3):621–644
Van Meijl H, van Tongeren F (2004) International diffusion of gains from biotechnology and the European Union's common agricultural policy. Agric Econ 31(2):307–316
Vigani M, Raimondi V, Olper A (2012) International trade and endogenous standards: the case of GMO regulations. World Trade Rev 11(3):415–437

Chapter 15
National Trade Interests

Guillaume P. Gruère

G. Gruere was a Senior Research Fellow at the International Food Policy Research Institute, Washington DC, when writing this chapter. He can be contacted by e-mail at ggruere@gmail.com.

15.1 Introduction

The coexistence between, and segregation of, genetically modified (GM), non-GM and organic crop production in supply chains is at the heart of the debates around the use and/or importation of specific GM products in a growing number of countries (Carter and Gruere 2012; Gruere and Sengupta 2009a). In this setting, the key question for policy-makers is how to manage negative market externalities induced by the introduction or use of GM products (Golan and Kuchler 2002; Moschini and Lapan 2006). Field testing and/or producing a GM crop may generate unintentional movements of pollen or seed to non-GM crops or fields. Introducing a GM product in a market chain (whether from the farm or via imports) may result in accidental comingling affecting non-GM supply chains. In a larger setting, adopting or importing GM crops may taint the reputation of non-GM marketing chain actors. In each of these cases, non-GM marketing chain actors may suffer economic losses due to market share restrictions or price decline.

These management issues are not trivial for decision makers. There are real economic risks from mishandling GM grains in the supply chain, as observed in the oft-cited cases of the StarLink corn, LL601 rice in the USA, and GM flax in Canada (e.g., Carter and Smith 2007; Carter and Gruere 2012; Ledford 2007; Ryan and Smyth 2012). At the same time, GM and non-GM crop coexistence has been successfully managed, mostly by private actors, in a number of countries without any reported economic loss for non-GM growers in domestic and/or international markets in the last 15 years (Carter and Gruere 2012). Moreover, an overcautious approach to market risks, defined as risks of affecting non-GM market opportunities, can be detrimental to regulatory development and technology use (Gruere and

G. P. Gruère (✉)
International Food Policy Research Institute, Washington, DC, USA
e-mail: ggruere@gmail.com

K. Ludlow et al. (eds.), *Socio-Economic Considerations in Biotechnology Regulation*, Natural Resource Management and Policy 37, DOI 10.1007/978-1-4614-9440-9_15,

Sengupta 2009a). For instance, claimed but unproven, negligible, or manageable market risks have been used as a political instrument by exporting industries to avoid potential costs at the detriment of progress in biosafety regulation in African countries, preventing these countries from making informed decisions on GM applications (Paarlberg 2008; Gruere and Sengupta 2009b; Gruere and Takeshima 2012). As with other risks, exposure, potential damage, and management options should all matter to rational and benevolent decision makers.

This chapter focuses on methods to assess market-related risks of GM crop production and marketing for non-GM production channels, using examples from the literature and existing regulatory practices. As a caveat, the discussion is limited to the implications of importing, testing, or using living modified organisms (LMOs) on conventional supply chain actors. This means that imports of processed products derived from GMOs are not considered.[1] Issues relating to possible gene flows on the ecology, including non-target organisms, will also not be addressed in this chapter. The focus is on the possible economic effects of GM imports and/or adoption of non-GM agricultural production, marketing, and trade. Lastly, the chapter focuses on economic considerations, for their prominence in this area, and because of the author's own expertise, but social considerations that arise from the described economic risks could be worth considering in a broader setting.

15.2 Methodologies

It is important to distinguish different categories of regulatory and marketing considerations that target countries may account for in their market-risk assessment. In particular, the incentive to separate non-GM from GM products varies from case to case. First, non-GM producers may need to satisfy implicit or explicit GM-free private standards in their domestic or international market (e.g., certified organic agriculture, fair trade, and GM-free standards). This constitutes the primary concern of non-GM market actors regarding GM production and use. Second, regulations may also play a role, especially if labeling of GM products is mandated in the target country. In such cases, non-GM producers, regardless of private standards, may have their production labeled as GM, which could result in market-access restrictions. Thirdly, firms will protect their intellectual property (IP) in instances where the technology is being illegally grown. These three modalities can be found together for IP-protected GM crops produced in a GM-labeling country, in the presence of non-GM private standards.

Another layer of complexity comes from the intended final use of the GM products. Most internationally traded GMOs are intended for food, feed, or processing.

[1] In the rest of the chapter, we use the GM/non-GM denomination to avoid confusion between non-GM market actors that do face risks and are concerned about living organisms that are not modified and their derivatives, from the case of non-LMOs, which would include processed grains and products derived from LMOs, but that are considered GM. Furthermore, GM and non-GM products are the common terms used in the literature.

Table 15.1 Typology of cases: sources of externality

Incentive for segregation	GM food	GM feed	Other GM products
Private standard	Maintaining organic or non-GM certification	Maintaining certification for GM-free or organic animal products	Maintaining certification for organic or fair-trade cotton
Labeling regulation	Risk of price decline if labeled as GM	No effect	
IP rights	Risk of paying GM technology fee		

The largest volume of grain is used for animal feed and non-food uses (Gruere 2006). Food uses of GM products remain marginal in volume and value. While all these categories may be subject to IP contracts and private standards, the two former classes do not generally face stringent labeling requirements. Table 15.1 provides a matrix of market-risk typology. Despite these differences, most cases raise risks regarding compliance with private standards.

To the author's knowledge, there is no standard methodology for assessing market risks and measuring the economic externality from GM crop production to non-GM supply chains. Instead, a number of approaches have been used focusing on different aspects of the question (e.g., Smale et al. 2008). Methods used in the literature and by regulatory agencies can be separated into two categories: those pertaining to market risks at the domestic level and those relating to risks at the international level. Each of these categories can then be divided into four types that gradually increase in difficulty (Table 15.2). At the domestic level, beyond rapid appraisal of a case (D1), a number of papers have provided benchmark economic analyses of coexistence and segregation options that can be used as support for decision makers (D2). A few studies have focused on effects of introducing a GM product on a supply chain (D3), and a number of studies have used advanced modeling of coexistence or segregation (D4). At the international level, regulatory agencies seem to have relied on qualitative information from traders or interest groups (I1), but researchers have used bilateral trade flow analyses (I2), buyer surveys (I3) that could be completed with a general market analysis, and *ex ante* economic simulations to measure the international effect of GM adoption on non-adopters (I4).

Most literature on domestic effects is concerned with assessing the cost and implications of management practices for coexistence and segregation.[2] This can be useful for countries concerned about keeping pre-existing markets for products under non-GM private standards or certification schemes. Risk assessors may use existing literature (D2), conduct their own analysis accounting for costs (D3), or

[2] For a list of estimates of segregation cost, see Gruere (2009). On coexistence, see for instance NRC (2010) on the US and Messean et al. (2006) on Europe, and the regulatory option discussion and references in Demont et al. (2009) and Beckmann et al. (2011).

Table 15.2 List of methodologies

Focus	Type	Methodology	Difficulty
Domestic market	D1	Fact looking (plant/other countries)	+
	D2	Review of literature on coexistence/segregation costs	++
	D3	Cost accounting and supply chain analysis	+++
	D4	Modeling coexistence/segregation options	++++
International market	I1	Basic macro market data review	+
	I2	Bilateral trade flow analysis	++
	I3	Survey of traders/market prospective	+++
	I4	Market simulations of adoption and segregations	++++

use modeling (D4). Because traders of non-GM products will always prefer avoiding any cost of segregation, they have an incentive to deter the introduction of a GM crop in a particular country (Gruere and Takeshima 2012). Policymakers uninformed about the feasibility of coexistence and segregation, and the potential cost they imply, that is, using only method D1, may believe these traders and bias their final decisions.

There are diverse methodologies for *ex ante* assessment of coexistence or segregation options. Yet, most focus on specific developed country context; all may not be adaptable to a developing country context. Articles focus on either determining optimal management decisions or assessing the cost of coexistence or segregation. The degree of complexity of models also varies from accounting costs in a supply chain (Huygen et al. 2003), to modeling their market implications (Wilson et al. 2008) or developing spatial models of coexistence (e.g., Demont et al. 2009).

A few studies evaluate the comprehensive effects of introducing a particular GM event in a supply chain (D3). Cost of segregation is once again a key variable to consider, but the required institutional changes that GM introduction may induce are also included. For example, Gruere and Cartel (2006) looked at the implications of Bt cotton introduction into the well-organized cotton supply chains in several West African countries. As part of their analysis, they identified key issues (Fig. 15.1), including coexistence with organic and fair trade cotton chains. Horna et al. (2013) in their report on Bt cotton introduction in Uganda, included specific management options for Bt cotton seed segregation to ensure that Ugandan organic cotton growers could continue certified production. The methods used were simple value chain analysis and identification of possible constraints and solutions, but could be extended to the use of primary collected data in a value chain analysis.

At the international level, a few studies (type I2) assess the possible exposure of countries to loss of market access or implications of new trade regulations (Paarlberg 2006; Gruere and Rosegrant 2008). Their focus is on GM-related trade flows, but they indirectly address concerns raised by non-GM actors in these countries. Paarlberg (2006) addresses concerns by African policymakers related to exports of non-GM products to Europe. Gruere and Rosegrant (2008) examine the implementation effects of information requirements on traded shipments of LMO-FFPs,

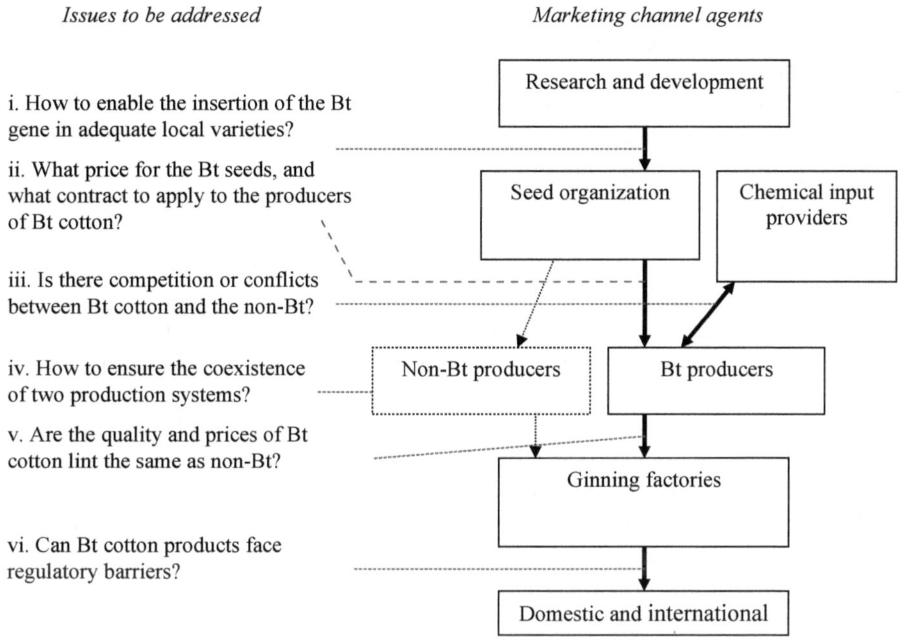

Fig. 15.1 Example of market chain analysis: key issues with the introduction of Bt cotton in West Africa. (Source: Cartel and Gruere 2006)

which would affect any shipment from GM adopting countries to Biosafety Protocol member countries. To do so, the authors compute share of relevant export volumes to countries with specific regulatory requirements using past bilateral trade data. The same method could be applied to determine the volume and value of non-GM products. The outcomes are not sufficient for a full determination of any risk, but give an indication of what could be at risk if GM was adopted.

A second subcategory of international studies (type I3) is based on the use of buyer surveys and market acceptance of GM and non-GM products (Bett et al. 2010; Knight et al. 2005, 2008). The possible introduction of a GM crop in a non-GM producing nation (like New Zealand) may impact buyers' preferences, and such studies, while requiring primary data collection and potentially facing some strategic biases from buyers, may provide useful indicators in an assessment of market risks.

The last subgroup of studies (type I4) uses simulation models to evaluate the effects of GM introduction in an international setting (Gruere et al. 2011). Partial equilibrium and computable general equilibrium models have been used to generate *ex ante* simulations. While market access for GM products is the primary focus, some of these studies explicitly look at the effects of adoption on non-adopters. For instance, Elbehri and MacDonald (2004) and Bouet and Gruere (2011) use computable general equilibrium models to assess the effect of adoption or non-adoption of GM cotton in African countries when other countries adopt it. Frisvold et al. (2006)

assess the price-dropping effect of Bt cotton on the international market. Berwald et al. (2006) assess the effect of adoption of GM wheat in Canada and/or USA when other countries adopt it. Gruere et al. (2009) compare the effect of adoption and non-adoption of GM rice in China and India. While these studies may not focus on market risk as defined here, they can be useful in assessing the full benefit of adoption compared to non-adoption, in a competitive international market. Establishing a valid counterfactual or baseline is necessary to avoid underestimating the benefits of adoption.

A few studies in the same category do look at segregation of non-GM, by simulating the effect of keeping two markets. To do so, they implement a cost of segregation or include various preference parameters. Nielsen et al. (2003) used shift in preference parameters to implement adoption and trade shocks in a global trade model. Gruere et al. (2009) derive an opportunity cost of segregation, giving a benchmark on the maximum cost of segregation for a country to break even with the adoption of a particular GM crop. This methodology could be used to include benchmark values for large non-GM exporting industries.

To complete the picture, some literature attempts to explain influence links between trading actors and non-actors around the introduction of a GM product (Gruere and Sengupta 2009a; Gruere and Takeshima 2012). These political economic analyses may be needed in the design of a regulatory mechanism but not necessarily in product development.

In terms of regulatory practices, few countries account for the economic effects of GM on non-GM market actors; most using explicit inclusion, or implicit representation of non-GM bodies in regulatory decisions, and a few have systems requiring studies. Assessors tend to have a very partial view based on rapid appraisal (D1).

- The USA has no specific requirement to incorporate coexistence in product regulatory management (Carter and Gruere 2012). But pursuant to the National Environmental Protection Act, the US Department of Agriculture Animal and Plant Health Inspection Service (USDA-APHIS), in charge of environmental risk management, has to consider the implications of production or marketing externalities for products with likely substantial risks. In cases that are found to have substantial risks, a full environmental impact statement must be completed by the agency, and is usually conducted by consultants, resulting in long reports.[3] It should be emphasized that the outcome of the statement plays no role in the final decision to deregulate (approve) a GM crop (Carter and Gruere 2012).
- In France, a specific socio-economic committee (Comité économique, éthique et social du Haut Conseil des Biotechnologies) has been included in the regulatory system with the mandate to discuss any concern relevant to an application for field trial or environmental release, including coexistence. No analytical action

[3] A number of recent court decisions against the USDA-APHIS have shown that the agency usually considers most cases not worthy of pursuing a full environmental impact analysis (including coexistence).

is conducted, but committee members can call upon studies to support their arguments. Still, because the committee is composed of few scientists, compared to a significant portion representing interest groups with pro or anti-GM positions, the committee has only published a few recommendations.
- In South Africa, the GMO executive council in charge of risk management is composed of different Departments, including the Department of Trade and Industry (DTI) (Gruere and Sengupta 2010). The GMO executive council only takes consensus decisions. According to various reports, DTI has been among the most proactive in supporting the voice of non-GM industries, in particular in cases that resulted in negative outcomes for the GM potato, GM sugarcane, and a GM vine (Gruere and Sengupta 2010).

In contrast, a number of countries have explicitly accounted for non-GM exports in discrete decisions in an ad hoc manner. The movement toward ban of GM products and GM food aid in Southern African countries, can be attributed at least partially to the voice of non-GM traders (Paarlberg 2008; Gruere and Takeshima 2012). The well-publicized case of non-GM exporter to Europe in the refusal to take GM food aid played a role in Zambia. Other cases exist where limited or negligible market risks blocked policy actions on GMOs (Gruere and Sengupta 2009a). There is no evidence that a formal assessment was conducted to support any of these decisions, and in one case (Namibia) a report of evidence was actually explicitly ignored by regulators (Gruere and Sengupta 2009b).

Similarly, decisions taken in Asia to restrict or avoid GM rice trials or GM papaya, among others, were reportedly influenced by non-GM interests (Gruere and Sengupta 2009a). Once again, there is no evidence that any analytical work was conducted to support these decisions, but it is likely that the voice of industry was sufficient to encourage decision makers in making precautionary decisions (Gruere and Takeshima 2012).

15.3 Critical Assessment

Assessing national trade concerns should take place within an overarching decision-making system. Such a system would first trigger the need for a trade risk assessment and then would guide toward the determination of plausible market risks. For instance, Gruere and Sengupta (2009b) suggest five discrete questions to determine whether there is a market risk or not (Fig. 15.2).

If market risk assessment is required, to answer Q1–Q4, the choice of methodology would follow. Not all described methodologies may be relevant for socio-economic assessment—the type of methods depends upon the specific question and need. Typically, market risk studies will need to be conducted *ex ante*, before introduction, and to save time, they could be conducted at the stage of likely commercialization. As noted above, there are gradual degrees of sophistication in existing studies. The most advanced techniques require more advanced knowledge, often

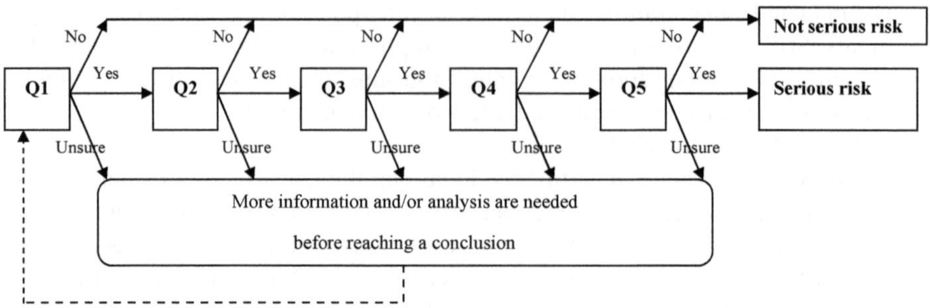

Fig. 15.2 Example of decision tree in the case of a market risk assessment. Notes: Q1. Is the alleged risk substantiated? Q2. Are export losses likely with the decision? Q3. Are presumed export losses non-negligible for the country? Q4. Is the risk unavoidable? Q5. Is the risk greater than the benefits? (Source: Gruere and Sengupta 2009b)

the collection of primary data and more time for completion. In contrast, a number of methods could be used with secondary data and less technical advancements in a relatively rapid fashion. Table 15.3 shows a possible three tier-system of degree of advancement in methodology with time and data collection requirements.

Even if the detailed decision tree in Fig. 15.2 is not followed, three critical stages may be necessary: (1) Determination of basic risks (Tier 1 methods); (2) If necessary, investigation with choice of method (Tier 2 or 3 methods) on market risk and management options; (3) Recommendation.

The basic determination of risks should be done rapidly, focusing on existing markets for non-GM, differentiation between GM and non-GM, and the likelihood of coexistence/segregation issues (Tier 1). A number of cases can be almost entirely dismissed, due to no market issue, unlikely comingling, or gene flow. Others may require more information before determination using more thorough methods (Tier 2), with data collection and simple analysis. In cases of critical possible risks, or where there are ambiguities not resolved by simple evaluations, Tier (3) methods may be needed.

The assessment process adopted in a specific country will depend on the degree of government involvement and types of regulations on the management of GM externalities. In the case of coexistence, government-prescribed guidelines should be followed and adapted to each case. In the absence of guidelines, basic practices may need to be provided to GM and non-GM farmers. In the case of segregation, a cost will be involved. The role of the government may be to decide whether this cost would be prohibitive or implementable, but it could also decide that only major market risks are worth considering in the risk assessment.

There are three possible recommendations from the assessment: (1) no evidence of risk; (2) presence of manageable risk; and (3) presence of risk with no easy management options. In the third case, more data from the applicants would be required. If the additional information does not provide a solution, the recommendation should be to reject the application.

Table 15.3 Categories of approaches, from rapid and inexpensive, to long and expensive

Tier	Methods involved	Category
Tier 1	Basic market data collection and literature review	D1-D2, I1-I2
Tier 2	Supply chain analysis, surveys of buyers, studies of exposure and possible costs of management options	D3, I3
Tier 3	Simulation methods for market effects of adopting and non-adopting for farmers and supply chain actors	D4, I4

15.4 International Arena

Many of the cases discussed above focus on market risks related to the domestic introduction of GM crops intended for environmental release and will not generally challenge WTO rules. But decisions regarding GM products intended for import that could affect non-GM supply chains may be more contentious. In such cases, requiring the importer to obtain more data on risk management options, and more importantly rejecting imports due to potential market risks, may be considered technical barriers to trade, if used improperly.

The rejection of an application to import corn-rootworm-resistant GM maize into South Africa provides a singular example (see Gruere and Sengupta 2010). South Africa produces, imports, and exports GM and non-GM maize, providing an example of relatively successful coexistence and segregation practices. Still, in 2006, the GM maize event approved in the USA was submitted for import and consumption use approval in South Africa. But it coincided with a new measure that made import approval dependent on planting approval. Given that the GM trait was not advantageous in the South African context, the corn rootworm being an inconsequential pest, the crop could not pass the test of increased agronomic performance and no planting application was submitted. As a consequence, import approval was rejected, resulting in a de facto ban of comingled GM maize imports shipments from the USA. In 2007, year of drought, South Africa imported significant quantities of maize, and the domestic animal feed industry requested the removal of the measure linking import to planting approval, noting that it resulted in increased grain prices. Yet the maize industry, that had supported it, opposed any change, as they felt that GM maize imports from the USA impeded their competitiveness. As explained in Gruere and Sengupta (2010), the case appeared to be a good example of non-tariff barrier to trade based on seemingly unjustified GM approval decisions. Ultimately, these GM products remained unauthorized for import partly because of presumed market risk reasons for other GM and non-GM growers, blocking unsegregated imports from the USA.

Another case with potential incompatibility with WTO requirements is the recent EU ruling on honey containing pollen from GMOs. The ruling argued that honey should follow the same rules as other GM products, including zero tolerance for unapproved events, labeling, and traceability. The claimants in the case before the European Court of Justice were German organic honey producers. They raised the risk of their honey being found to contain GM pollen, with loss of certification and imposition of labeling requirements, if GM crops were introduced in their vicinity. The objective cause was the introduction of more stringent requirements for GM maize producers. While the case centered on domestic market risks, it was bound to affect imports of honey from all GM-producing countries. If those countries allowed GM crops not approved in the EU, they faced a ban on export of their honey into the EU. De facto, this introduced a new import restriction based on non-product-related process and production methods, whose compatibility with WTO rules is questioned. The ruling's implication was that a market-risk-related standard could imply the ban of honey imports only from nations that do not have the same list of approved GM products as the EU.

15.5 Administrative Consequences

Two opposite approaches could be adopted for socio-economic assessments: it could be left to applicants and reviewed by authorities, or conducted by regulatory agencies (or their external consultants).[4] Naturally, intermediate solutions exist, with responsibilities shared by the applicant and the regulatory agency.

In the case of market risk assessment, as defined above, the possible involvement of multiple external chain actors, from farmers to traders, manufacturers, and retailers means that an applicant-only approach is unlikely to be appropriate. While the applicant may be asked to provide justification or data showing that the product is not likely to create risks for non-GM market actors, and/or points toward management options, it may be difficult for applicants to provide a credible and complete assessment of the risk for all actors involved in a particular market chain.

Instead, other intermediary approaches may be considered, representing different degrees of involvement for applicants and regulatory agencies. In all cases, the regulatory agents would be responsible for triggering a more complete assessment and the final recommendation to the biosafety institutional body. Table 15.4 suggests a possible list of options.

Interestingly, existing examples of regulatory decision-making in this area have used option O3, even if they did not use any rigorous method to arrive at their outcome. The more involved the regulatory agency, the higher capacity it may require

[4] This discussion for risk assessment bears some similarities with issues related to Smyth et al.'s (2006) analysis of options for managing liabilities from GM crops. While liabilities occur after risk realization, the same contrast between private and public resolution occurs. Smyth et al. (2006) propose three options: one private, one scientific and regulatory, and one based on market strategies. Of these three, the scientific option may not be directly relevant for an application ready to advance, but could enter into discussions of management options.

Table 15.4 Implementation options

Option	Basic market data/case study analysis	Basic risk determination	Deeper investigation of risk and management options (if needed)	Recommendation
O1	Applicant	Regulatory agency	Applicant	Regulatory agency
O2	Applicant	Regulatory agency	Consultant/Regulatory agency	Regulatory agency
O3	Regulatory agency	Regulatory agency	Consultant/Regulatory agency	Regulatory agency

either internally or externally. But at the minimum, the regulatory authority should have personnel competent to review basic market and trade data, and relevant analyses.

15.6 Summary/Synthesis

- The use of GM crops (LMOs) can generate negative externalities on non-GM supply chains.
- Potential market risks can be significant, but most presumed risks are manageable or avoidable; they should be assessed objectively and rigorously on a case-by-case basis.
- Regulators should consider adopting a decision tree, including a first layer of basic market analysis using rapid market analysis that could trigger further inquiry into potential market risk and management options.
- Objective market risk assessment, if conducted by a regulatory agency, requires a minimum internal expertise to analyze market and regulatory data, with calls for outside expertise on a case-by-case basis.
- Decisions to reject LMO-FFPs imports due to potential market risks, while seemingly in line with the Biosafety Protocol, may be incompatible with WTO requirements, especially if they favor domestic producers, and/or specific trade partners. Care should be taken to ensure that rejection decisions occur only in cases where management options are not feasible or prohibitively costly.

References

Beckmann V, Soregaroli C, Wesseler J (2011) Coexistence of genetically modified (GM) and non-modified (non-GM) crops: are the two main property rights regimes equivalent with respect to the coexistence value? In: Carter C, Moschini G, Sheldon IM (eds) Genetically modified food and global welfare. Frontiers of Economics and Globalization, vol 10. Emerald, Bingley

Berwald D, Carter CA, Gruere GP (2006) Rejecting new technology: the case of genetically modified wheat. Am J Agric Econ 88(2):432–447

Bett C, Okuro Ouma J, de Groote H (2010) Perspectives of gatekeepers in the Kenyan food industry towards genetically modified food. Food Pol 35(4):332–340

Carter CA, Gruere G (2012) New and existing GM crops: in search of effective stewardship and coexistence. Northeast Univ Law J 4(1):169–207

Carter CA, Smith AD (2007) Estimating the market effect of a food scare: the case of genetically modified starlink corn. Rev Econ Stat 89(3):522–533

Demont M, Dillen K, Daems W et al (2009) On the proportionality of EU spatial ex antecoexistence regulations. Food Pol 34(6):508–518

Elbehri A, MacDonald S (2004) Estimating the impact of transgenic Bt cotton on west and central africa: a general equilibrium approach. World Dev 32:2049–2064

Frisvold GB, Reeves JM, Tronstad R (2006) Bt cotton adoption in the United States and China: international trade and welfare effects. AgBioForum 9(2):69–78

Golan E, Kuchler F (2002) Labeling biotech foods: implications for consumer welfare and trade. In: Krissoff B, Bohman M, Caswell J (eds) Global food trade and consumer demand for quality. Springer, New York

Gruere GP (2006) An Analysis of Trade Related Regulations of Genetically Modified Food and their Effects on Developing Countries. EPT Discussion Paper 147, Washington DC: International Food Policy Research Institute.

Gruere GP (2009) Asynchronous approvals of GM products, price inflation, and the codex annex: what low level presence policy for APEC countries? Conference paper presented at the 2009 IATRC summer symposium. Pullman, WA: International Agricultural Trade Research Consortium. http://iatrc.software.umn.edu/activities/symposia/2009Seattle/seattle-Gruere.pdf

Gruere GP, Cartel M (2006) Trading hot potato from farm to port: the case of Bt cotton in West Africa. Selected paper presented at the winter meeting of the IATRC 2006. International Agricultural Trade Research Consortium

Gruere GP, Rosegrant MW (2008) Assessing the implementation effects of the biosafety protocol's proposed stringent information requirements for genetically modified commodities in countries of the Asia Pacific economic cooperation. Appl Econ Perspect Policy 30(2):214–232

Gruere GP, Sengupta D (2009a) The effects of GM-free private standards on biosafety policymaking in developing countries. Food Pol 34(5):399–406

Gruere GP, Sengupta D (2009b) Biosafety and perceived commercial risks: the role of GM free private standards. IFPRI Discussion Paper 00847. Washington, DC: International Food Policy Research Institute. http://www.ifpri.org/sites/default/files/publications/ifpridp00847.pdf

Gruere GP, Sengupta D (2010) Reviewing South Africa's marketing and trade related policies for genetically modified products. Dev South Afr 27(3):333–352

Gruere GP, Takeshima H (2012) Will they stay or will they go? The political influence of GM averse importing companies on biosafety decision makers in Africa. Am J Agric Econ 94(3):736–743

Gruere GP, Mevel S, Bouët A (2009) Balancing productivity and trade objectives in a competing environment: should India commercialize GM rice with or without China? Agric Econ 40(4):459–475

Gruere G, Bouet A, Mevel S (2011) International trade and welfare effects of biotechnology innovations: GM food crops in Bangladesh, India, Indonesia and the Philippines. In: Carter C, Moschini G, Sheldon IM (eds) Genetically modified food and global welfare. Frontiers of Economics and Globalization, vol 10. Emerald, Bingley

Horna D, Zambrano P, Falck-Zepeda J (2013) *Socioeconomic considerations in biosafety decisionmaking: Methods and implementation* IFPRI Research Monograph 180, IFPRI, Washington, DC. IFPRI, Washington DC

Huygen I, Veeman M, Lerohl M (2003) Cost implications of alternative GM tolerance levels: non-genetically modified wheat in Canada. AgBioForum 6(4):169–177

Knight JG, Holdsworth D, Mather DW (2008) GM food and neophobia: connecting with the gate keepers of consumer choice. J Sci Food Agric 88:739–744

Knight JG, Mather DW, Holdsworth D (2005) Impact of genetic modification on country image of imported food products in European markets: perception of channel members. Food Pol 30:385–398

Ledford H (2007) Out of bounds. Nature 445:132

Messean A, Angevin F, Gómez-Barbero M et al (2006) New case studies on the coexistence of GM and non-GM crops in European agriculture. European Commission Joint Research Centre (DG JRC). Institute for Prospective Technological Studies: Madrid, Spain

Moschini G-C, Lapan H (2006) Labeling regulations and segregation of first- and second-generation GM products: innovation incentives and welfare effects. In: Just RE, Alston JM, Zilberman D (eds) Regulating agricultural biotechnology: economics and policy. Springer, New York

National Research Council (NRC) (2010) The impact of genetically engineered crops on farm sustainability in the United States. National Academy Press, Washington DC.

Nielsen CP, Thierfelder K, Robinson S (2003) Consumer preferences and trade in genetically modified foods. J Policy Model 25:777–794

Paarlberg R (2006) Are genetically modified crops a commercial risk for Africa? J Tech Glob 2(1–2):81–92

Paarlberg R (2008) Starved for science. How biotechnology is being kept out of Africa. Harvard University Press, Cambridge

Ryan CD, Smyth SJ (2012) Economic implications of low-level presence in a zero-tolerance European import market: the case of Canadian triffid flax. AgBioForum 15(1):21–30

Smale M, Zambrano P, Falck-Zepeda J, Gruere GP et al (2008) The economic impact of transgenic crops in developing countries: a note on methods. Int J Biotechnol 10(6):519–551

Smyth S, Phillips PWB, Kerr WA (2006) Managing liabilities arising from agricultural biotechnology. In: Just RE, Alston JM, Zilberman D (eds) Regulating agricultural biotechnology: economics and policy. Springer, New York

Wilson WW, DeVuyst EA, Taylor RD et al (2008) Implications of biotech traits with segregation costs and market segments: the case of roundup ready wheat. Eur Rev Agric Econ 35(1):51–73

Chapter 16
Producer Choice

Ari Novy and Latha Nagarajan

16.1 Introduction

Several respected bodies and stakeholders have begun to identify socioeconomic considerations (SECs) relevant to producer choice. For example, the Netherlands COGEM report has identified freedom of choice, for both producers and consumers, as a key theme in consideration of SECs surrounding genetic modification. Producers should possess the freedom to deploy GMOs or be GMO-free and also be free to innovate and research such technologies (COGEM 2009). The Third Word Network is an international network of organizations concerned with various issues, including agricultural development. They have identified several key areas of SECs that could affect producer choice including income security, control over production by poorer farmers, contamination of organic agriculture, and farmers' rights to save seeds (Third World Network 2008).

The heart of any producer choice consideration must be an understanding of factors that producers weigh when making decisions. In agriculture, these factors can be diverse and heterogeneous in both time and space. With regard to GMOs, producers only have a choice if there are desirable GM crop varieties available. Choice is therefore a function of each country's national decision-making process to allow, or not allow, GMOs. In addition, specific GM crops must be available, in appropriate genetic backgrounds (i.e., varieties), for producers to exercise full choice.

Few producers base their decision to utilize GM technology simply on the classification of this technology as GM or non-GM. Producers make decisions based

A. Novy (✉)
United States Botanic Garden, Washington, DC, USA
e-mail: anovy@aoc.gov

Smithsonian Institution, National Museum of Natural History, Washington, DC, USA

L. Nagarajan
Department of Agriculture, Food and Resource Economics, Rutgers University, New Jersey, USA

International Fertilizer Development Center (IFDC), Washington, DC, USA

K. Ludlow et al. (eds.), *Socio-Economic Considerations in Biotechnology Regulation,* Natural Resource Management and Policy 37, DOI 10.1007/978-1-4614-9440-9_16,

on a variety of expectations including, but not limited to, yield, profit, risk perception, inputs (labor, fertilizer, water, pesticides, etc.), health effects, environmental stewardship, market factors, tradition, and culture. Furthermore, these concerns are likely to be ranked differently by each producer. Therefore, evaluation frameworks regarding producer choices need to be relevant to each application situation. The specific choices that producers wish to exercise, with regard to the GMO in question, often require assessment on a case-by-case basis. In order to do so, a detailed understanding of the factors likely to impact producer choice must be generated and explored.

16.2 Methodologies

Fransen et al. (2005) generated a list of research methodologies that would be useful to assess SECs. They noted that a comprehensive assessment may require the combination of several methodologies. They identified economic modeling, cost–benefit analysis, social impact assessment, sustainable livelihoods framework, systemic "relevance assessment," and participatory research as key assessment approaches for examining producer choice issues. To this list, we would add social audit and biological assessment as additional useful approaches with regard to producer choice. Table 16.1 lists examples of several of the methodologies discussed below to facilitate access to examples from the literature.

16.2.1 Economic Modeling

The basic question posed at the producer (farm) level of an economic study involves quantifying the advantages of GM crop varieties with respect to yield, pesticide use, input cost, revenue, profit and/or any other variable of interest, by producer type and geography (Smale et al. 2006). Many studies utilize standard economic (statistical, empirical, and modeling) methodologies to explore the impacts, both *ex ante* and *ex post,* of GMO adoption or non-adoption. In fact, the methodologies of such studies have been extensively reviewed (e.g., Smale et al. 2008, 2009; Qaim 2009; Raney and Matuschke 2011, and specifically for cotton in Smale et al. 2006; Nazli 2010). Therefore, we will briefly summarize various methodologies and highlight several examples.

Most studies examine GMO producer effects using already deployed farm data (*ex post*). For these studies, researchers generally choose from two main approaches. The first is farm accounting, also called partial budgets. Farm accounting is essentially a type of cost–benefit analysis comparing the economic gains of GMO adopters and non-adopters at the farm level and is discussed in more detail in the section on cost–benefit analysis. The second approach uses an econometric or quantitative model to describe field data empirically based on a theoretical economic

Table 16.1 Various studies on the impact of GM crops on producers

Crop	Country	Publication year	Authors	Data type	Study type	Approach	Methods
Banana	Uganda	2006	Edmeades and Smale	Statistical survey, key informant	*Ex ante*	Economic modeling	Farm survey analysis, *ex ante* adoption model, simulation
Banana	Uganda	2010	Kikulwe	National agricultural statistics, statistical survey	*Ex ante*	Cost–benefit analysis, economic modeling	Maximum incremental social tolerable irreversible costs, willingness to pay, choice experiment, real option
Cotton	Burkina Faso	2007	Vitale et al.	Field trials	*Ex ante*	Economic modeling	ANOVA model
Cotton	China	2003	Huang et al.	Statistical survey	*Ex post*	Economic modeling	Farm survey analysis, multivariate pesticide use model
Cotton	China	2004	Hossain et al.	Farm survey	*Ex ante*	Economic modeling	Health-production function
Cotton	China	2006	Kuosmanen et al.	Statistical survey, leaf tissue	*Ex post*	Biological assessment, economic modeling	Damage control production function
Cotton	India	2003	Qaim and Zilberman	On-farm trials	*Ex ante, ex post*	Cost–benefit analysis	Trial data analysis, yield-density function, logistic damage control function
Cotton	India	2007	Crost et al.	Statistical survey	*Ex post*	Economic modeling	Farm survey analysis, fixed effects, panel data, selectivity bias, Cobb-Douglass production function
Cotton	Pakistan	2010	Ali and Abdulai	Farmer survey	*Ex post*	Economic modeling	Treatment effect model, propensity score matching
Cotton	South Africa	2005	Shankar and Thirtle	Statistical survey	*Ex post*	Economic modeling	Farm survey analysis, damage control production function, tests for endogeneity of pesticide use and Bt choice, model tests, value of marginal product analysis
Cotton	South Africa	2005	Morse et al.	Company data	*Ex post*	Economic modeling	Farm records analysis

Table 16.1 (continued)

Crop	Country	Publication year	Authors	Data type	Study type	Approach	Methods
Cotton	South Africa	2005	Shankar and Thirtle	Statistical survey	*Ex post*	Economic modeling	Farm survey analysis, damage control function, tests for endogeneity of pesticide use and Bt choice, model tests, value of marginal product analysis
Cotton	South Africa	2006	Bennett et al.	Farm records	*Ex post*	Economic modeling	Farm records analysis, farm survey analysis, Gini coefficient
Cotton	South Africa	2007	Shankar et al.	Farm records	*Ex post*	Economic modeling	Stochastic dominance model and stochastic production function
Cotton	South Africa	2008	Morse and Bennett	Farm survey	*Ex post*	Sustainable livelihoods framework, participatory research	Farm survey analysis
Eggplant	India	2007	Krishna and Qaim	Statistical survey, field trials	*Ex post, ex ante*	Economic modeling	Farm survey analysis, trial data analysis, double-bounded dichotomous choice, choice experiment, public–private partnership, open-pollinated variety hybrid comparison
Eggplant	India	2008	Krishna and Qaim	Statistical survey, field trials	*Ex post, Ex ante*	Economic modeling	Farm survey analysis, trial data analysis, economic surplus, farmers stated preference
Eggplant and tomato	Spain and Italy	2011	Groenveld et al.	National agricultural statistics	*Ex ante*	Cost–benefit analysis	Coexistence compliance costs, pesticide costs
Maize	Africa	2009	Ezezika et al.	Interview, observation, focus groups	*Ex ante*	Social audit	Open and closed-ended questionnaire analysis
Maize	Kenya	2011	De Groote et al.	Field trials	*Ex ante*	Economic modeling, biological assessment	Economic surplus model, regional assessment, insect resistance consideration
Maize	Mexico	2007	Birol et al.	Farmer survey	*Ex ante*	Economic modeling	Latent class model, choice experiment

Table 16.1 (continued)

Crop	Country	Publication year	Authors	Data type	Study type	Approach	Methods
Maize	Philippines	2006	Yarobe and Quicoy	Statistical survey	*Ex post*	Economic modeling	Farm survey analysis, Cobb-Douglass production function, Heckman model, selectivity bias, producer surplus
Maize	South Africa	2012	Regier et al.	Farm survey	*Ex post*	Economic modeling	Stochastic dominance model
Rice	Uruguay	2006	Hareau et al.	Statistical survey	*Ex ante*	Economic modeling	Economic surplus model
Rice (golden)	Philippines	2004	Zimmermann and Qaim	Health data	*Ex ante*	Cost–benefit analysis	Disability adjusted life years, scenario approach, efficacy model
Soybean	Argentina	2005	Qaim and Traxler	Pilot survey, key informant	*Ex post*	Economic modeling	Farm survey analysis, economic surplus, large open economy, three regions, institutional analysis
Sweet potato	Kenya	2001	Qaim	Pilot survey, key informant	*Ex ante*	Economic modeling	Farm survey analysis, economic surplus, closed economy, cost–benefit analysis, sensitivity

framework. These models are considered to represent a more rigorous approach to hypothesis testing.

Identifying the specific research question(s) is key to these approaches. Since each GM trait and variety may have advantages in specific cropping systems, it is important to carefully evaluate the model's appropriateness for each situation. For example, Shankar and Thirtle (2005) used a damage control production function to explore farmer choices surrounding pesticide use and Bt cotton adoption in a region of South Africa. Due to the differences in cropping systems and pesticide usage patterns in this region relative to other well-studied Bt cotton regions (e.g., China and Argentina), the authors considered their approach to be more appropriate since producers in their study viewed the technology as damage-abating, as opposed to output-enhancing. Distinctly, Huang et al. (2003) were interested in cost savings that producers could achieve through selection of Bt cotton in China. They used multivariate analysis to determine that the main economic benefit of Bt adoption came from decreased pesticide and labor expenditures in their study area.

Predictive (*ex ante*) approaches are less common than *ex post,* but often use similar economic modeling techniques. Nevertheless, several examples of *ex ante* studies are included in Table 16.1 since they may be of interest to policy-makers in countries that have not yet deployed commercial GM agricultural technology. *Ex ante* studies are useful to estimate the economic benefits that may result from producers' adoption of a GMO before its release (Smale et al. 2009). For example, Krishna and Qaim (2008) assessed the potential impact of Bt eggplant in India using an economic surplus model based on field trial and farm survey data. Interestingly, they estimated future adoption rates by directly surveying potential adopters. Birol et al. (2007) used a choice experiment and latent class model to determine preference heterogeneity among Mexican maize farmers. This approach proved valuable in determining which farmers were likely to desire and/or benefit from GM maize introduction. De Groote et al. (2011) used an economic surplus model to assess the potential economic impact of Bt maize in Kenya. Their approach allowed for differing GMO adoption rates among farmers based on the success of GMO resistance to specific insect species and the estimation of benefits based on producer agroecology. Hareau et al. (2006) used stochastic simulation to explore the economics of multinational corporation involvement in a small market country by examining the potential of GM rice in Uruguay.

16.2.2 Cost–Benefit Analysis

Cost–benefit analysis (CBA) is a method to quantify an action's costs and benefits so as to facilitate a decision when considering more than one option, often in the context of regulatory approval. CBA requires the translation of the options under consideration into a common unit, often an economic valuation. This kind of analysis presupposes that a decision should not be taken unless: (1) the benefits of that decision outweigh the costs; (2) it is useful to quantify all the costs and benefits relating to the decision; and (3) the value of the analysis itself warrants the resources necessary for completion (Kelman 1981). Virtually all farm-level economic-impact assessments of GM agriculture consider the effects of adoption vs. non-adoption, and are therefore CBA variations. These include the economic modeling examples listed earlier. Additionally, there are many other statistical methods for determining costs and benefits of GM agriculture. For example, Bennett et al. (2006) examined, *ex post,* the costs and benefits of adoption vs. non-adoption among Bt cotton smallholder farmers in South Africa relative to yields, pesticide use, labor, and gross margins. They also gathered data from hospitals to consider the health benefits (through reduced pesticide poisoning) resulting from the technology's application. Groeneveld et al. (2011) estimated the costs and benefits of contained (i.e., inbreeding) GM crops in Italy and Spain to quantify the potential of a specific class of agronomic species (i.e., solonaceous vegetables).

In addition to strictly economic comparisons, CBA can be applied to compare social, environmental, or other issues of potential concern to producer choice. Essentially, any decision that can lead to distributional equity issues, winners and losers, can be approached (Fransen et al. 2005). For example, Zimmermann and Qaim (2004) applied a commonly used public health metric (disability-adjusted life years) in a cost–benefit framework to evaluate the public investment potential of utilizing golden rice in the Philippines to improve health outcomes. Similarly, Hossain et al. (2004) used an econometric method to predict the impact of reduced pesticide use, resulting from Bt cotton adoption, on farmer health in China.

16.2.3 Social Impact Assessment

Social impact assessment (SIA) is an outgrowth of environmental impact assessment. Recognizing that alteration of environmental systems leads to alteration of socioeconomic systems, many development scholars have called for a more comprehensive consideration of such processes through an SIA. Burdge and Vanclay (1996) define SIA as an advance estimate of the social consequences that are likely to result from a policy action.

SIAs involve determining future impacts by examining social impacts of previous events. Analysis is based on a variety of social science data, including demographics, institutional and community structure, community resources, lifestyle, beliefs, sociopolitical context, unequal distribution of benefits, power structures, and racial and cultural diversity. Both qualitative and quantitative data are used in SIAs, leading to a potential lack of objectivity (Stabinsky 2000). SIAs may be most relevant when a regulatory process requires an environmental impact assessment. In that case, SIAs would be appropriate for examining relevant SECs (Fransen et al. 2005), including those likely to impact producer choice. We could not find any currently agreed upon specific methodologies for implementing SIAs. Instead, it may be more convenient to conceive of SIAs in the context of a core set of values, principles, and guidelines. Such a context is available through the International

Association for Impact Assessment (Vanclay 2003, http://www.iaia.org/publicdocuments/special-publications/SP2.pdf). A participatory example examining tobacco and rice is available through the Participatory Assessment of Social and Economic Impacts of Biotechnology project (http://www.agecon.vt.edu/biotechimpact/index.htm).

16.2.4 Sustainable Livelihoods Framework

Sustainable livelihoods framework, as it is applied to agricultural biotechnology, seeks to enhance understanding of food security and poverty. This is accomplished by examining people's lives in rural communities through analysis of the relationships between relevant factors at the household, community, and regional levels. This framework goes beyond conventional measures such as income or nutrition and includes concepts such as vulnerability, assets, and empowerment (Falck-Zepeda et al. 2002). Methodologies for this framework include surveys, focus groups, key informant interviews, household case studies, and examination of various secondary sources. Such research has the potential to clarify issues related to distribution of benefits among producers and can lead to technical research priorities sensitive to small-scale farmers' needs, including labor issues (Fransen et al. 2005). Similar to SIAs, sustainable livelihoods framework may best be thought of as a set of core principles and approaches. Guidelines for the framework can be accessed through ELDIS (http://www.eldis.org/), an information clearing house for development issues.

Morse and Bennett (2008) assessed the livelihood impacts of Bt cotton farmers in South Africa. This approach allowed the authors to measure the proportion of farmers reporting benefits from the technology and to test whether realization of benefits was related to gender and farm size. In addition, the authors were able to explore how farmers utilized income gain from the Bt crop. They found that farmers used the additional income to educate their children, invest in agricultural resources, and pay off debt.

16.2.5 Systemic "Relevance Assessment"

The systemic "relevance assessment" is a unique approach to SECs in that it considers the problem that a technological solution addresses rather than the technology itself (Vanloqueren and Baret 2004, cited in Fransen et al. 2005). In addition, this methodology relies on a systems approach to consider the entire production system rather than only a single part of that system. In practice, this requires a two-pronged approach. The systems approach is used to determine the problem that the biotechnology-based solution addresses, examine all existing farmer practices that could also meet the need addressed by the biotechnology, and then analyze the socioeconomic implications of implementing these various strategies. In parallel, the relevance assessment aims to identify relevant stakeholders, circulate information among the stakeholders, seek input on technology implementation, research policy interventions from the stakeholders' perspectives, and carry out evaluation of the stakeholders' opinions relative to the problem in order to generate recommendations (Fransen et al. 2005).

Although we could not find any published studies utilizing this methodology to examine producer choice in specific cases of GMOs, this technique could prove quite valuable. First, it has the potential to put farmer needs into focus since the agricultural problems that farmers face must be explicitly identified and examined. Second, it has the potential to deemphasize the most ideologically controversial aspect of GMOs (i.e., the legitimacy of GMOs *per se*) by placing a specific GMO technical solution in context relative to other technical solutions.

16.2.6 Participatory Research

In recommending that producers have as much choice as possible with regard to new crop varieties, including GMOs, the Nuffield Council on Bioethics (2004) specifically called for farmers to be included in deciding which traits should be incorporated by plant breeders. This would allow producers themselves to play a primary role in setting research priorities. Furthermore, the Nuffield Council recommended that farmers should also be involved in the breeding research so that farmers are "informed about the technological potential and management requirements of GM crops" (Nuffield Council on Bioethics 2004, p. 76). Ashby and Sperling (1995) take the more expansive view that participatory research is not sufficient to meet the needs of diverse producer groups. They argue that participatory research must be institutionalized through consideration of producers in client-driven research. Consequently, the relevant client-driven agendas are likely to emerge only when the clients are given some control over research funds. This would include decentralization of technology development and careful inclusion of all relevant producers. Fransen et al. (2005) note that many research frameworks should be participatory, as it is important to include the concerns and experience of those most affected by a new technology. Several of the studies listed in Table 16.1 use participatory techniques within the framework of other methodologies including Birol et al. (2007), Krishna and Qaim (2008) and Morse and Bennett (2008).

16.2.7 Social Audit

Social audits are similar to SIAs in that both seek to increase social awareness of regulatory processes or development activities. They are included as a separate category because literature on the subject has developed out of for-profit managerial and accounting concerns, as opposed to SIAs, which have developed out of regulatory and policy frameworks. As such, social audits may be more appropriate for private foundations working in the agricultural biotechnology sector, while SIAs may be more appropriate for those seeking a more anthropological point of view.

Ezezika et al. (2009) proposed the social audit model as a method to mitigate the risk of project failure associated with agricultural biotechnology projects. Although definitions vary, social audit has been interpreted in the development context to facilitate increased organizational transparency and accountability, enhance stakeholder interest representation, and improve social and organizational performance. This is achieved through performance measures relative to organizational core aims, specific stakeholder aims, and societal aims (Dawson 1998). For agricultural biotechnology projects, a social audit system is promising because it relies heavily on consulting stakeholders (including producers) so that their concerns and preferences can be addressed.

Ezezika et al. (2009) developed a social audit model specifically to counter skepticism and resistance to biotechnology in order to advance the Water Efficient Maize for Africa (WEMA) project, a public–private partnership, funded by the Bill and Melinda Gates Foundation. WEMA seeks to use biotechnology to provide east African countries with drought tolerant maize. In this case, the social audit model yielded several principles that may be applicable to other development projects: (1) creation of a framework based on stakeholder needs and project goals, (2) identification of stakeholders, (3) engagement with stakeholders when developing assessment criteria, (4) ensuring transparency, and (5) accountability of project management to funders.

16.2.8 Biological Assessment

In any system based in the physical world, human understanding of natural laws will be of paramount importance in understanding the socioeconomic consequences of human stewardship of such systems. Although the biophysical characteristics of agricultural systems have been under investigation since the dawn of empirical science, certain biological aspects of agriculture are under particular focus in the context of agricultural biotechnology. Here, we briefly summarize some of the main biological aspects of GM agriculture that are likely to be of importance to SEC determination.

The issue of gene flow, whether through pollen or seed admixture, is of fundamental importance to producers choosing not to utilize biotechnology, especially when those producers serve a market requiring GMO-free products. These issues are also important for many other producers, especially seed producers (GMO or otherwise), who need to protect the genetic integrity of their product. Since each crop, in each agroecological environment and production system, may be subject to differing rates of gene flow and seed admixture, decisions to protect the purity of crop genetics will rely on various biological factors on a case-by-case basis. The biophysical parameters that govern gene flow are well studied and reported in the coexistence literature, especially in Europe (reviewed in Devos et al. 2009), but may need to be revisited for new crops and new environments. From a policy perspective, decisions relating to acceptable, or tolerated, rates of gene flow and admixture will need to be informed and balanced against both the biophysical realities of cropping systems and other concerns, such as economics (e.g., Demont and Devos 2008). The setting of tolerance limits should not be taken lightly, as they are likely to have profound implications on the ultimate quantification of SECs. For example, differing approaches to admixture limits in the United States and EU present a host of legal and economic problems (Endres 2005).

A farmer's choice to save seeds is also highly dependent on specific crop biology. It may also be dependent on intellectual property issues, but we will not deal with those here. Of most relevance in current GMO usage is whether or not a GM crop is available as a hybrid or open-pollinated variety (OPV). Though the function of the inserted gene should not change based on the GMO's designation as hybrid or OPV, hybrid crop biology necessitates a different crop production model than OPVs. In the simplest terms, a farmer planting OPVs is able to save seed to plant for the next season without a major change in the agronomic properties of the crop. In contrast, hybrid seeds (which are more complicated to produce) are usually bought at a price premium each season but offer some agronomic advantage that compensates for the additional cost. Often, hybrid seeds require more inputs and are thus more attractive to resource rich farmers. OPVs are often more attractive to resource poor farmers. Therefore, a GM crop's hybrid or OPV status may be an important feature when considering producers' desire to save seeds, and/or engage in higher or lower input agriculture.

Finally, because of the prevalence of insecticidal GM crops (i.e., Bt), the issue of refuges, and their SECs, should be considered. In order to maintain insect susceptibility to GM insecticidal proteins, resistance management strategies, especially the high dose/refuge approach, have been used to maintain the public good of pesticide susceptibility (Bates et al. 2005). Refuges, and other management strategies, are important to producer choice because they may require producer participation in the management program. A refuge program may impact producer decision making because it requires compliance with spatial or temporal planting recommendations, including planting a portion of crop as non-GM, or intercropping (e.g., Yenagi et al. 2011). Well-defined refuge plans are therefore requisite for farmers to make informed decision ahead of their decision to adopt GM crops. The ability of farmers to effectively carry out a resistance management plan may be dependent on the organizational structure of regional farm management. SECs affecting such farm management practices may therefore be impactful on producer decision making.

16.3 Critical Assessment

Table 16.2 lists each assessment method and briefly summarizes its advantages, disadvantages, and typical uses. By far, economic analyses (including both CBA and modeling approaches) have been the most explored methodologies used to

Table 16.2 Major methods of evaluating SECs relevant to producer choice

Method	Advantages	Disadvantages	Typical use
Economic modeling	Often utilized, many examples, highly quantitative and objective	Potential biases	Any situation
Cost–benefit analysis	Often utilized, many examples, highly quantitative and objective	Potential biases	Any situation
Social Impact Assessment	Utilizes both qualitative and quantitative data, complementary to environmental impact analysis	Used infrequently, subjective	Where there is concern over social consequences for producers
Sustainable livelihoods framework	Goes beyond conventional metrics to include vulnerability, assets and empowerment	Used infrequently, subjective	Where producers are susceptible to food insecurity and poverty
Systemic "relevance assessment"	Can put producer needs into focus and deemphasize the most controversial aspects of GMOs	No precedent for use in GMO consideration, subjective	Where GM solutions should be contextualized among other technological solutions
Participatory research	Inclusive of producers	Not stand alone	Any situation
Social audit	Developed out of for-profit managerial and accounting concerns	Used infrequently, subjective	Where GM solutions are a part of a privately financed development effort
Biological assessment	Much of the relevant data may already be generated during biosafety assessment	Not stand alone	Any situation, but especially when considering socioeconomic consequences of gene flow, hybrid vs. OPV seeds, and insecticidal refuge plans

determine which factors influence producers' decisions to utilize GMOs; however, there are problems associated with these studies. Although there is an emerging consensus in the economic literature that GM agriculture, on average, provides benefit for producers, the generalizability of this conclusion must be approached with caution. Not all producers benefit and the magnitude of benefits vary widely, reflecting the heterogeneity of farms, farmers, and markets. Furthermore, biases must be carefully considered. Many existing studies may be subject to measurement bias, estimation bias, and endogeneity (Smale et al. 2009).

The great advantage of economic assessment methodologies is that they are highly quantitative and objective. These methodologies will remain incredibly important since the polarized debate on GMOs has prompted certain groups to claim that producers are harmed while others contend that producers benefit. In dealing with such competing claims, it will remain necessary to deploy objective and rigorous methodologies. Continuing to develop and refine such methodologies will remain a vital and challenging area.

The majority of the non-economic methodologies mentioned earlier are less quantitative and more subjective in nature. They are best leveraged where there are specific principles or qualitative concerns at stake. SIAs will be most useful where there is concern over the social consequences of GMO regulation on producers. Sustainable livelihoods framework will be useful where the producers likely to be affected by GMO introduction are susceptible to food insecurity and poverty. Systemic "relevance assessment" will be most useful where a contextualization of the GMO as a technological solution is most appropriate. Social audits will be most appropriate where the GMO is specifically a part of a privately financed development effort.

Biological assessment and participatory research are not stand-alone methodologies. We have included them here because of their importance, together with economic assessment, in creating the foundational knowledge necessary to properly consider producer choice as an SEC. Biological assessment is often conducted as a part of the biosafety regulatory process, which can be independent of SECs. Nevertheless, the biological information gathered during biosafety assessment is often relevant to SECs and may be considered in a different light during review of SECs.

All of the methodologies described can be used in both *ex post* and *ex ante* analyses. However, *ex ante* studies (with the exception of studies wholly based on surveys of producer preference) require input data to calibrate forecast models. These data (e.g., yields, input costs) must be obtained from either test studies or from other countries that have already planted similar GM products. There is therefore a problem for countries seeking to gather preliminary data that are unwilling to allow field trials or research to proceed, especially if no other country with an analogous agroecological environment is testing that GMO. This point may be particularly important since the benefits of GM agriculture may increase for early adopters (e.g., Falck-Zepeda et al. 2008).

Compared to the economic approaches, the other methodologies mentioned earlier have been utilized relatively infrequently, or not at all, in examining SECs relevant to producer choice. Therefore, all examinations of SECs using unproven approaches will have to be approached carefully and rigorously. Any study design must be as explicit as possible about the key research question(s). Since producers are likely to make their choices differently for each GM trait, crop, and agroecological environment, each assessment must be tailored to the concerns that make the most sense for the affected producers.

Not all of the methodologies mentioned earlier are in use for GMO considerations. Specifically, we could find no example of systemic "relevance assessment" in the literature. Of the applied methodologies, all are used by academics, research groups and government agencies. If governments choose to adopt producer choice as an SEC to be examined for regulatory approval of a GMO, it is critical to determine the appropriate regulatory decision point (Falck-Zepeda and Zambrano 2011). For example, in order to best calculate likely economic impacts on producers, it would be helpful to design contained field trials in a way that can inform such studies. Therefore, SECs relevant to producer choice would need to be identified and considered during, or before, the approval process for field trials.

More generally, in order to ensure the desired expression of producer choice, the producers must be consulted at the appropriate stages in the GMO pipeline. Given the highly controversial nature of GMOs and the general lack of knowledge about the technology, it will be most beneficial to include stakeholders in decision making as early and often as possible. This will be facilitated if producer organizations (including commercial farmers, NGOs, and other farmer-centric stakeholder groups) are involved in all steps of the regulatory process from research to approval.

The importance of using multiple methodologies cannot be stressed enough. Virtually all studies of GMO adoption are sufficiently complicated as to warrant a complex understanding of the variables affecting producer choices. Consequently, models used to forecast adoption consequences need to be similarly complex. Furthermore, producer choice is not an SEC that can be considered in a vacuum. As an example, when examining the potential adoption of GM bananas in Uganda, Kikulwe (2010) recognized that farmers are both banana consumers and producers, as is the case for farmers of many crops. Therefore, he chose to link farm-level decisions and their propensity of purchase to the propensity to adopt new technologies. Doing so required multiple methodologies that included comparing several different economic models while integrating an understanding of the producers' knowledge, attitudes, and perceptions relative to GM banana.

When considering SECs, producer choice ought to be of fundamental importance. However, once an understanding of producer choice is obtained, regulatory decisions that ignore or downplay producer desires should be carefully considered. If a GMO provides a significant advantage, it may be difficult to impose a ban or moratorium. The case studies of stealth GM seeds in India and Brazil serve as important examples. As Herring (2007) chronicled, producers in Brazil and India obtained and circulated stealth seeds (i.e., GM seeds that had not been vetted by existing domestic regulatory processes) because producers perceived advantages to these seeds, yet were impeded from legally obtaining them by regulatory and/or economic factors. The unavailability of desired legal GM seeds, combined with the inability of regulators to suppress illegal plantings, resulted in widespread illegal deployment of GMOs despite existing regulatory frameworks. Such occurrences emphasize the importance of realistic and proportional regulatory oversight of SECs. As Herring (2007, p. 130) noted, farmers are "active, creative and autonomous…." If their needs and preferences are not properly considered, vis-à-vis GMOs, established regulatory frameworks may not have their desired effect.

As a final point, it should be noted that issues of "political economy" surrounding GMO adoption are likely to be important decision-making factors for any SEC, including producer choice. In spite of the sound methodologies present in the research literature, at the end of the day it is often political will that is the most meaningful determinant of GMO regulatory processes. Yet, there are few studies that examine the effects of "political economy" (e.g., lobbying methods) on GMO regulatory processes, and none of the abovementioned methodologies specifically accounts for them. Therefore, a careful evaluation of the "political economy" surrounding GMO producer choice will be necessary to fully consider how SECs may impact the regulatory approval process in each country.

16.4 International Arena

The most relevant international agreements relative to producer choice are the International Treaty on Plant Genetic Resources for Food and Agriculture (ITPGRFA), the Agreement on Trade-Related Aspects of Intellectual Property Rights (TRIPS), and the Union for the Protection of New Varieties of Plants (UPOV). These treaties are relevant to producers since farmers are both consumers and producers of seed. The relationships between the Cartagena Protocol on Biosafety (CPB) and the WTO are also important in this context but are discussed in Part I of this book. In the simplest terms, ITPGRFA can be characterized as an agreement that channels compensation arising from novel agricultural products back to the producers, including traditional farmers, who have conserved and developed contributing genetic resources while also recognizing farmers' rights to save, use, exchange, and sell their seed. TRIPS and UPOV, on the other hand, seek to establish intellectual property in an effort to spur investment in research and development of agricultural products, including GMOs. While it may seem that these agreements are in opposition, they are not in any specific legal conflict since protection can simultaneously be offered to indigenous and improved crops, although there may exist challenges to avoid future legal conflicts (Gerstetter et al. 2007).

We should note that there has been general concern that the treaties granting intellectual property rights to GM varieties could have the effect of decreasing the number of varieties available on the seed market due to consolidation in the seeds sector, thereby impacting producer choice. Evidence from the United States indicates otherwise. Between 1997 and 2008, the number of maize hybrids available annually grew from 3,060 to 4,300 and soybean varieties grew from 650 to 1130 (Kalaitzandonakes et al. 2010).

16.5 Administrative Consequences

While we could find no quantitative analyses detailing the institutional and administrative resources required to assess producer choice as an SEC, it is reasonable to assume that expenditures in time, human capital, and financial resources would be substantial. As has been noted by others, biotechnology adoption is limited by research, regulatory, and law enforcement capacity in developing countries (e.g., Takeshima and Gruere 2011). Building the capacity to conduct the multi-year studies required to fully utilize many of the methodologies mentioned in this chapter may represent a significant challenge in many countries, especially considering that many of the resources required to assess SECs are distinct from those required for biosafety assessment. Many countries may also lack experts trained to perform the relevant analyses. Therefore, any decision to mandate inclusion of producer choice is likely to require many years and considerable outlay of expense in order build and maintain the capacity to administer the institutions that would analyze the effects of GMO adoption on producer choice.

16.6 Summary/Synthesis

- SECs relevant to producer choice include freedom of choice, income security, control over production, contamination of organic agriculture, and farmers' rights to save seeds; however, producer concerns are heterogeneous in time and space.
- There are a variety of research methodologies available for assessing SECs including economic modeling, cost–benefit analysis, social impact assessment, sustainable livelihoods framework, systemic "relevance assessment," participatory research, social audit, and biological assessment.
- Economic analyses have been the most explored methodologies, generally showing that GM agriculture does provide a benefit to producers on average.
- Non-economic methodologies are appropriate where there are qualitative concerns.
- In many cases, the usage of multiple methodologies will be necessary to fully explore SECs such as producer choice.
- Understanding issues of "political economy" and involving producers in all stages of research, development, and regulatory processes will be necessary to ensure that regulatory frameworks have their desired effect.

References

Ali A, Abdulai A (2010) The adoption of genetically modified cotton and poverty reduction in Pakistan. J Agric Econ 61:175–192

Ashby JA, Sperling A (1995) Institutionalizing participatory client-driven research and technology development in agriculture. Dev Chang 26:753–770

Bates SL, Zhao J-Z, Roush RT et al (2005) Insect resistance management in GM crops: past, present and future. Nat Biotechnol 23:57–62

Bennett R, Morse S, Ismael Y (2006) The economic impact of genetically modified cotton on South African smallholders: yield, profit and health effects. J Dev Stud 42:662–677

Birol E, Rayn E, Smale M (2007) Farmer preferences for milpa diversity and genetically modified maize in Mexico: a latent class approach. IFPRI discussion paper 00726

Burdge RJ, Vanclay F (1996) Social impact assessment: a contribution to the state of the art series. Impact Assess 14:59–86

COGEM (2009) Socio-economic aspects of GMOs: building blocks for an EU sustainability assessment of genetically modified crops. COGEM Report CGM/090929–01

Crost B, Shankar B, Bennett R et al (2007) Bias from farmer self selection in genetically modified crop productivity estimates: evidence from India data. J Agric Econ 58:24–36

Dawson E (1998) The relevance of social audit for Oxfam GB. J Bus Ethics 17:1457–1469

De Groote H, Overholt WA, Ouma JO et al (2011) Assessing the potential economic impact of *Bacillus thuringiensis* (Bt) maize in Kenya. Afr J Biotechnol 10:4741–4751

Demont M, Devos Y (2008) Regulating coexistence of GM and non-GM crops without jeopardizing economic incentives. Trends Biotechnol 26:353–358

Devos Y, Demont M, Dillen K et al (2009) Coexistence of genetically modified (GM) and non-GM crops in the European Union. Rev Agron Sustain Dev 29:11–30

Edmeades S, Smale M (2006) A trait-based model of the potential demand for a genetically engineered food crop in a developing economy. Agric Econ 35:351–361
Endres AB (2005) Revising seed purity laws to account for the adventitious presence of genetically modified varieties: a first step towards coexistence. J Food Law Policy 1:131–163
Ezezika OC, Thomas F, Lavery JV et al (2009) A social audit model for agro-biotechnology initiatives in developing countries: accounting for ethical, social, cultural, and commercialization issues. J Tech Manag Innov 4:24–33
Falck-Zepeda J, Cohen J, Meinzen-Dick R et al (2002) Biotechnology and sustainable livelihoods—findings and recommendation of an international consultation. International Service for National Agricultural Research, briefing paper no. 54
Falck-Zepeda J, Horna D, Smale M (2008) Betting on cotton: potential payoffs and economic risks of adopting transgenic cotton in West Africa. Afr J Agric Res Econ 2:188–207
Falck-Zepeda JB, Zambrano P (2011) Socio-economic consideration in biosafety and biotechnology decision making: the Cartagena protocol and national biosafety frameworks. Rev Policy Res 28:171–195
Fransen L, La Vina A, Dayrit F et al (2005) Integrating socio-economic considerations into biosafety decisions: the role of public participation. World Resources Institute, Washington DC
Gerstetter C, Gorlach B, Neumann K et al (2007) The international treaty on plant genetic resources for food and agriculture within the current legal regime complex on plant genetic resources. J World Intellect Prop 10:259–283
Groeneveld RA, Ansink E, van de Wiel CCM et al (2011) Benefits and costs of biologically contained genetically modified tomatoes and eggplants in Italy and Spain. Sustainability 3:1265–1281
Hareau GG, Mills BF, Norton GW (2006) The potential benefits of herbicide-resistant transgenic rice in Uruguay: lessons for small developing countries. Food Policy 31:162–179
Herring RJ (2007) Stealth seeds: bioproperty, biosafety, biopolitics. J Dev Stud 43:130–157
Hossain F, Pray CE, Lu Y et al (2004). Genetically modified cotton and farmers' health in China. Int J Occup Environ Health 10:296–303
Huang J, Hu R, Pray CE et al (2003) Biotechnology as an alternative to chemical pesticides: a case study of Bt cotton in China. Agric Econ 29:55–67
Kalaitzandonakes N, Magnier A, Miller D (2010) Market power and dynamic efficiency in the US seed industry. Working Paper, EMAC, University of Missouri, USA
Kelman S (1981) Cost-benefit analysis: an ethical critique. Regulation 5:33–40
Kikulwe EM (2010). On the introduction of genetically modified bananas in Uganda: social benefits costs, and consumer preferences. PhD Thesis. Wageningen University, Netherlands
Krishna VV, Qaim M (2007). Estimating the adoption of Bt eggplant in India: who benefits from public-private partnership? Food Policy 32:523–543
Krishna VV, Qaim M (2008) Potential impact of Bt eggplant on economic surplus and farmers' health in India. Agric Econ 38:167–180
Kuosmanen T, Pemsl D, Wesseler J (2006) Specification and estimation of production functions involving damage control inputs: a two-stage, semiparametric approach. Am J Agric Econ 88:499–511
Morse S, Bennett R (2008) Impact of Bt cotton on farmer livelihoods in South Africa. Int J Biotechnol 10:224–239
Morse S, Bennett RM, Ismael Y (2005) Bt-cotton boosts the gross margin of small-scale cotton producers in South Africa. Int J Biotechnol 7:72–83
Nazli H (2010) Impact of Bt cotton adoption on farmers' wellbeing in Pakistan. PhD Thesis, University of Guelph, Canada
Nuffield Council on Bioethics (2004) The use of genetically modified crops in developing countries: a follow-up discussion paper. Nuffield Council on Bioethics, London. Available at www.nuffieldbioethics.org/gm-crops-developing-countries
Qaim M (2001) A prospective evaluation of biotechnology in semi-subsistence agriculture. Agric Econ 25:165–175
Qaim M (2009) The economics of genetically modified crops. Annu Rev Resour Econ 1:665–693

Qaim M, Traxler G (2005) Roundup ready soybeans in Argentina: farm level and aggregate welfare effects. Agric Econ 32:73–86
Qaim M, Zilberman D (2003) Yield effects of genetically modified crops in developing countries. Science 299:900–902
Raney T, Matuschke I (2011) Current and potential farm-level impacts of genetically modified crops in developing countries. Frontiers of Economics and Globalization 10:55-82
Shankar B, Thirtle C (2005) Pesticide productivity and transgenic cotton technology: the South African smallholder case. J Agric Econ 56:97–116
Shankar B, Bennett R, Morse S (2007) Output risk aspects of genetically modified crop technology in South Africa. Econ Innov New Tech 16:277–291
Smale M, Zambrano P, Cartel M (2006) Bales and balance: a review of the methods used to assess the economic impact of Bt cotton on farmers in developing economies. AgBioForum 9:195–212
Smale M, Zambrano P, Falck-Zepeda J et al (2008) The economic impact of transgenic crops in developing countries: a note on the methods. Int J Biotechnol 10:519–551
Smale M, Zambrano P, Gruère G et al (2009) Measuring the economic impacts of transgenic crops in developing agriculture during the first decade: approaches, findings and future directions, Chap. 3—Impacts on farmers. International Food Policy Research Institute
Stabinsky D (2000) Bringing social analysis into a multilateral environmental agreement: social impact assessment and the biosafety protocol. J Environ Dev 9:260–283
Takeshima H, Gruère GP (2011) Pressure group competition and GMO regulations in Sub-Saharan Africa—insights from the Becker model. J Agric Food Ind Organ. doi:10.2202/1542-0485.1325
Third World Network (2008) Assessing the socio-economic, cultural and ethical impacts of GMOs. Third World Network briefings for MOP 4–3
Vanclay F (2003) Social impact assessment: international principles. International association for impact assessment, special publication series no 2
Vitale J, Boyer T, Uaiene R et al (2007) The economic impacts of introducing Bt technology in smallholder cotton production systems of West Africa: a case study from Mali. AgBioForum 10:71–84
Yenagi BS, Patil VC, Biradar DP et al (2011) Refuge cropping systems for *Helicoverpa armigera* (Hubner) resistance management in BT cotton (*Gossypium hirsutum*). Acad J Entomol 4:102–107
Yorobe JM Jr, Quicoy CB (2006) Economic impact of Bt corn in the Philippines. Philipp Agric Sci 89:258–267
Zimmermann R, Qaim M (2004) Potential health benefits of golden rice: a Philippine case study. Food Policy 29:147–168

Chapter 17
Culture and Religion

Alexandra Coe

17.1 Introduction

This chapter commences by referencing definitions of religion and the understanding of culture. However, the scope of these two concepts in the global arena is daunting. It is difficult to provide a comprehensive definition of religion or culture since different definitions are used for different purposes. When looking at religious and cultural relationships with genetically modified organisms (GMOs), the most important thing to understand is that the beliefs, habits, and rituals attached to religion and culture are so deeply rooted, and that these undercurrents of human thought possess the power to decide if something is acceptable or unexceptionable, in an instant. The speed at which religion and culture can deem something welcomed or unwelcomed is why it is critical to understand the potential religious and cultural interpretations of agricultural biotechnology (agbiotech) before the opposition begins. It is prudent, and preferable, to engage in this understanding on the front end of research and development. The future of agbiotech rests in true cooperative engagement across all sectors of the agricultural network and this requires a strong religious and cultural understanding of how biotechnology might play into a region's agricultural landscape.

One advantage agbiotech has in the area of religion and culture is that the technology is so new, only in the last 15 years have large religious organizations gathered together to debate the religious acceptance or rejection of agbiotech (Agbioworld, n.d.). The debate still continues but generally religious acceptance has occurred across major religious organizations. The greatest challenge of religious and cultural assessments is in determining whether a particular biotechnology is being

A. Coe (✉)
Goucher College, Department of Cultural Sustainability, MD, USA
e-mail: ajwcoe@mac.com

Livable Planet Consulting, Gentle Storm Productions, Sarasota, FL, USA

K. Ludlow et al. (eds.), *Socio-Economic Considerations in Biotechnology Regulation*, Natural Resource Management and Policy 37, DOI 10.1007/978-1-4614-9440-9_17,

evaluated within the dialogue of true religious and cultural paradigms or if it is being influenced by the current wave of anti-GMO propaganda.

One cannot ignore the current debate around GMOs when considering religious and cultural interpretations. Interestingly, most of the opposition comes from the developed world, while most of the perceived need is in the developing world. One might say it is easy to oppose GMOs on a full stomach but there are other factors that have fueled this current opposition. Some of this is influenced by the Green Revolution of the late 1960s, which significantly increased agricultural yield for many, but not all regions of the globe. Promises of feeding the planet were not realized as poverty and famine continued, especially in Africa which was not viably a part of the Green Revolution.

Agbiotech advocates have consistently articulated that GM crops are essential "to feed a hungry planet"; yet, the 2008 food crisis proved to many anti-GMO groups that 15 years of growing GMOs was not the solution. The agriculture of GMOs is viewed in a similar light as the Green Revolution and often compared to it in having similar shortcomings. This image of agriculture dominated by farm corporations using excessive amounts of chemical fertilizers and pesticides and various forms of monoculture is promoted as the enemy of sustainable agriculture as many opponents believe growing GMO crops contributes to environmental degradation and disease.

This view that agbiotech is just a continuation of the Green Revolution limits the potential benefits GMOs could provide for many crops threatened by current agricultural challenges, such as climate change. To further complicate the potential, well-funded NGOs fuel the controversy and increase the numbers of the "passionately misinformed" (Coe 2009). Increasing this controversy, religious and cultural aspects can be highly emotionally fueled on their own, but those influenced by the current misinformation surrounding agbiotech will create barriers, unless different methodologies and assessments for the introduction of GM crops are considered outside of the current methods. Key to the assessments of the potential impact of the introduction or import of GMOs is the creation of thoughtful, educational information to balance the anti-GMO message. The prevailing global myths about GMOs must be dispelled prior to entering an understanding of regional religious or culturally rooted challenges.

If agbiotech is going to become a vital tool for sustainable agriculture, something must shift the global GMO myth paradigm. Shifting the focus to regional, small-scale agricultural challenges that support traditional food-ways and small-scale farmers could be one critical way to achieve this. It is within this more regional focus that the importance of religious and cultural aspects of agriculture will have the greatest significance and also within this focus, better application of the science of plant biotechnology can occur and impact these aspects in a positive way.

Why do religion and culture play such an important role in the future developments of GMOs? Looking back on human history, food supply has always been related to the divine or weaved within the cultural fabric of societies. Religion and culture mix and separate within modern human paradigms at an ever-changing rate. To attempt to grasp this, it is important to simplify the concepts of religion and culture and the dynamics that fueled these aspects of human existence, within a framework that can apply specifically to the issues that need to be addressed when looking at their relevance to the regulation, production and imports of GMOs. It is important to try to differentiate between religion and culture by stating that religion primarily addresses concerns over whether or not something violates the laws set forth by God, while culture addresses those behaviors and practices that people tenaciously cling to as part of their individual and collective identity, as well as daily habits and rituals. It serves this book well, at this point, to define ethics as that barometer of morality that exist within religion and culture equally; however, to separate this chapter from Chap. 7, it is important to present core concepts that exist in religion and culture, yet are completely outside the scope of ethics.

It can be extremely difficult to separate ethics, religion, and culture, as so many aspects of them are intertwined. Yet, in order to move forward, with an understanding of how to evaluate agbiotech in respect to religion and culture, it is necessary to find a place where religion and culture share themes that can be applied to almost any society. Those themes are the concepts of sacredness and the desire for happiness and well-being. Although the true understanding of individual religious and cultural relationships with the introduction of agbiotech cannot be distilled completely to those two concepts, some of the answers to the big questions about agbiotech can truly be answered when evaluated within the scope of sacredness and happiness. This chapter examines sacredness and happiness as the common values all religions and cultures share, as the lens of weighing and balancing the perceived risk and benefits associated with agbiotech.

17.2 Methodologies

Current methods to evaluate the risk and benefits of GM crops are not looking at culture or religion. They function under a premise that GMOs will increase yield, that GM crops are safe to eat, and there is no significant difference between them and their conventional counterparts. Also within this assumption, it is held that GMOs pose no threat to human or animal health and are environmentally safe. These factors may not matter, when looking at the religious and cultural assessment. Simple facts such as seeds being handed down generation after generation, of a crop looking just like the crop that the grandparents grew, or the simple fact that a new GMO is not grown following the traditional farming methods of a region, or that it might be rejected by a protective god, may instantly negate all of these current methodologies that have generally become accepted within the industry and within governments.

Although there is some *on the ground* data collection by such organizations as FAO, who have a large collection of case studies from agricultural regions across the globe, this impressive and important data overlooks important cultural markers as they pertain to regional agriculture. Understanding the culture of agriculture is one of the most critical data collection area for the future of agbiotech.

Culture is the realm of the anthropologists and many of them focus on food production and regional agriculture. Agbiotech companies would benefit by building partnerships with anthropologists. Many are already involved in developing countries and can easily gather culturally relevant data on the relationship between culture, religion and agriculture in any particular region. This understanding will aid in determining which crops are best suited for the current agricultural landscape that exists in any individual region and form a cost-effective route for data collection.

Prior methodologies fall short of gathering this valid data. The most common methodology is some variation of democratic engagement, involving summits and other formal conferences of stakeholders said to be representative of particular topics of the events. Although this form of interaction does create a certain type of data, it fails to have the ability to consider cultural or religious aspects of agriculture. Many of the democratic engagement methodologies seem more like business deals as closed door treaties are signed, policy written and companies and countries negotiate agriculture's future. Unfortunately many people affected by major agricultural issues cannot even afford to attend these expensive summits and conferences. Therefore how can they be truly democratic? This is why direct regional engagement is critical to future assessment methodologies.

Conferences do provide a successful method for assessing broad-based analysis, particularly with the bodies of the world's major religious authorities. They allow the gathering of key minds within religious authority to evaluate agbiotech under a single lens, religion. Most of these conferences look at agbiotech's role in feeding the hungry with only some consideration as to the effects GMOs have on human health and on the environment. Religious acceptance by the major religious sectors can be the first step toward regional partnerships and opening doors to engagement that starts the cultural dialogue.

It is important to note that, just as some religious opposition is obvious, such as religious groups that do not eat swine would universally oppose the use of swine genes in agbiotech, once cultural data is collected, patterns will emerge that will reveal obvious barriers. It is not just all about risk. Interestingly, under Jewish dietary laws, called Kasrut, the safety and healthiness of food is not necessarily an overriding factor when determining if something is kosher. One would think health and safety would be part of some cultural role in connection to happiness and well-being, but it might not be. An Islamic evaluation of agbiotech tends to view halal food under a much broader scope than Christianity or Judaism, using the objectives of the Shari'ah, which are the benefits of protection and preservation of the religion, life, intellect, progeny, property, and the environment (Abu-Sway 1998).

The current methodology that has the most negative impact on considering religious and cultural aspects is the typical business model for agbiotech to gain access to new markets. Being extremely top-down, corporations submit scientific data directly to government, basically supporting the industry assumptions previously stated, to regulators with little dialectic engagement with the communities directly affected. If access is approved, governments inform their farmers that this crop is available to plant. Often these GMO crops are non-indigenous crops and are

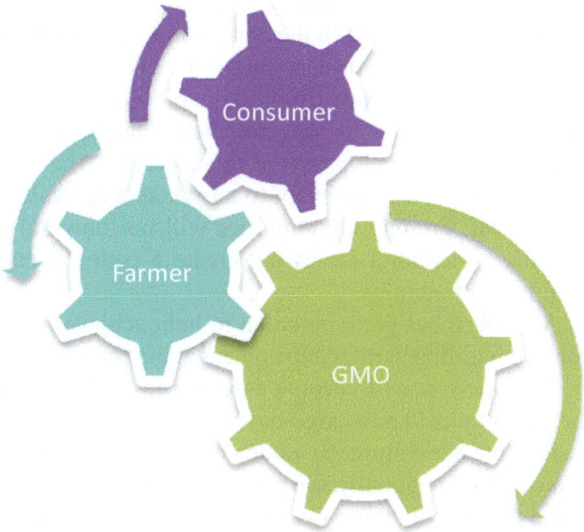

Fig. 17.1 The farmer link

brought in and planted by foreign farmers managed by corporate interests. Rarely is the local farmer growing the crop.

The most important problem arising from democratic engagement methodologies, which needs to be addressed in new methodologies, is that the real stakeholders, the farmers, feel left out. In many regions, farmers feel that the introduction of agbiotech is, in some way, a deal between corporations and governments to surrender regional farmers to corporate interests. Equitable inclusion of farmers in research and innovation is essential to future assessments. Embracing regional farmers and allowing traditional knowledge to influence the technology will help agbiotech understand the essential place it has in regional sustainable agriculture. The farmer is the link to acceptance by the consumer (Fig. 17.1). By engaging regional farmers in educating the industry on the regional culture of agriculture, companies can learn how the technology can honor the sacred and provide happiness and well-being for the farmer and consumer alike.

17.3 Assessment

Current assessments, such as methodologies, focus on safety, productivity, and risk to health and environment. Few of these assessments include the informed consent of farmers. Even though some may be involved in the industrialized methods used to produce the crop, they are not equal partners in creating their own agricultural future.

Small farmers have the greatest potential for feeding a hungry planet. The future of agbiotech lies in understanding the needs of regional farmers and in supporting agriculture that is consistent with promoting social equity; this is how people feed themselves as opposed to the current concept that large corporations must *feed the world*. Feeding the world should no longer be the goal but providing communities with tools and pathways for sustainable agriculture should be. The future of GM crops lies in discovering how the science can be applied in such a way that it allows for the regional development of food security. The main purpose of gathering these data is to reveal the pathway to sustainable agriculture for regional communities. Equity with farming communities will not only solve some of the world's most pressing agricultural issues but create a pathway out of poverty for many small farming communities.

Agbiotech has great potential to help achieve this success but it must be in the spirit of building partnership and engaging farmers as equals. During this data gathering process, it will become essential to evaluate current agbiotech products and the current plant knowledge-base to determine how they fit in a region's models for sustainable agriculture. By understanding the culture of agriculture and how people relate to food regionally, the potential of the science, being placed more in the hands of the people with this assessment approach, will find its natural place.

The type of data that builds these critical partnerships can only be collected through a participant style of engagement and not just interviews and short interactions. The agriculture must be observed because it is only through this type of engaged observation that the nuances of religious and cultural aspects of agriculture reveal themselves. Possible methodologies for data collection are as follows:

1. Engage regional farmers in the gathering of data on agricultural systems including crop data. Any farming study should cover at least a three-year cycle in order to understand regional systems as well as environmental and other influences.
2. Create lists of potential threatened culturally significant crops.
3. Record regional agriculture challenges and current methods used to address them.
4. Conduct environmental impact studies on regional agriculture methods.
5. Expanded studies of any specific religious or culturally significant crop.
6. Look for social inequity and distribution factors that prevent true implementation of new technologies.

Most importantly, gathering useful data for religious and cultural understanding of a region's agriculture requires asking the right questions. Whether assessing the regional agriculture or evaluating the technology, the essential religious/cultural question is: *Can agbiotech improve agriculture in a region while respecting that which is sacred and promoting health and well-being within the community?* Some initial questions that help in gathering the right type of data to learn the answer to the essential question might be as follows:

> What are the regional challenges of agriculture and what is being done currently to address them? Does a possible threat to regional and culturally significant foods exist by the introduction of GM crops? What can a specific biotechnology contribute to the improvement and sustainability of the regional agriculture? Are there similar cultural regions that have benefitted from a particular GMO? How will the region, community, and farmers benefit

from the introduction of a particular GMO? What are the current agricultural methodologies and techniques utilized? Are any of these incompatible with GMO agricultural systems? Are there other means of genetic introduction that improve access or may have lower cost before embarking on R&D for GM solutions? Might a mixture of approaches create the desired result? Are there other changes in the process and systems of regional agriculture that should be implemented first? The most difficult question to answer honestly, Is the technology needed?

Biotechnology is science in its purest form and must distinguish itself as a science apart from industrialized agriculture, as presently many believe that agbiotech is industrialized agriculture. Agbiotech must be viewed as a tool which can be used to improve agriculture and preserve the future of the food supply. Until agbiotech changes the media and world-wide perception of the science, pathways for approval and importation of GM products will continue to be delayed and rejected, indeed with negative consequences for many agricultural products and the people that could benefit from the science.

Hawaii provides an excellent case study for how the collection of data, based on the methodologies presented, could create greater acceptance and understanding of potentially beneficial crops. This example looks at both levels of stakeholder engagement, the farmer and the consumer, and shows how the culture of agriculture affected the acceptance of one GM crop, papaya, and the rejection of another, taro.

17.3.1 GM Papaya

The potential threat of the spread of papaya ring spot virus (PRSV) was identified in 1978. If the virus reached Puna (Gonsalves and Gonsalves, forthcoming), the major agricultural region for papaya, the spread of this disease would cripple the papaya industry in Hawaii and affect the lives of many farmers and the future of Hawaiian papaya farming. A group of scientists embarked on proactive research to find a solution. The scientists first tried classic breeding methods using cross-pollination but could not get a resistant strain. PRSV is spread by aphids; yet, increased pesticide use was not proving to be effective and was exposing farmers to increased risk. This research was being conducted at the dawn of modern agbiotech and the concept of pathogen-driven resistance, which states that a transgenic plant that expressed a transgene of a pathogen would be resistant to that given pathogen, was an emerging science (Gonsalves and Gonsalves, forthcoming). Genetic modification was attempted with the resulting production of one papaya strain showing resistance to PRSV. To lower costs and speed success, one resistant plant was cloned for field testing. Successful field testing resulted in the distribution of free seeds to farmers. Scientist then collected data from surveys of papaya farmers regarding their satisfaction and adoption of these new GM varieties. GM papaya was successfully adopted by Hawaiian farmers, and papaya (*Carica papaya*) became the first horticultural fruit crop on the market that was produced by agricultural biotechnology. GM papaya has been grown in Hawaii since the mid-1990s with little opposition. Hawaiian papaya is sold and eaten by millions of people across the USA.

17.3.2 GM Taro

As early as 1980 and confirmed in 2009, documentation states that the Hawaiian taro plant is susceptible to no less than 23 pathogens. The most serious of these is the fungal disease caused by *Phytophthora colocasiae,* commonly known as leaf blight. Dithane-M45 is the fungicide recommended to deal with these outbreaks. The material safety data sheet issued by Dow AgroSciences on Dithane-M45 fungicide (Dow AgroScience, n.d.) states clearly that not only is this fungicide toxic to aquatic organisms, but it also causes cancer and birth defects in laboratory testing. It would appear on the surface that not only the taro farmers would immediately benefit from the introduction of GM leaf blight-resistant taro by not having to use this fungicide, but in addition, the Hawaiian aquatic ecosystem would benefit by minimizing the use of this fungicide. Looking at the success of papaya why the resistance to taro?

Obviously risk, safety, and biosecurity assessments have been completed on GM taro yet resistance exist. Why? Is this rejection fueled by the global anti-GMO movement? Is it deeply rooted in the agricultural practices of Hawaii and the sacred relationship to the taro plan? Is it possible that the apparent religious and cultural resistance to taro is simply a convenient reason for what is truly an expression of anger because of the way other GM crops have invaded the landscape of Hawaii? In taro's case, it is all three.

The main resistance to GM taro rests in the sacred relationship between the plant and the native Hawaiian's belief that taro is the incarnation of their ancestors. To change the genetics of Hawaiian's taro is to alter that which is divine. Taro is a sacred gift to the people and as a gift it must be unchanged. This is the core belief system both religiously and culturally between the Hawaiians and taro, and for some, nothing will ever change this.

Additional resistance rests within the farming methods and traditions with taro, which do not apply to papaya. Taro is planted by almost all families in Hawaii. Taro saplings are shared among neighbors and families. Hawaiian children are taught how to cultivate taro as part of understanding their culture. They are taught how to plant, nurture, harvest, and pound it and even to make "poi" out of it. This closeness to taro cannot be separated from larger scale production of taro either because the family farmer and the production farmer have the same responsibility to the nurturing of the ancestors through taro farming. There is also a core element of farming that is critical; when a plant suffers, it is speaking. This is a language only farmers and those connected to plants understand. The plant is communicating something is out of balance. For the indigenous farmer, this is a sacred communication that must be honored. GM intervention at this juncture is seen as a band-aid and not a solution to the underlying problems that address the future security of taro in Hawaii. The failure to understand the totality of the agricultural systems of Hawaii created an environment where the benefits of GM taro could not even be considered.

The final aspect of opposition to GM taro rests in the fact that GM taro is not the first GM crop to upset the Hawaiians. Hawaii is blessed with some of the most fertile soil on the planet and this, to the Hawaiians, is their source of life. Religiously, the Hawaiian Islands themselves are sacred; the mountains, the plants, and even the rocks contain the souls of the ancestors. To not understand and respect this is to threaten the very existence of every native Hawaiian.

The island of Kauai has been a hotbed for GMO conflict in recent years. Plots of land on the west side of the Island have been used for seed production for GM corn and other crops for years. Pesticide and herbicide runoff and airborne spraying have been blamed for illness and environmental degradation. Lawsuits filed against agbiotech companies continue and outrage over the use of Hawaiian land for corporate profit fuels opposition. Taro is rather the final sacred straw. This failure to understand the religious and cultural beliefs surrounding the agriculture of the native Hawaiians was very short-sighted. Many Hawaiians feel biotech companies overlooked culture in favor of profit. Hawaii is now a hotbed for the rejection of GMOs that could positively contribute to the environmental sustainability of Hawaii. Farmers refuse to plant it and consumers refuse to eat it.

The importance of regional dialogue and regional understanding of agriculture for the future implementation of GMOs cannot be too greatly emphasized at this point. In today's world, the ability of agbiotech to hold the keys to creating sustainable regional agriculture, and solve some of the most pressing issues that threaten agriculture today, decreased water supplies, salt intrusion, soil degradation, to name a few, cannot be neglected as a critical part of the puzzle for feeding a hungry planet. This future potential will only be realized through embracing agricultural knowledge of smallholder regional farmers and engaging their equal participation in solving regional agricultural challenges. Otherwise, there will be many "Hawaiis" that jeopardize the introduction of a potential technology that can preserve culturally significant foods and provide food sovereignty and security for many people.

17.4 International Arena

One recent document that supports the equitable engagement of farmers is the adoption of the International Treaty on Plant Genetic Resources for Food and Agriculture (Planttreaty.org). The original document entered into force in 2004, but at a recent high-level roundtable at the 2012 Earth Summit in Rio+20, the governing parties for the Treaty established six points of action. Two of these points which apply to earlier discussion follow: first, to raise awareness of the actual and potential value of underutilized species of local and regional importance for food security and sustainable developments, and second to sensitize policymakers and other key stakeholders about the importance of fully implementing the Treaty, not only for food and agriculture, but also for food security, nutrition, and the resilience of agricultural systems, particularly in the context of climate change. Within the second of these points of action, one could easily replace *the Treaty* with agbiotech and create what would become a very powerful statement.

The Treaty was established to recognize the enormous contribution farmers have made to the diversity of crops that feed the world. The Treaty seeks to create a global system to provide farmers access to plant genetic material and it seeks to ensure there are shared benefits in the use of these genetic materials particularly in countries where they have originated. Although biotech is not specifically mentioned in the Treaty, any guidelines regarding plant genetic material would logically apply.

There was a distinct shift in the dialogue in Rio. It suggests that governments and other authoritative agencies were finally listening to the voices of farmers. There were discussions about representing farmers' rights and allowing them access to biotechnology and benefit sharing. Utilizing species of local and regional importance and understanding the added value of how diversity contributes toward food security. There was also affirmation that plant diversity is the answer to many agricultural issues of climate change and that we "must realize what nature has already given us." Additional language included a call for corporate transparency, especially in issues of land grabbing and the right of farmers to save seeds. Most importantly, plant diversity should be seen as a way to respect and guarantee cultural diversity.

An additional document that should be considered in the assessment process is *The Future We Want,* which is the outcome document from Earth Summit Rio+20. Within this final document is a variety of needs and visions, contributed by stakeholders across every region of the planet.

17.5 Administrative Consequences

Information dissemination and implementation rests largely on biotech companies to take the initiative in creating the platform for this dialogue. Cultural agricultural knowledge and techniques need to be communicated to the research and development sector and applicable biotechnology needs to be accessed according to cultural practices to maximize acceptance and benefit. There exist two key cultural stakeholders in this process: the farmer growing the food and the consumer eating the food. Each of these requires a different process for assessment and information gathering and each require separate outreach and educational engagement. Barriers may lie in one or both of these groups but understanding where and why they exist is critical for successful introduction and application of agbiotech. The future of agbiotech rests in addressing the most pressing regional challenges as they relate to hunger, poverty, biodiversity and regional diets. Through culturally sensitive education and public outreach utilizing regionally focused media-driven campaigns that seek to involve, inform and educate about the importance of GMOs ability to contribute to food sovereignty and security challenges.

Agbiotech is a product and just like any product, it needs to be evaluated and rebranded to reach its greatest marketing potential. Products that support sacredness, happiness and well-being are the products that will be part of the sustainable future. Development of new crops must answer the critical question: Can agbiotech improve agriculture in a region while respecting that which is sacred and promot-

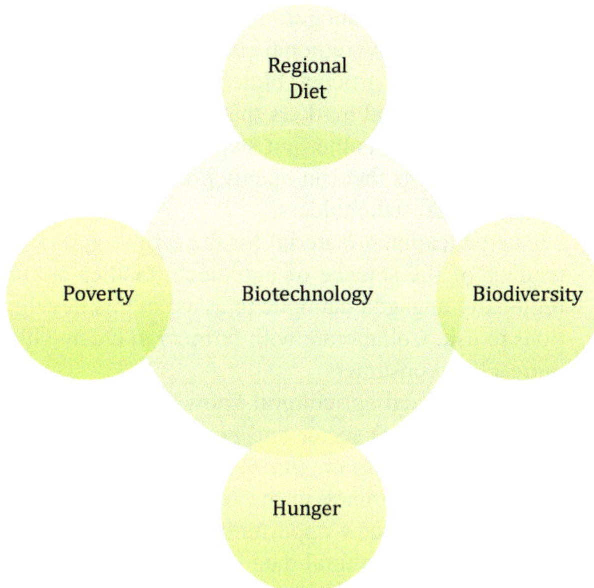

Fig. 17.2 The future of agricultural biotechnology

ing health and well-being within the community? If the answer is no, then what has been created needs to be abandoned and a renewed focus must ensure that the power that rests in plant biotechnology is a form of blessed knowledge bestowed to do that which is the highest and greatest good. If the planet is truly sacred, and is here so that we may create happiness and well-being for all of its inhabitants, why are toxic things still made, known carcinogens put in the environment? This is not our sustainable future. Whatever the image of agbiotech is right now, it is the result of the behavior of many individuals who have neglected to ask if their actions are honoring that which is sacred and that which promotes the health and well-being of the people. To continue down this path is to deny science its highest and greatest good. The future of agbiotech (Fig. 17.2) lies in developing crops that satisfy the preservation of regional diets, decrease regional poverty, eliminate hunger and support biodiversity.

17.6 Summary/Synthesis

- Respect and document traditional regional farming methods. Acknowledge that farming is deeply rooted in the very fabric of culture. Agbiotech must be presented as a tool for sustainable agriculture that can assist food security for threatened regionally significant crops. As a tool, it must engage the regional knowledge and practices already applied to agriculture. GMO introduction should enhance the cultural dynamics of farming in the region.

- Through cultural dialogue, the agbiotech industry must seek to gain a full understanding of how regional culture and religion might influence the acceptance or rejection of a particular biotechnology. Skilled cultural experts with the ability to identify cultural markers that will influence introduction need to be utilized. This information is the first step toward developing better regional public relations campaigns that can openly present the benefits of agbiotech through engaging essential stakeholders.
- Create educational material for the farmer and consumer that aid in the understanding of the science of agbiotech. Gather culturally relevant data based on regionally engaged dialogue to assist in this development. Know the right questions to ask. Collaborate with farmers to create culturally relevant public affairs outreach to consumers.
- Culturally based agricultural knowledge and techniques need to be communicated to the R&D sector and biotechnology needs to be accessed according to cultural practices to maximize acceptance. Corporations and governments must have full transparency in the review of GMOs. New business models need to be evaluated, as new opportunities for smaller regional biotech companies will emerge as more cultural data is collected that deal with regional food problems and focus on culturally significant foods.
- Finally, is the answer yes to the critical question? Can a biotech improve agriculture in the region while respecting that which is sacred and promoting health and well-being within the community?

References

Abu-Sway M (1998) Fiqh al-Bi'ah fil-Islam. Towards an Islamic jurisprudence of the environment. Paper presented at Belfast Mosque, Northern Ireland, February

Agbioworld (n.d.) Playing God or improving human lives? Religious, moral and ethical perspectives on food biotechnology. http://www.agbioworld.org/biotech-info/religion/. Accessed 22 Apr 2013

Coe A (2009) Public affairs education and the rethinking of the dialogue surrounding genetically modified crops. Unpublished paper presented at Food, Famine and Future Technologies: Ethical Dilemmas in a Hungry World. Appignani Bioethics Center, under the auspices of the United Nations Headquarters, New York City, USA

Dow AgroScience (n.d.) http://www.dowagro.com/usag/prod/060.htm. Accessed 8 November 2013.

Gonsalves C, Gonsalves D (forthcoming) The Hawaiian papaya story. In: Smyth SJ, Phillips PWB, Castle D (eds) Handbook of agriculture, biotechnology and development. Edward Elgar, Cheltenham

Planttreaty.org (n.d.) International treaty on plant genetic resources for food and agriculture. Accessed 22 Apr 2013

The Future We Want Outcome (2012) Earth Summit Rio+20. http://www.uncsd2012.org/thefuturewewant.html. Accessed 22 Apr 2013

Chapter 18
Animal Welfare

Leslie J Butler

18.1 Introduction

Consumer attitudes toward the food they purchase and consume have changed radically in the last 50 years. Healthfulness and food quality have become increasingly important in their decisions. Globalization, innovative changes in technology, and increased product differentiation are responsible for the wider variety of foods available, with increasingly broader quality attributes and dietary health characteristics. With the advent of animal biotechnology, consumers and others are becoming increasingly interested in the welfare of the animals that contribute to food products (Thiermann and Babcock 2005). It is not just a social concern for animal welfare, although that is certainly an important aspect, but increasingly consumers are aware of, and concerned about, the important links between animal health and animal welfare, the safety and quality of food products and their broader implications for biodiversity, the environment and for human, plant and animal health, and safety within that environment. In essence, consumers everywhere are increasingly demanding their right to make more informed choices about the food products they purchase and consume, including how animals are bred, raised, kept, used, transported, and slaughtered (Mitchell, 2001, The Boyd Group, 1999).

The question this chapter addresses is: what measures, obligations, and implications a country faces *if* animal welfare may be taken into account? It is important to understand that arguments for and against regulations pertaining to animal welfare fall into two major categories: social or ethical concerns associated with animal welfare, and those aspects of animal welfare that can be considered to be scientifically acceptable and objective assessments. The focus in this chapter is on the latter category of objective measurability, evidence based and scientific acceptability.

L. J. Butler (✉)
Department of Agricultural & Resource Economics
University of California, Davis, CA, USA
e-mail: ljbutler@ucdavis.edu

K. Ludlow et al. (eds.), *Socio-Economic Considerations in Biotechnology Regulation*, Natural Resource Management and Policy 37, DOI 10.1007/978-1-4614-9440-9_18,

That is not to say that social or ethical arguments about animal welfare are unacceptable or irrelevant. However, the overall focus of consideration of SECs in these chapters requires adherence to science-based assessments. Social or ethical arguments would require quite different models of assessment, and much more space.[1]

Animal welfare is the physical and psychological well-being of animals, which can be measured by indicators of behavior, physiology, longevity, and reproduction. The scientific concern for animal welfare is based on the awareness that nonhuman animals are sentient beings and therefore consideration should be given to their health and well-being. Many of these concerns are especially important in the consideration of animals used by humans, whether they are slaughtered for food, used for scientific research, or kept as farm animals or as pets. How human activities affect the survival of animal species, especially endangered species, and the maintenance of genetic diversity is also relevant in this context, but is probably more appropriately addressed under the auspices of a wider ranging discussion of biodiversity.

Most countries have some form of animal welfare regulations, ranging from basic anticruelty laws, to laws and regulations concerning the rearing, handling, transportation, and slaughter of animals. Most of these are based on cultural, religious, political, and/or social values of the relevant jurisdiction. The broad array of values, and stakeholders, has created an equally diverse array of laws and regulations in individual countries (NRC 2012).

Animal welfare should *not* be confused with animal rights or animal liberation. While some aspects of the animal rights movement are associated with animal well-being, animal welfare does not imply that the use of animals by humans is unacceptable, or that animals cannot be regarded as property, as is often argued by those who advocate animal's rights. We are also *not* advocating any particular position on animal welfare in this chapter—rather we are considering those aspects of animal welfare for which society has deemed necessary to provide minimum standards of well-being by enacting certain rules, regulations, and laws that ensure that animals are afforded freedom from thirst and hunger, discomfort, pain, injury and disease, fear and distress, and the freedom to express normal behavior.

Finally, it is important to separate the circumstances under which animal welfare in conjunction with animal biotechnology is being discussed. There are two distinctly different situations discussed in the literature, both of which have implications for this chapter. First, where animals are used for research, animals may be used for medical research or the development of new or productivity-enhancing technologies that enhance their food and fiber production abilities. Second, the welfare of on-farm animals that have been subjected, in some way, to genetic modification, and are used for food and/or fiber production. In each of these cases, different issues arise. Discussion is made more complex by the recognition that, in each case, animal welfare issues arise at different stages of end product development.

[1] An excellent discussion of assessing the ethics of animal welfare in animal biotechnology can be found in (Kaiser 2005).

As an SEC, animal welfare also has two distinct levels at which it could be evaluated. The first is whether the production or import decision is based on *current law* in the importing country. The second is the impact of the LMO's introduction on human, plant or animal health, biodiversity or the environment that is not addressed by existing state law.

18.2 Methodologies

18.2.1 Potential Impacts of Animal Biotechnology

The genetic modification of animals has immense implications for the advancement of science and for the future of agriculture and food production. Genetic modification can enhance productivity or quality traits, improve animal health and disease resistance, produce products for human medical and therapeutic use, advance medical research through the development of animal models and produce industrial products. The most common development of genetically modified (GM) animals has been for research purposes, and some are beginning to produce commercial products. In the USA in 2009, the FDA announced the first approval of a drug from a GM animal. The drug is a human anti-clotting agent produced in the milk of GM goats (CAST 2011). However, to date, no GM animals destined to enter the human food chain have been approved for commercial use in the USA, or as far as we know, in any other country.

Despite the obviously large potential benefits of advances in animal biotechnology, there are a number of welfare concerns associated with the genetic modification of animals that need to be considered. Some aspects of gene transfer have the potential to create infectious disease hazards and/or impaired reproduction. There are questions about whether genetic modification, and other technologies, stress animals unnecessarily, subject them to higher rates of disease and injury, and hasten death. For example, ruminants produced by in vitro culture often result in higher birth weights and longer gestations. This leads to large-offspring syndrome (LOS). LOS animals usually have more congenital malformations and higher perinatal mortality rates than "normal" farm animals. Abnormalities may include skeletal malformations, incomplete development of vascular system and urogenital tract, immune system dysfunction, brain lesions, double-muscling, leg and joint problems, hydroallantois,[2] heart failure, enlarged organs, cerebellar dysplasia,[3] and abnormal limbs and spinal cords. LOS also leads to dystocia (difficult calving) and often involves delivery via caesarian section. Repetition of these procedures increases the welfare concerns.

[2] *Hydroallantois* is primarily the result of dysfunction of the placenta, resulting in an increased production of fluid within the allantoic sac.

[3] Dysplasia refers to an abnormality of development and is typically used when the cellular abnormality is restricted to the originating tissue, as in the case of an early, in situ neoplasm.

Nuclear transfer techniques to propagate GM animals may also increase risks to the reproductive health and welfare of both the surrogate female animals and their GM offspring. For example, animal welfare problems can arise because of poorly controlled expression of the introduced gene. The most frequently cited example of welfare problems arising from inappropriate transgene expression in the USA is that of the so-called Beltsville pigs (National Academy of Science 2002). These were engineered with a gene for human growth hormone in an attempt to improve growth rate and decrease carcass fat content. The pigs were plagued by a variety of physical problems including diarrhea, mammary development in male pigs, lethargy, arthritis, lameness, skin and eye problems, loss of libido, and disruption of estrous cycle. Other examples include sheep born with diabetes, and fish with growth abnormalities of the head and jaw (National Academy of Science 2002; CAST 2011; Devlin et al 1995; Beardmore and Porter 2003). Other evidence of problems such as anatomical, physiological, or behavioral abnormalities have been reported in GM animals (National Academy of Sciences 2002). Most scientists appear to agree that animals originating from some forms of genetic modification usually require closer observation and care.

Having pointed out some of the potential problems and concerns of GM animals, it needs to be emphasized that the concerns noted here pertain only to concerns about animal welfare, and should not be construed as general concerns for the broader practice of genetic modification of animals. Indeed, there is a plethora of reasons to recognize that while there may be challenges associated with genetic modifications of animals, it is expected that the research will produce unprecedented breakthroughs in medicine and the animal sciences. These include gene therapy, disease models, improved test system development, gene discovery, life-span extension, and xenotransplantation. And in the animal sciences, enhanced production attributes for food, improved animal health, pharmaceutical products for both humans and animals, and other industrial or consumer products.

18.2.2 Issues in Pain, Suffering and Distress

According to the authoritative *Animal Biotechnology: Science Based Concerns* (National Academies Press 2002), animal welfare concerns associated with LMOs are their potential to

1. Cause pain
2. Cause distress (both physical and psychological)
3. Result in behavioral abnormalities, and
4. Result in physiological abnormalities and/or health problems.

As discussed in Sect. 18.1, two broad aspects of animal biotechnology need consideration in a discussion of the role of animal welfare in biotechnology regulation. The first involves the use of animals in medical research, while the second focuses on animal agriculture and food production. Both have important, but differing, implications for animal welfare. At the same time, regardless of the situation in which

GM animals are involved, all cases entail the potential for animals to experience pain, suffering, and distress.

The fundamental welfare concern about the use of animals in research and testing is the potential for pain, suffering, or distress. But the creation and use of GM animals presents additional problems for effective monitoring of these issues because of the potential for unexpected adverse phenotypes[4] and the large numbers of animals involved. With respect to GM animals, or their off-spring, that are used for food and fiber production, this concern extends to the potential that inadvertent, or purposeful, mutations and deformities may cause pain and suffering for the animals created.

According to most commentators and authorities, the underlying principles of the ethical care and use of animals (and particularly laboratory animals) are provided in what is commonly referred to as the 3 Rs—Refinement, Reduction and Replacement.[5] Alternatively, the more modern interpretation of the basic principles is made up of guidelines known as the Five Freedoms.[6] Broadly speaking, the 3 Rs are generally referred to when addressing the ethical care and use of animals in a research/laboratory setting, whereas the Five Freedoms are generally used in the context of animal welfare for farm animals.

---

[4] Unexpected adverse phenotypes are animals that are born or created with unexpected and undesirable and usually debilitating deformities, and usually have a negative impact on animal welfare. See for example http://www.marquette.edu/researchcompliance/research/documents/IACUCUnexpectedOutcomes.2012.pdf

[5] *Refinement* of experimental procedures to reduce or eliminate pain and distress. Where the use of animals is unavoidable, minimize pain, distress, lasting harm, or other threats to animal welfare. For example, researchers should ensure that accommodation meets animals' needs; use pain treatment drugs; and specify humane endpoints—that is, when a study design should be changed or a study ended early due to concerns about animal pain, distress, or welfare.

Reduction in the number of animals being used. Use methods that enable equivalent information to be obtained from fewer animals or more information from the same number of animals, such as through the use of advanced imaging techniques.

Replacement of animals with other reliable models. For example, use alternative methodologies, such as computer modeling, or replace higher order animals with those of a lower order (such as using amphibians or invertebrates instead of mammals).

[6] *Freedom from thirst and hunger*—by ready access to fresh water and a diet to maintain full health and vigor.

Freedom from discomfort—by providing an appropriate environment including shelter and a comfortable resting area.

Freedom from pain, injury, and disease—by prevention or rapid diagnosis and treatment.

Freedom to express normal behavior—by providing sufficient space, proper facilities and company of the animal's own kind.

Freedom from fear and distress—by ensuring conditions and treatment which avoid mental suffering.

18.2.3 The Modalities or Measurement and Therapeutic Approaches Associated with the Concerns

Many researchers have examined the nature of nociception (the experience of pain in sentient beings), pain, and suffering and whether animals are capable of experiencing any or all of them. The current consensus is that all vertebrates, and many invertebrates, are capable of experiencing pain. Most professionals therefore agree that

- Animals experience (i.e., suffer) distress, discomfort, and pain
- Analgesics or other interventions are available and should always be administered if animals are experiencing discomfort or pain, unless there is a compelling veterinary justification not to
- Humane endpoints should always be employed and re-evaluated regularly; and
- Prevention of pain, suffering or distress is the ideal and if this is not possible then prompt recognition and treatment are essential.

However, it is also recognized by almost all professionals that pain is subjective, and the evolution and function of a sense of pain (nociception) in animals is relatively obscure mainly because:

- There is a lack of research on nociceptors in many animals
- The identification of pain relies mainly on anthropomorphic judgments by humans
- There are very few, if any, objective scientific procedures available to identify pain, and, in particular, procedures that can be easily implemented.

Despite this, most professionals still agree that the assessment of animal pain through human judgment is valid and that once identified, pain and suffering should be evaluated and alleviated (Hawkins 2002; NRC 2000; 2009).

Techniques have been devised to assist with animal monitoring and recognition of discomfort, pain, and distress, but their effectiveness have not been widely evaluated in practice, and the extent to which they are used is not widely known. Current techniques for assessing animals and recording observations range from clinical observation sheets and "score" sheets, data management systems, to phenotype assessment protocols (e.g., SHIRPA[7]) and visual analog scales (Hawkins 2002). Recent scientific research has considerably advanced our understanding of animal pain; however, there are still few scientifically validated pain-assessment techniques. Therefore, in most circumstances, pain is assessed based on the appearance of an animal and its overall behavior. Current best practice is to combine a structured clinical examination with a good knowledge of the normal appearance and behavior of the animals involved (NRC 2000; 2009).

Many organizations around the world are advocating animal welfare assessments that move beyond the mere presence of the positive aspects of "good" animal welfare practices, to a system that recognizes what are commonly called "iceberg

[7] SHIRPA is a standardized set of experimental procedures used by scientists to characterize the phenotype of GM laboratory mice. The protocols are designed to test muscle function, cerebellar function, sensory function, and neuropsychiatric function.

Table 18.1 Development of EU welfare quality assessment

Welfare criteria	Welfare principle	Example of assessment measure
Good feeding	Absence of prolonged hunger	Body condition score
	Absence of prolonged thirst	Quality of water provision
Good housing	Comfort around resting	Cleanliness, abnormal rising
	Thermal comfort	Panting
	Ease of movement	Tethering, slipping
Good health	Absence of injuries	Injuries, lameness
	Absence of disease	Mastitis, diarrhea
	Absence of pain from management	Dehorning, tail docking
Appropriate behavior	Expression of social behaviors	Agonistic behaviors
	Expression of other behaviors	Stereotypical behaviors
	Good human–animal relationship	Avoidance distance
	Absence of general fear	Reaction facing novel situation

indicators". The "iceberg" concept recognizes that there are key physiological, behavioral, and psychological indicators of animal welfare such as body condition, normal behavior, and alertness. Just as the sighting of an iceberg signals that 90 % of its bulk is below the water line, and we ignore its tip at our peril, so iceberg indicators are critical signs of welfare.

One example of the application of animal welfare principles combined with an attempt to design an "objective" measure of animal welfare is the European Union (EU) Welfare Quality© Project (Blokhuis 2007). The Project's aim is to produce practical methods to assess the welfare of cattle, pigs, and chickens. Based on the Five Freedoms, researchers have developed a set of welfare "principles" based on four welfare criteria (Table 18.1). Assessment is focused on welfare outcomes and the protocols are being turned into reliable tests for use on the farm and at the abattoir (see for example Table 18.1, which describes some of the development of the Welfare Quality project). They are currently being trialed on 570 farms (pig, cattle, and poultry) in 15 EU countries and could form the basis of a common EU scheme of animal welfare assessment. However, they have not yet been accepted by the scientific community or farmers and will need to be closely appraised by EU Governments and Commission officials for their suitability.

Despite the fact that the EU Welfare Quality assessment principles have yet to be accepted and approved by the scientific community as a "scientifically acceptable and objective assessment" of animal welfare, there appears to be growing agreement among professionals that scientific and objective assessments of animal welfare are moving in the right direction. According to many veterinary researchers, the search for objective measures of pain is ongoing. For example, in one recent New Zealand study (Johnson 2007), two approaches to the quantification of animal pain appear to offer significant advances over previous techniques. The paper elucidates the use of spectral analysis of electroencephalograms (EEGs) and the ethological analysis of behavior. According to Johnson (2007), while the methods have very different applications and limitations, when used in conjunction with each other, they can give a complete and objective picture of the pain felt by groups of animals and the impact of different analgesic regimes upon that pain. Thus while it is still very much accepted that pain is subjective, the use of objective measures of pain felt by animals and the conditions under which they are felt can be measured and recorded in a scientifically and objective manner, and may eventually lead to further research in approaches to measuring pain and suffering in animals.

18.3 Critical Assessment

If rules on a biosafety risk assessment with respect to animal welfare are to be specified in the context of the Cartagena Protocol, then they need to be:

1. Objectively measurable
2. Evidence based, and
3. Identifiable in advance.

18.3.1 Objective Measurement of Animal Welfare Impacts

As stated previously, the measurement of pain and suffering in animals is usually acknowledged as being subjective, and the evolution, function and recognition of a sense of pain (nociception) in animals is relatively obscure, mainly because of the reasons set out above. Despite this, most professionals agree that the assessment of animal pain through human judgment is valid and that once identified, pain and suffering should be evaluated and alleviated. In addition, new and recent research indicates that the objective measurement of pain and suffering in animals is plausible and near to fruition.

18.3.2 Evidence of Pain and Suffering

While there is little in the way of objective measurement, there are many examples of evidence of pain and suffering in animals as a result of GM procedures. The previously mentioned report *Animal Biotechnology: Science Based Concerns* (National Academies Press 2002) is probably the most comprehensive listing of actual and potential pain and suffering of animals as a result of GM procedures. However, the most revealing point about the discussion surrounding animal welfare is that while the discussion has been fruitful and concerning, it has not led to any conclusive actions with respect to collecting evidence of pain and suffering in animals subject to GM procedures. For example, there is no established protocol for recording information with respect to pain and suffering which could be entered into a database so that researchers and practitioners could share the information. Such a database could build on the concepts developed, and establish, over time, the requisite "scientific" assessments necessary to give any resulting rule-making the objectivity required to put the rules in place and have them objectively enforced.

18.3.3 Advance Identification of the Impacts on Animal Welfare

The creation and use of GM animals presents additional and unique problems for effective monitoring of pain, suffering, and distress, because of the potential for unexpected adverse phenotypes. While many effects can be predicted within a limited range of certainty, more subtle effects are also possible that may not be detected for several generations. Of major concern are the phenotypic changes that can occur (and have occurred in mice) which can be transmitted to offspring, including retarded growth and abnormal DNA methylation patterns. Inadvertent transfers of DNA may result in an array of defects such as severe muscle weakness, missing kidneys, seizures, behavioral changes, sterility, and disruptions of brain structure, neuronal degeneration, inner ear deformities, and limb deformities. Mutations can vary enormously and because many mutations are recessive, they can result in generations of animals with mutations and deformities that are welfare concerning.

18.4 International Arena

18.4.1 Domestic Laws

While most countries have some form of animal welfare regulations in place, the use of domestic policy to impose restrictions on LMO imports based on animal welfare concerns must take into account the economic repercussions. For example, animal welfare regulations can have the effect of raising production costs, which may in turn cause higher food prices for domestic consumers. Furthermore, if a country's domestic animal welfare regulations are more restrictive than those of a competing country, domestic producers can be priced out of the export market for that product because they are unable to match the price competition of producers in other country(s). Thus, it is incumbent on countries to carefully weigh the economic outcome of imposing restrictions on other countries based on animal welfare regulations (Mitchell 2001; Gaisford and Lau Chui-Ha 2000).

18.4.2 General Agreement on Tariffs and Trade (GATT)

Apart from the economic impacts of these restrictions, a number of international regulations are also relevant if a country imposes animal welfare–based standards on other nations wishing to export similar products. Among the most important of these international rules governing world trade are the WTO Agreements. Articles I and III of the GATT require identical treatment for "like" products, regardless of the country of origin. That is, production measures that do not alter the final product in any material or differentiable way must be treated as "like" products. Any decision to prevent the import of a product must be based on the characteristics of the product itself and not on the process or production method (PPM)[8] by which the product was manufactured. Thus, production measures that might ensure animal welfare standards must demonstrate that the product is materially different to similar products that do not use the same (animal welfare enhancing) production measures.

Based on Article XX (a) of the GATT, trade could also be restricted "to protect public morals". Presumably, one would have to argue that the product itself was offensive and against "public morals" and not the production process which, as pointed out above, would probably contravene Articles I and III. In addition, in the case of a product produced outside of the importing country's jurisdiction, the question arises as to whether this exception could apply. Finally, it is not clear that if one were to argue that public moral concerns for, say, animal cruelty would constitute a legitimate argument for imposing restrictions based on those public morals.

Article XX(b) provides for trade restrictions to protect human, plant, and animal health. While many might argue that animal welfare is an important aspect of animal health, it is not, in general, strictly an issue of animal health. Alternatively, animal health can be included as an important aspect of animal welfare, but reliance on this issue would require a scientifically established link between the two. That is, if scientific research can show that there is a direct connection between animal health and a specific animal welfare standard, and that maintaining that standard is a method of protecting animal health, then restricting trade based on such a standard should be acceptable under WTO rules (Thiermann and Babcock 2005). However, again, in the case of a product produced outside of the importing country's jurisdiction, the question arises as to whether this exception could apply, particularly if Articles I and III of GATT are invoked.

[8] PPM stands for "process and production method". An important technical distinction is made within WTO rules between product-related PPMs (PR-PPM) and non-product-related PPMs (NPR-PPM). The distinction between the two relies on the how the PPM affects the final product. If the final product of PPM-A is used, handled or disposed of essentially the same way as PPM-B, then it is a NPR-PPM. If the final product of PPM-A is used, handled or disposed of in a completely different way than PPM-B, then it is a PR-PPM.

Article XX(g) allows measures that are "related to the conservation of exhaustible natural resources". In 1989, the US embargoed shrimp from countries who did not adopt the policy of protecting sea turtles by adopting turtle excluder devices to their nets while fishing for and catching shrimp. One of the arguments employed by the USA (and which was recognized under US Conservation Policy and applied to US fishermen) was that sea turtles were an exhaustible natural resource that should be protected from inadvertent snaring while fishing for other species (WTOa 2013). Similar claims are also made in the Tuna/Dolphin case (WTOb 2009). The basic argument is that under GATT XX(g), "exhaustible natural resources" are not limited to resources that are susceptible to exhaustion or extinction in a static sense, but may be extended to natural resources that are "evolving". Thus, it is possible that certain animals could be interpreted as being "exhaustible natural resources[9]". In other words, it is conceivable that an argument could be made under GATT XX(g) that if a GM animal was released into the wild and its release impacted the conservation of an exhaustible natural resource, then it could be listed as an exclusion under GATT XX.

18.4.3 The Sanitary and Phyto-sanitary (SPS) Agreement

The SPS Agreement specifically recognizes the OIE[10] (Office International des Epizooties or World Organization for Animal Health) which addresses issues of animal health measures. The OIE's Terrestrial Animal Health Code and the Aquatic Animal Health Code both have individual sections (Section 7 in both reports) relating to animal welfare.[11] Measures based on international standards, guidelines or recommendations developed by the OIE are presumed to be consistent with the SPS Agreement, and Members who base their measures on them can be confident of compliance with the SPS Agreement. International standards are sometimes described as providing a "safe harbor" for governments. Clearly, however, Members have the right to challenge all SPS measures, and particularly if they believe that the claim of being based on an international standard is ill-founded (Kogan, 2007).

[9] While the Shrimp/Turtle case was considered legitimate by the WTO Appellate Body, the USA lost the case because it discriminated between WTO members http://www.wto.org/english/tratop_e/envir_e/edis08_e.htm The Tuna/Dolphin case http://www.wto.org/english/tratop_e/envir_e/edis04_e.htm report was never adopted and therefore is not a legal interpretation of GATT law. Nevertheless it still attracts lot of attention because of its implications for environmental disputes.

[10] See their website at http://www.oie.int/en/ for more information on the activities and recommendations of the OIE. Two other international bodies are of relevance—The IPPC (International Plant Protection Convention) and the Codex Alimentarius (Codex). It should be noted that the SPS Agreement makes no legal distinction between the "standards," "guidelines" and "recommendations" of these three organizations. All three types of norms have equal status under the SPS Agreement. *Source:* World Trade Organization documents, http://www.wto.org.

[11] See http://www.oie.int/international-standard-setting/terestrial-code/access-online/ and http://www.oie.int/international-standard-setting/aquatic-code/access-online/

18.4.4 The Technical Barriers to Trade (TBT) Agreement

A plausible alternative to restrictions on trade based on animal welfare concerns is the use of labeling of products according to production standards, with the main objective of allowing consumers to be informed. Labeling issues come under the auspices of the TBT Agreement. However, it is uncertain whether labeling of imports according to production methods is possible within the framework of the TBT. The TBT was designed to ensure that labeling and other technical requirements do not create unnecessary obstacles to trade. As such, the TBT Agreement encourages countries to develop standards and technical regulations based on international standards. However, the problem is that the TBT Agreement fails to identify the relevant standard setting bodies for international standards, unlike the SPS Agreement discussed previously. (One could assume in this case that since the OIE is specifically identified as a "standard-setting body" for animal welfare under the SPS Agreement that it would also apply under the TBT Agreement.)

The TBT also allows governments to choose measures based on their national requirements if international standards do not meet their needs. The objectives may include the prevention of deceptive practices, protection of human, plant or animal health or safety, or protection of the environment. Whatever the case, the TBT Agreement requires advanced notices (similar to those required for SPS measures), and also notices of bilateral technical agreements and compliance by national standard setting bodies with a CGPPAAS.[12]

Finally, all WTO Members are required to establish national enquiry points and to keep each other informed through the WTO. The TBT Committee is the main clearinghouse for sharing information among Members and the primary forum for discussing concerns about the regulations and their implementation.

In summary, the main question with respect to *animal welfare and trade compliance* is that domestic (national) rules, regulations, and laws pertaining to animal welfare are likely to be viewed as trade barriers unless there is demonstrable (scientific) product-related evidence that the use of such standards can be shown to be necessary for the protection of human, plant or animal health, biodiversity, or the environment.

Some qualifications are appropriate:

- From a scientific point of view, what is safe in one country should be safe in another. However, the same level of risk is not perceived the same way everywhere and differs significantly across countries (e.g., the legality of the production and sale of cheese made from raw or unpasteurized milk vs pasteurized milk in various countries).
- Science is not always conclusive.
- It is almost impossible to separate scientific considerations from economic and political ones. The perception is that science is often used for profit and is subject to the influences of industry and special interest groups.

[12] Code of Good Practice for the Preparation, Adoption and Application of Standards.

- The substantial debate over the extent to which the SPS and TBT Agreements allow (or disallow) trade restrictions based on specifications related to the PPMs reveals such inconsistencies that could actually diminish a nation's ability to uphold its own environmental, animal welfare, or public health principles.

18.5 Administrative Consequences

By far, the most important issue associated with the consideration of animal welfare in biotechnology regulation is the appropriate balance between the costs of regulation and the benefits associated with them. The economic consequences of imposing animal welfare regulations in an international context were discussed in Sect. 18.4. While the passage of animal welfare laws may offer individual and social benefits to consumers, it may also result in higher prices for all consumers, and result in trade policy concerns for consumers and producers, both domestically and internationally.

Administrative concerns are also at the forefront for organizations and researchers who use animals for research. For example, in one academic facility that used large numbers of GM animals, over 90 % of them appeared to be clinically normal, but some abnormal phenotypes did occur. Because each technician was responsible for checking around 500 cages of 4–5 mice every day, if 6 hours a day was spent monitoring animals, this allowed 43 seconds at most for each cage, or 9–11 seconds for each mouse. Objective measures such as body mass were rarely taken due to lack of time! Indeed, one of the conclusions from the International Animal Research Workshop held in Buckinghamshire, UK, in July 2011 (NRC 2012), concluded that there was a need to recognize that regulations create administrative burdens, and that the costs of administering regulations need to be (1) balanced, or in line with, the benefits and (2) justified on the basis of improving animal welfare, and not just for the sake of ensuring animal welfare.

18.6 Summary/Synthesis

- Consumers, worldwide, are demanding their right to make more informed choices about the food products that they purchase and consume, including how animals are bred, raised, kept, used, transported, and slaughtered.
- While the creation and use of LMOs in agriculture and the food system harbors great advances in science and the future of the food system, it is also responsible for an increased awareness of, and concern for, the important links between animal health and animal welfare, and the safety and quality of food products, and their broader implications for biodiversity, the environment, and for human, plant, and animal health and safety within that environment.
- Despite the fact that the development of risk and quality assessments of animal welfare has yet to be accepted and approved by the scientific community as "scien-

tifically acceptable and objective," there is growing agreement among professionals that the concepts used and the technologies developed mean scientific and objective assessments of animal welfare are not only plausible, but near fruition.

- The main question with respect to animal welfare and trade compliance is that domestic (national) regulations pertaining to animal welfare are likely to be viewed as trade barriers unless there is demonstrable (scientific) product-related evidence that the use of such standards is necessary for the protection of human, plant or animal health, biodiversity, or the environment.
- Animal health and animal welfare should be at the forefront of ALL production, processing and development of animal food products that result from modern biotechnology, domestic or international, regardless of its status as a SEC, provided that an appropriate balance exists between the costs of regulation and the benefits associated with them.

References

Beardmore JA, Porter JS (2003) Genetically modified organisms and aquaculture. FAO Fisheries Circular No 989, Rome, p. 35

Blokhuis HJ (2007) International cooperation in animal welfare: the welfare quality project Acta Veterinaria Scandinavica, 2008, 50 (Suppl 1):S10 http://www.biomedcentral.com/content/pdf/1751–0147-50-S1-S10.pdf

Council for Agricultural Science and Technology (CAST) (2011) The science and regulation of food from genetically engineered animals. CAST Commentary QTA2011–2. CAST, Ames, IA, USA

Dennis MB (2002) Welfare issues of genetically modified animals. ILAR Journal/ National Research Council, Institute of Laboratory Animal Resources 43:100–109

Devlin RH, Yesaki TY, Donaldson EM et al (1995) Production of germline transgenic Pacific salmonids with dramatically increased growth performance. Can J Fish and Aquat Sci 52:1376–1384

Gaisford JD, Chui-Ha L (2000) The case for and against import embargoes on products of biotechnology. The Estey Cent J Int Law Trade Policy 1(1):83–98

Hawkins P (2002) Recognizing and assessing pain, suffering and distress in laboratory animals: a surveyof current practices in the UK with recommendations. Laboratory Animals 36(4):378–395

Johnson C (2007) New approaches to identifying and measuring pain. Australian animal welfare strategy science summit on pain and pain management, May—Proceedings, pp 1–10

Kaiser M (2005) Assessing ethics and animal welfare in animal biotechnology for farm production. Rev Sci Tech Off Int Epiz 24(1):75–87

Kogan LA (2007) World Trade Organization biotech decision clarifies central role of science in evaluating health and environmental risks for regulation purposes. Glob Trade Cust J 2(3):149–155

Mitchell L (2001) Impact of consumer demand for animal welfare on global trade. In: Regmi A (ed) Changing structure of global food consumption and trade. USDA:ERS, Washington DC, p. 111

National Academy of Science (2002) Animal biotechnology: science based concerns. Committee on defining science-based concerns associated with products of animal biotechnology, The National Academies Press, USA

National Research Council (2000) *Definition of pain and distress and reporting requirements for laboratory animals: proceedings of the workshop held June 22, 2000*. The National Academies Press, Washington DC
National Research Council (2009) *Recognition and alleviation of pain in laboratory animals*. The National Academies Press, Washington DC
National Research Council (2012) *International animal research regulations: impact on neuroscience research: workshop summary*. The National Academies Press, Washington DC. http://www.nap.edu/catalog.php?record_id=13322
Sendashonga C, Hill R, Perini A (2005) The Cartagena protocol on biosafety: interaction between the convention on biological diversity and the world organization for animal health. Rev Sci Tech Off Int Epiz 24(1):19–30
The Boyd Group (1999) Genetic Engineering: animal welfare and ethics. A discussion paper from the Boyd Group http://www.boyd-group.demon.co.uk/genmod.htm accessed 7 March 2012
Thiermann AB, Babcock S (2005) Animal welfare and international trade. Rev Sci Tech Off Int Epiz 24(2):747–755
World organization for animal health. http://www.oie.int/international-standard-setting/terestrial-code/access-online/ accessed 16 March 2012
World organization for animal health. http://www.oie.int/international-standard-setting/aquatic-code/access-online/ Accessed 16 March 2012
World Trade Organization (WTOa) (2013) India etc versus US: 'shrimp-turtle'. http://www.wto.org/english/tratop_e/envir_e/edis08_e.htm accessed 18 June 2012
World Trade Organization (WTOb) (2013) Mexico etc versus US: 'tuna-dolphin'. http://www.wto.org/english/tratop_e/envir_e/edis04_e.htm accessed 18 June 2012

Part III
Navigating the Challenges

Crystal-ball gazing is fraught with peril and while we are cognizant of such perils, offering some poignant synthesizing observations is essential to frame the preceding discussions. Other forms of knowledge in addition to science are necessary to understand the values embedded in and transmitted through the social, economic and cultural dimensions of the environment, and how these things may be affected by or in turn affect a particular use of a technology. The key issue is how to relate these other forms of knowledge to scientific knowledge in the decision-making process. This policy milieu is complicated as the examination is, in fact, of complex interactions between democratic societies' right to decide how to proceed, producers' freedom to operate and consumers' rights and freedom to know. These issues are extremely important for regulatory design as they help define a functional system and deal with relevant issues such as individual privacy and confidential business information filed with regulatory dossier applications. This section contains a practical synthesis of the practicalities and options relevant to the assessment of SECs and their introduction and implementation into the agricultural GMO decision-making process.

Chapter 19
Ensuring Functional Biosafety Systems

Karinne Ludlow, Stuart J. Smyth and José Falck-Zepeda

19.1 Introduction to a Three-Dimensional Problem

Were regulation a living matter, without a doubt, it would be called a complex creature. While domestic regulations are complex enough on their own, elevating regulation to the international level adds an even greater degree of complexity. Complying with international regulations is frequently a matter that requires considerable attention, but even with detailed levels of scrutiny, disagreements pertaining to interpretation are commonplace. International governance is a growing area of complexity, with numerous institutions, organizations, and agreements, all proclaiming some degree of legitimacy and authority to oversee the diverse area of activities regarding regulation in specific jurisdictions. Developing domestic regulations will frequently fall under the oversight of several of these international bodies, making it challenging for countries with limited regulatory development resources to be compliant in all aspects. As well, countries will give a greater, or lesser, degree of importance to particular international governance bodies, creating varying interpretations, responses, and legislation, all pertaining to the same jurisdiction. As has been demonstrated in Part II, the international regulation and governance of agbiotech is rife with complexity.

As with the CPB itself, which provides no guidance on the matter, the various authors of the SEC chapters were not provided with detailed definitions of the particular consideration they were invited to address. This was deliberate—providing such a definition would of itself influence how an expert responded to the SEC. Instead

K. Ludlow (✉)
Monash University, Clayton, Australia
e-mail: karinne.ludlow@monash.edu

S. J. Smyth
University of Saskatchewan, Saskatoon, Canada
e-mail: stuart.smyth@usask.ca

J. Falck-Zepeda
International Food Policy Research Institute, Washington, DC, USA
e-mail: J.FALCK-ZEPEDA@cgiar.org

K. Ludlow et al. (eds.), *Socio-Economic Considerations in Biotechnology Regulation*, Natural Resource Management and Policy 37, DOI 10.1007/978-1-4614-9440-9_19,

experts in particular fields were invited to write on a nominated consideration, which they were to interpret in light of their expertise. The experts were drawn from different disciplines, including economics, humanities, science, law, and management. They were also deliberately drawn from diverse jurisdictions including Europe, North America, Africa, and Australia. This meant that even with a common chapter template, the 15 chapters show the breadth and range of interpretations of, and responses to, SEC assessment in biotechnology regulation. This part of the book brings together those interpretations and responses and finds commonalities that can be used to inform future discussion and developments relevant to the assessment of SECs and their introduction into the agricultural GMO decision-making process.

SEC has been used in this book to refer to those considerations not currently addressed in what are commonly referred to as the science-based risk assessment frameworks considered in Chap. 2, used to evaluate risk to human, animal and plant health and safety, and the environment. Concerns outside of this have been treated in this book as SECs. As has been highlighted in Chap. 3, the present application of SECs varies substantially, with little to no consistency in implementation strategy or range of regulatory oversight by countries that have incorporated SECs into their domestic biosafety regulatory frameworks. However, it must be remembered that even for those risks considered in science-based risk assessment frameworks, there can be closely related impacts that are not addressed. Thus, for example, as Novy and Nagarajan discuss in Chap. 16 on producer choice, the ability of producers to effectively carry out resistance management plans designed to respond to identified environmental risks may depend on the typical organizational structure of farm management in the jurisdiction concerned. Subsistence farmers may, for example, be unable to afford to comply and this in turn will be relevant to the assessment of the SEC, producer choice.

This issue of definitions or boundaries for SECs included in a decision-making process is a central one to address and will be particularly difficult to resolve. Resolution could be sought by conceptualizing each SEC as being an individual color in a rainbow. Therefore, while the precise boundary between any two adjacent colors may be a matter of dispute, because all colors are displayed and "considered" this blurring does not matter. But conceptualization of SECs in that linear fashion is not accurate. First, even a linear model can have gaps. Mugwagwa (2012, p. 43) explains that "[o]ften this is not because the information or other attributes to fill these gaps are not there, but because of a lack of obligation among the various players to take forward what the player at the other level (lower or higher) has done. This issue is best explained within the social arena of problems…, where multiple perspectives may not overlap enough to cover the issue area adequately."

Second, SECs have greater "3D" implications than a rainbow. They are more like a cluster of colored helium-filled balloons. The perception of the color of any individual balloon is greatly influenced by the color of its neighbors and by whether that balloon is seen by looking through a neighbor. Returning to SECs, the content of any particular SEC is greatly influenced by surrounding factors such as who is defining it and the context in which the definition is taking place. Context may mean the jurisdiction the SEC is being assessed in, but also the other SECs identified as most relevant to that jurisdiction which are also being assessed as part of the same process.

The objective of this chapter is to highlight the methodological issues that arise and to provide a sense of the fundamental challenges that will need to be resolved prior to the introduction of a regime where SECs are included as part of the regulation of the import and production of GMOs and that are in compliance with the implementing countries' international obligations.

19.2 Structuring the SEC Discussion

A classical illustration of the 3D nature of SEC assessment is the SEC of environmental impact. In Chap. 6, Wesseler and Smart posit two starting points that are relevant to all SECs: regional aspects are crucial, and agriculture, regardless of whether it is GM or otherwise, always has an environmental impact. These points are relevant to the baseline used to measure any SEC against, an issue we will return to in the final chapter. In relation to environmental impacts, Wesseler and Smart address how the broad array of potential positive and negative impacts of agricultural biotechnology on the environment could be assessed. This discussion illustrates the inevitable intertwining of the SECs. For example, one identified impact is a change in pesticide use. This impact is identified as an issue relevant to environmental impact assessments, as Wesseler and Smart have done, because a decrease in pesticide use lowers pesticide residues in the environment, residues in the resulting crop, and exposure to the farmer. It may also potentially increase on-farm biodiversity. But these spill-over effects mean changes in pesticide use may also need to be included in an SEC assessment of impact on biodiversity, labor impacts, and food safety.

In Chap. 10, impacts on biodiversity from GM crops are discussed as a component of an SEC risk assessment. Impacts on the environment are part of science-based risk assessments, but the concept has been expanded slightly and included as a SEC. Falck-Zepeda, Zambrano, and Smale identify that one of the methodologies used to assess biodiversity impacts is known as an ecosystems service evaluation, which comprises evaluations of the biophysical, sociocultural, and monetary. In essence, it is a stocktaking of all goods and services, tangible and intangible that are of value, and could be of value, to humans. The potential scope of such an evaluation is limitless and the data requirements are quite staggering for an assessment of this nature.

Addressing the strongly tangential SEC of labor impacts (or balloon to continue with the visual cue), reductions in farmers' exposure to pesticides and other chemicals, as well as the time required to undertake the chemical application, would be central to any assessment. Like Wesseler and Smart in Chap. 6, Gouse (in Chap. 13) highlights the importance of recognizing regional differences, which for the SEC of labor impacts is particularly apparent regarding the introduction of new agricultural technology. Developed nations, where mechanized production systems are used, and developing nations, where agriculture is often the sole source of livelihood and employment, will assess a decrease in pesticide usage differently.

Gouse focuses on agricultural production in sub-Saharan Africa (SSA) where agriculture is labor intensive and notes that in such agricultural systems there is often no incentive to expand production because of limited profitable marketing opportunities. He notes that a labor-saving technology, such as herbicide tolerant (HT) GM crops which require less weeding, can be beneficial to some but detrimental to others. Where labor is a limiting factor, HTGM crops may be beneficial, allowing farmers to reallocate their labor to other activities, such as expanding their production to increase yields. In contrast though, where landless unemployed depend on weeding for their livelihood, and where market limitations restrict expansion, the introduction of HTGM crops could have a negative impact. This is an interesting conundrum because the research issues define the methodology which in turn can and does define the results from an assessment. This state of research clearly shows the high level of expertise and experience needed to conduct such assessments so that they become a valuable tool to support decision-making.

The opportunity to trade surplus yields is thus important in the assessment of the SEC of labor impact but is also an SEC of its own. Chap. 15 addresses the SEC of national trade interests, focusing on the assessment of possible domestic economic effects of GM imports and/or adoption, particularly segregation and coexistence options, on non-GM agricultural production, marketing, and trade. Market access and assurance that international trade will not be disrupted have become dominant socio-economic issues for many developing nations and are judged to be a major reason for the lack of GM crop adoption in Africa (Paarlberg 2008; Smyth et al. 2013). Gruère notes that whether the introduction or use of GM crops causes economic loss to non-GM supply chain actors is driven by many considerations, including the incentive for separating non-GM from GM products—GM-free private standards in the domestic or international market, labeling regulations, and intellectual property (IP) protection.

In Chap. 14, Viju further considers domestic regulatory measures but in the context of their impact on trade policy and market access at a global level. She extends Gruère's discourse at the international level, positing that three methods can be used to determine if the commercialization of a GM crop would be welfare enhancing in specific trade markets—partial-equilibrium models, general equilibrium models, and gravity models—each of which offering somewhat different results. She points out that a focus point for literature in this field is the important regional differences between GMO regulatory frameworks wherein particular countries are divided on the basis of whether the country of interest is a major exporting nation that has adopted GMOs, a major developed nation that imports GMOs but has strict regulatory frameworks, or a developing country that may be a potential importer. Importantly, particularly on the issue of equity the focus of Chap. 7, Viju notes that welfare consequences for the world economy of current and prospective GM crop adoption by some countries and of trade/domestic policy responses by other countries need to be estimated. Thus, for example, changes in the policy of some developed nations toward GMO adoption could greatly impact global welfare with most of those gains going to the developing countries. This chapter also considers the "bigger picture" outcomes for global trade of consistency across jurisdictions in GMO regulation.

Chapter 16 focuses on the SEC of producer choice, to which the issue of reduction of pesticide usage mentioned earlier will be relevant, among many other matters. Novy and Nagarajan note that producers make decisions based on a variety of expectations including yield, profit, risk perception, inputs (labor, fertilizer, water, pesticides etc.), health effects, environmental stewardship, market factors, tradition, and culture. Many of these are also considered as individual SECs in this book. Indeed, Novy and Nagarajan point out that the assessment of producer choice will need to address many factors, including freedom of choice, income security, control over production, contamination of organic agriculture, and farmers rights to save seeds that will be included in other SECs. They also point out that real choice will often require case-by-case assessment, tailored for each application situation and, as is common in the assessment of SECs, will be influenced by the regional jurisdiction of the assessment.

For example, farmers' choice to save seed, which informs farmers' decisions whether to adopt GM crops, depends on IP but is also highly dependent on specific crop biology of the crop concerned: seeds from open-pollinated varieties, which are often more attractive to resource poor farmers, can be saved, subject to IP constraints; seeds from hybrid varieties, which cannot be saved, often require more inputs and usually have a price premium because of some offered agronomic advantage, are less attractive to resource poor farmers. One methodology for assessment considered by Novy and Nagarajan is the sustainable livelihoods framework with its focus on food security and poverty. Food security in this context then is taken as meaning access to sufficient food although it goes beyond income and nutrition to include concepts such as vulnerability, assets and empowerment, a different interpretation to that given to food security in Chap. 8. A final commonality to be drawn from Chap. 16 is Novy and Nagarajan's observation that if a tolerance limit for GM comingling is set, that limit will have profound implications for the ultimate quantification of other SECS.

IP is the focus of Lawson in Chap. 12. Lawson, like Wesseler and Smart in Chap. 6 regarding environmental impacts, observes the influence of private rights in SEC assessment. Some argue that the operation of IP could have major impacts on developing countries by biasing agri-food production and development toward modern improved varieties. Moreover, it is possible that owners of IP may effectively appropriate or marginalize traditional knowledge in ways that jeopardizes landraces or traditional farming practices. The resulting dislocation of indigenous people and practices could, in some circumstances, be linked to effects on biodiversity. IP sits within a matrix of rigid regulation through international agreements, but Lawson describes the various national laws on IP as a patchwork, each attempting to articulate at least the minimum standards in the context of national and regional standards. He also notes that in addition to the multi-party international agreements, there are important entrenched obligations under bilateral and regional trade agreements which often impose significantly higher IP obligations. This means, Lawson concludes, where decisions about SECs are being made, the particular IP landscape of each nation will need to be determined according to their particular commitments. Further, the IP content of the particular GMO being assessed and related materials, processes, etc. will have to be identified and the relevant owners determined. While the data can be found in various online IP databases in developed countries, such as the European Patent Office, the Canadian Intellectual Property Office, and the US Patent and Trademark Office, few developing countries have such developed systems. Lawson suggests that such owners may also need to be involved in the decision-making process because of their private rights and where a decision adversely affects their IP, a mechanism of review and compensation will be required.

Concern for the individual is also addressed in Chap. 5, where Moses and Fischer discuss consumer choice. While the EU provides the background for this chapter, much of the discussion is relevant whatever the jurisdictional context. They point out that assessment of consumer opinion and the reasons for those opinions is difficult. Consumer attitudes also change over time although they note that nutritional or health benefits seem more attractive to consumers than price advantage. While food security and safety issues are reviewed in Chap. 8, with health impacts of GMOs addressed in Chap. 9, Moses and Fischer conclude that consumers' expressed attitudes to GM food products may well differ from their actual choices when such products are available in the stores. They also point out that GM food sometimes raises issues of the "unnatural" challenging traditional perspectives of nature, agriculture, and humanity's place in nature, which may bring about moral objections. These issues are more thoroughly addressed in Chaps. 7 and 17. Further, they note that in some countries growing skepticism about information from industry, governmental and other official sources about food safety were the precursors of the GM food debate which remains partly unresolved today. They state that lay skepticism about GM foods may be influenced by a lack of trust in the institutions and actors responsible for the new technology or by a lack of a sense of agency.

These issues are considered further in Chap. 8 on food security and safety. In this chapter, Strauss addresses food security and safety in the context of the developed world. Like Moses and Fischer in Chap. 5, Strauss considers the impact of the GM food debate in SEC assessment. She addresses the SEC of food security from the viewpoint of one of choice and consumers' trust in government regarding assessment of food safety, particularly government's choice of acceptable level of risk. Strauss considers that food safety as addressed in the science-based risk assessment framework does not go far enough. Like Moses and Fischer, Strauss explores concerns raised by consumers regarding scientific uncertainty over risks to safety and biodiversity and consumers' ability to respond to their concerns through their food purchasing decisions, which in turn links to labeling issues, raised by Gruère in Chap. 15.

Chapter 4 integrates these discussions even further by focusing on the impacts of GM crops and foods on producers and societies. Assessments of this nature are typically made by gathering on-farm data and determining what the impacts of GM crops have been on the adopting farmers, be it in India with GM cotton or in Canada with GM canola, and then aggregating the results to be applied to the larger economy. While studies of this type can be done with a single year's data, confidence in the results greatly increases with the ability to interpret the results based on multi-year data. This of course takes time and considerable investments of financial resources. Falck-Zepeda and Smale highlight that the results of applying this SEC to biosafety regulations are valuable to policy makers; however, they also caution that no one methodology is capable of providing the complete assessment and that each method applied has its own advantages and disadvantages. This same conclusion can actually be extended to all the different research issues and approaches discussed in this book.

Newell-McGloughlin in Chap. 9 also pursues concerns focused on a particular societal segment but this time in the context of the potential impact on the health and well-being of humans and animals by GM plants, focusing on the value-added output traits of improved nutrition and food functionality. While acute safety concerns raised by nutritional changes to GM plants and livestock are addressed by current risk assessments, Newell-McGloughlin points out that assessment of this SEC will require detailed measurement at the individual human or animal level given that nutrition intervention from a functionality perspective has a personal dimension. She also posits that the "cost" of precautionary inaction needs to be addressed from an ethical viewpoint relative to the "cost" of failing to adopt a new product that could significantly benefit sizable world populations. In addition, the commonality of the importance of regional context is raised by her observation that some impacts may be dependent upon certain variable conditions, such as regional differences in processing, consumption, and health affections.

Phillips, in Chap. 11, addresses the SEC of traditional knowledge (TK), which includes within it concern regarding protection of genetic resources, so important for biodiversity. Phillips discusses a number of important hurdles that need to be met for legitimate assessment of TK to occur. He notes the definition of TK is itself contentious and that valuing it is hotly contested. Difficult aspects include that TK and related genetic resources have intrinsic cultural or spiritual values that cannot be quantified in monetary terms, often do not enter modern markets and are often only a component, albeit an important one, in developing other products. Defining the ownership of TK and the rights and rules associated with it (of which the formal IP laws are only part), which in turn impacts its overall "value" and what infrastructure is necessary to realize that value, are further difficulties. Returning to the endemic issue of regional differences, claims by different groups to TK and genetic resources often overlap so that national boundaries and specific indigenous communities are spanned, which in itself can give rise to regional conflicts. Despite this, ultimately what is protected and how that is done and exploited, issues so closely related to the economic value of TK, are matters for domestic law. Phillips points out that this means international benchmarking and standardization will be difficult and the spectrum of models, methods, and metrics for valuing TK allows any viewpoint to be supported. He predicts this is likely to lead to complicated and contested solutions. This may be in fact the "state of nature" for all SECs that are being discussed at the international level and of the assessment approaches that may be used to evaluate them. In fact, we may have to limit the discussion to those elements that characterize implementation of models, methods, and metrics as there is no ideal approach which may necessitate approach triangulation, and thus the need to consider inevitably the possibility of contradictory results. Lawson, in Chap. 12, also raises the issues of TK and equality of access to genetic resources and sharing of benefits from those accessed resources, in his consideration of the SEC of IP.

Thompson, in Chap. 7, addresses the SECs of ethics and equity. This chapter also considers the issue of fairness, raised in Chap. 11, regarding TK. In the context of ethics though, Thompson points out that while ethics may be described in its broadest sense as addressing whether the "right thing" is being done, the answer to this requires more than a consideration of fairness. He notes that in principle, all considerations relevant to answering how one should act are potentially relevant to an assessment of ethics. However, practically ethical considerations generally include criteria other than those usually considered part of environmental, safety or economic assessments. Valuing an ethical impact or outcome is made more difficult according to Thompson because generally if a consideration is categorized as ethical, it has been acknowledged as being controversial. Further, and yet again, returning to the endemic issue of regional differences, Thompson points out that the human ability to recognize equality in an ethical sense is subject to cultural, historical, and experiential modifications and further, and that even agreeing on what it is that should be equally shared is itself contentious. We note that it is our opinion that we may have to agree to devolve the question of the issues and approaches that may be important for countries to focus on to internal (national) discussions. This may help reduce the number of issues which may be considered at the international level, but still leaves a state of contention and thus tension between stakeholders. A final key observation from the chapter and applicable to all SECs is that the metric used to measure any SEC is essentially an economic one when in fact, well-being or happiness or some other non-economic value may be what is really of interest.

Coe addresses both of these final observations of Thompson, well-being and happiness, in her chapter (Chap. 17) on religion and culture as an SEC. While ethics is often included with religious and cultural aspects pertaining to biotechnology, Coe acknowledges that it can be problematic in trying to separate ethics due to it being interwoven with religious and cultural aspects. Coe suggests that two key themes set religious and cultural aspects apart: "the concepts of sacredness and the desire for happiness and well-being." Given that discussions involving religious and cultural aspects of an innovative technology can be emotionally laden and subject to misinformation, numerous countries have utilized citizen consultations. Coe highlights her chapter by providing insights into how the sacredness of taro in Hawaii has contributed to the rejection of biotechnology applications to taro, while, at the same time, GM papaya has been adopted by an estimated 80 % of Hawaiian papaya producers. Clearly engagement with stakeholders is a crucial part of the process regarding the commercialization of innovative agricultural technologies, but even full consultation is no guarantee of acceptance. In fact, the need may arise of pursuing qualified regulatory outcomes where specific areas of a country are deemed as "not approved" for commercialization and use of a specific GM crop. This introduces a higher level of post-release monitoring but may be the only way to address cultural concerns pertaining to traditional and ancestral knowledge codified in varieties and agricultural practices.

Chapter 18, in which Butler addresses the SEC of animal welfare, also concerns an issue not generally thought of in economic terms. Animal welfare also raises issues of food safety and quality, and links between animal health and welfare and biodiversity, the environment and plant and human health and safety within that environment. Butler points out that the development of risk and quality assessments of animal welfare has yet to be accepted by the scientific community as objective, but such agreement is close. However, he also acknowledges that there is a second category of concerns regarding animal welfare, the social and ethical concerns, which he does not address in the chapter which will require different models of assessment. As with Thompson's chapter (Chap. 7) on ethics and equality, Butler notes that the form of animal welfare regulations used in a particular nation is shaped by the cultural, religious, political and/or social values of the jurisdiction and creates a broad array of approaches. He also points out that there are economic repercussions of imposing restrictions on trade with other countries based on their animal welfare regulations that would need to be considered. There are also domestic economic repercussions caused by animal welfare regulations such as increased food prices. These economic repercussions then tie back to SECs such as consumer choice, producer choice, and market access and trade.

Clearly, applying SECs to a domestic regulatory framework is a diverse and daunting challenge. Some of the challenges exist at the methodology stage as there is a lack of quantifiable methods available to undertake the assessment of the particular SEC, while at the other end of the scale, gathering the data required to make an assessment is time consuming and expensive. Challenges such as this would strain the ability of regulators in developed countries, let alone those in resource stressed developing countries. A constant theme across the chapters of Part II is that while valuable information can be provided to policy makers, it comes with a cost. This cost can be measured in the additional years it would take to gather the data to make a complete risk assessment or fiscal allocation of scarce resources to additional levels of regulatory requirement. The greatest cost, and arguably the most difficult to quantify, is the loss of the technology to a country if the developer decides that the regulatory environment contains too high a degree of uncertainty, thereby seeking other jurisdictions to commercialize the technology.

19.3 The Cost of Increasing GMO Regulations

In addition to the difficulty of agreeing on what SECs are, or should be, included in the decision-making process and the boundaries or content of those SECs, is the important issue of the future costs of enhanced regulatory oversight. The value of that "cluster of balloons" can be assessed today but, assuming they continue to exist into the future, re-assessment will be needed. Future values may also affect the value given to the balloons today. Another side to the cost projection problem is that what is done (or not) today will impact on the future value of the SECs. Perron-Welch (2012) argues that intergenerational equity requires the present generation not to introduce a technology that irreparably harms the environment or the socioeconomic situation left for future generations, but also requires the present generation not to deny future generations the possibility of benefiting from biotechnology and its socioeconomic gains. These potential costs and benefits should then be included in the assessment process.

However, the approach offered by Perron-Welch raises a common concern—a trade-off is required between the benefit and the adverse outcome. Take, for example, the labor reallocation impacts from GM crop adoption in South Africa as identified by Gouse in Chap. 13. Gouse's research has determined that female smallholders that adopt GM crops spend 10–12 fewer days hand weeding their fields, with

Table 19.1 The typology of errors

Decision	Product is safe	Product is unsafe
Accept as safe	Correct	Type 1 error
Reject as unsafe	Type 2 error	Correct

this time being reallocated to caring for their children or tending their garden plots. Obviously, this is a significant benefit from GM crop adoption. Conversely, Gouse identifies that the reduced need for hand weeding has resulted in a decline in the need to hire labor to undertake the necessary weeding. This loss in income opportunity may have significant impacts on the laborer's domestic situation. Therefore, the ethical dilemma faced by posing Perron-Welch's argument is who decides whether the benefits gained by a reduction in hand weeding by the female smallholder outweighs the loss in employment opportunity to the for-hire farm laborer? The greater and more practical challenge for technological innovations though is how impacts such as these can be quantified in any meaningful sense prior to the point of commercialization? While cost-benefit analysis can be done by any accountant, the results can be manipulated as needed to rationalize any foregone policy decision.

The point that is raised by Perron-Welch is symbolic of the second of two types of error in risk assessment systems—a so-called type 2 error. Ultimately and obviously, risk-assessment systems ought to be designed to make the right decisions; that is accepting safe products and rejecting unsafe products. As with any human system, there is potential for error, especially when a new class of products is being considered where there is no empirical evidence, such as in commercializing a GM crop in a country for the first time. While the system is and should be designed to avoid accepting something that is not safe (a so-called type 1 error), it also has to be mindful of the trap of making type 2 errors, that is, rejecting safe products and activities (Table 19.1). While we can tally up the cost of type 1 errors in lost lives or damaged ecosystems, we cannot convincingly estimate the cost of foregone opportunities and all of the attendant benefits that could flow from them. An important difficulty raised by including SECs in agricultural biotechnology regulation is that social amplification of risk significantly raises the potential of making a type 2 error, thereby diminishing the flow of new and innovative products and progress in a science-based economy.

Innovation adoption is addressed by a number of authors in this book. For example, Novy and Nagarajan, in Chap. 16, discuss systemic "relevance assessment" whereby a systems approach is used to determine the problem the biotechnology-based solution addresses, examines existing practices that could also meet that need, and then analyses the socioeconomic implications of implementing the various strategies. In this perspective they argue, needs (of at least producers) are focused on, and the specific GMO technical solution is placed in context relative to other technical solutions. This perhaps mirrors Wesseler and Smart's, in Chap. 6, suggested approach of comparison of the GMO under consideration with the second best alternative. Novy and Nagarajan also suggest that assessing a GMO in this way has the potential to deemphasize the most ideologically controversial aspects of GMOs, their legitimacy per se. Thus the approach taken to assessment of one SEC, in this case producer choice, impacts the "value" of another SEC, in this case ethical concerns regarding GMOs.

The further regulations move away from being science based, the greater the opportunity for subjectivity to become part of the decision-making process. In a science-based system, it is conceivable that two parts of the regulatory risk assessment might produce something at odds with each other. For example, suppose the novel protein in a new GM crop variety had a low level of toxicity for human consumption, but took longer to break down in the gut, suggesting an increased potential for allergenicity concerns. Under a science-based risk assessment, the two findings could be measured against other proteins to determine if the combined risk assessment falls within the category of safe. Hence, the outcome is likely to be more transparent and unlikely to be subject to manipulation. Conversely, with two SECs that render competing results, such as addressing an identified need, but in an ethically unacceptable way, results in an opaque, subjective decision-making process. Regulators making these decisions can be subject to numerous forms of manipulation and bias with no way to defend their decision, and, hence, face attack from both sides of the biotech debate, regardless of the decision.

The time and the place where an innovation is introduced are obviously important in shaping the possibilities and expectations of those considering the innovation. Any assessment of an SEC will therefore only be a snapshot of that particular framing. This fragmented reality causes many difficulties for innovation assessment. Smyth and McDonald (2012) have highlighted that innovation requires investment but that investment requires a commitment of resources at a point in time with the benefits flowing in the future. This dynamic nature means that it is not possible to do a simple comparison of benefits and costs, and that the time value of money must be considered when appraising an investment. They point out a further significant confounder—that research often creates a new process or new germplasm that provide a very important base onto which subsequent research is built. "Thus, innovations in the form of new varieties contribute to the stock of knowledge or germplasm, which continue to play a role long after the particular innovation has been supplanted by newer innovations" (Smyth and McDonald 2012, p. 7).

Smyth S, McDonald J also explain (2012, p. 6) that because investors want the cash flow to begin as soon as possible, and delays in adoption decreases the eventual return (because there is no cash return until the project is completed and that in the meantime, other varieties are developed in competition with the investor's variety), uncertainty regarding time delays in a regulatory approval process means, at best, an increase in uncertainty as to the amount of return or at worst decreases in the return, and therefore may cause a change in willingness to invest.

Uncertainty, as it pertains to a jurisdiction's regulatory system for GMOs, can be expected to impact the commercialization of innovative technologies. Technology developers have strong preferences for regulatory systems that deliver timely, repeatable decisions. In most cases, those systems that do not use this approach have an elevated degree of uncertainty. However, even science-based systems have some level of uncertainty—and subjectivity—associated with them, such as is witnessed by the challenge of GM sugar beets in the USA or mutagenic low-phytate barley in Canada. Both of these varieties experienced unanticipated regulatory decision-making processes, causing delays in the commercialization of the varieties. Regulatory systems that have incorporated SECs as part of the regulatory framework will enhance the uncertainty beyond a science-based system. SEC-based regulatory systems, as is established above, will return risk assessments that will be diverse and potentially at odds with each other. Deciding on how to deal with the inconsistencies, let alone who deals with them, is not anticipated to be an expeditious process. An unfortunate consequence of the movement away from science-based regulation is that uncertainty may become the largest risk and be used to rationalize the decision of technology developers to forego investment decisions in technologies or products within some jurisdictions.

19.4 Consistency with International Obligations

As most chapter authors in Part II note, the WTO Agreements, particularly the SPS Agreement (and the role of Codex within that), TBT Agreement, and TRIPS Agreement, are particularly relevant to any discussion about SEC decision-making regarding agricultural biotechnology under the CPB and CBD. The important issue of the possible clash between free trade as pursued under those Agreements and inclusion of non-economic factors in decision-making has been addressed in Part I and will not be repeated here. However, one point should be noted—the different conclusions reached by authors on what that potential overlap or clash means for norm setting in SEC assessment in agricultural biotechnology regulation. There is no agreed commonality.

Focusing instead on additional parts of the international governance matrix identified by the authors, Thompson points out, in Chap. 7, that any international agreement arguably articulates the international community's values and is therefore relevant to an assessment of SECs. The editors of this book note though that where such agreements have few members, the international consistency of that value is doubtful. Strauss, in Chap. 8, in regard to food security and safety also discusses the International Covenant on Economic, Social and Cultural Rights (CESCR), which she believes mandates members to ensure the right to the highest attainable standard of health care and other socioeconomic rights, including the right to food. Phillips, in Chap. 11, discusses the wide range of international agencies, treaties, and committees relevant to TK and the many proposals, guidelines, and commitments created by them, but notes that there is little operational or judicable structure. He also describes the international inconsistency to attitudes and approach to TK. Phillips details recent attempts for consistency such as that by the International Labour Organization, UN and Inter-American Draft Declaration on Rights of Indigenous Peoples, the UNDP/UNCTAD and European, Asian, and African Development Banks, and the Nagoya Protocol. He concludes though that ultimately TK protection and benefit sharing is left to domestic laws and policies. This makes it difficult to benchmark international norms for this SEC.

Lawson, in Chap. 12, on IP provides examples of the numerous IP agreements, both multiparty international agreements, but also bilateral and regional IP agreements. He points out that this means the IP landscape of each nation will need to be determined according to its particular commitments because of the lack of international consistency. Gouse, in Chap. 13, sees the WTO framework as the most relevant to international approaches to the SEC of labor impacts, but notes there are other international agreements that may be relevant to particular aspects of labor impacts, including the International Labor Organisation conventions such as Indigenous and Tribal People's Convention. Novy and Nagarajan, in Chap. 16, raise the International Treaty on Plant Genetic Resources for Food and Agriculture and Union for the Protection of New Varieties of Plants (UPOV) in addition to the usual WTO Agreements as international agreements impacting on the SEC of producer choice.

In terms of international compliances, the WTO is likely to predominantly be the benchmark. If the implementation of an SEC risk assessment can be proven to be compliant with the WTO, and more particularly the SPS Agreement, the more likely it will be compliant with international obligations. However, if an SEC risk assessment cannot document compatibility with the WTO, the implementing country will have implemented a barrier to trade and should expect to be included in a case of non-compliance to the WTO's Dispute Settlement Body. This observation is reinforced by Isaac et al. (2002), who posit that:

> [h]owever, as with most aspects of international law, the legal subtleties of a given situation are less significant than the political realities. Politically, WTO agreements (and the obligations therein) are generally taken more seriously by states than obligations incurred under other instruments perhaps due to the compulsory jurisdiction of its dispute settlement body.

The potential for long-lasting and far-reaching impacts to be found in non-compliance of WTO obligations is likely to be of greater concern to many governments contemplating the inclusion of SECs into regulatory frameworks than is the SEC that could trigger the state's non-compliance.

19.5 Governance Constraints

The CPB has led to institutional fragmentation for countries because of the very specific and technical nature of the Protocol's subject matter (Morgera and Tsioumani 2011). This inevitably means that countries will need to support independent sub-processes separate from the usual biosafety risk-assessment processes. Continuing this, some authors in Part II consider the issue of the most appropriate agency to undertake SEC assessments. Strauss, in Chap. 8, for example, focuses on the use of an international agency and suggests that possibly a new international body could be created to focus on global food supply. Gouse, in Chap. 13, suggests any study required for assessment be paid for by the applicant and then reviewed by the regulatory authority. However, Gruère, in Chap. 15, concludes that an applicant only approach is unlikely to be successful because of the involvement of multiple external chain actors, from farmers to traders, manufacturers and retailers which makes it difficult for applicants to provide credible and complete assessment of risk for all actors in a particular market chain. So, he suggests, at a minimum, a regulatory agency should be involved in basic risk determination. This means such an agency will need personnel competent to review basic market and trade data and relevant analyses.

The need to improve human capacity before assessments can be undertaken is also pointed out by Phillips, in Chap. 11. He notes that there is little evidence that international institutions are being vested with authority to administer policies and adjudicate disputes regarding TK, meaning nations or their indigenous people will need to devise their own approach and negotiate with those seeking to use it or who object to it. This individual tailoring of assessment for each application is also pointed to by Lawson, in Chap. 12. Lawson observes that assessment of the SEC of IP will probably need the relevant IP owners to be involved in the decision-making process. If IP is adversely affected by the decision-making process, mechanisms of review and compensation would also need to be created. Novy and Nagarajan, in Chap. 16, also note the need for substantial expenditures in time, human capital and financial resources and the particular need to build capacity to undertake multi-year studies "especially considering that many of the resources required to assess SECs are distinct from those required for biosafety assessment."

Thompson, in Chap. 7, along with Coe, in Chap. 17, observes that because issues arise regarding the inherent characteristics of the technology itself, its products and the consequences of the use of such products, the administrative body responsible for assessment will need to take a broad approach to the issues it addresses. It will also need to continually monitor these issues because of the rapid pace of change in societal values. Similarly, Butler, in Chap. 18, notes that risk assessment regarding animal welfare will need to balance the costs of regulation of animal welfare against benefits associated with that regulation, noting that the passage of animal welfare laws may offer individual and social benefits to consumers, but may also result in higher prices for all consumers and result in trade policy concerns for consumers and producers, domestically and internationally. Gouse, in Chap. 13, on the other hand suggests that assessment of labor impacts would only really be needed where an application is made for general release and not for import or contained use applications because these products do not enter the production system.

A second common message consistent across the Part II chapters is that many SECs require more and better data collection. Viju, in Chap. 14, makes this point regarding the assessment of market access and trade concerns where she notes that better data collection and more work on modeling are needed so that costs of segregation and identity preservation can be properly included. Similarly, Wesseler and Smart regarding assessment of environmental impacts, in Chap. 6, observe that countries will need to document both positive and negative environmental effects and have institutional processes in place to maintain the information. They also note that there needs to be more international investigation of the impact of yield increases on land use. Inclusion of pest and weed resistance issues in this SEC will require the monitoring of pests and their properties in the specific relevant sites. Similarly, Moses and Fischer, in Chap. 5 on consumer choice note that any decision that labeling of GM food be required should be based on clear information on the purpose of the labeling. Gathering such information cannot, they conclude, be done on the basis of polls conducted in an artificial "what if" scenario. Creation of this data will be made more difficult in some cases because, as Novy and Nagarajan, in Chap. 16, point out data may sometimes need to be obtained either from test studies or from other countries seeking to gather preliminary data, and this will be particularly difficult if no other country with an analogous agro-economical environment is testing the particular GMO.

19.6 Conclusions—Commonalities for All SEC Assessments

Innovation, with all its various attributes, impacts, and effects, is inevitable. While the science of innovation may in some theory of scientific progression be predicted or even measurable, the ultimate determinant is how society responds to a commercialized innovation. Innovation will be more readily embraced by societies which have a high degree of trust in their regulatory authorities, believe that the innovation will produce more benefits than costs, and accept or can address the reality that not everyone in their society will equally benefit from the technology. Dissenters will exist for any innovation and today's ability to communicate, particularly via social media, can lend credence to the appearance that the dissenting voice is the dominant voice. However, upon closer examination, there is not necessarily an accurate correlation between the size of the voice of dissent and the number of dissenters. For example, Canada and the USA have some very vocal opponents to biotechnology, yet in a market economy, their demands are muted by the demands of the silent majority.

A poignant question for societies facing rapid innovation to plant and animal agriculture then, is how to appropriately, economically, and efficiently manage the advancing state of change? Is it appropriate, economic and efficient to have a strong science-based regulatory system or is it better to install trade regulations or policies concerning SECS, which some nations may use as disguised trade barriers? Strong science-based regulatory systems require fiscal resources to invest in such a system, but also fiscal resources to invest in a system of monitoring and compliance. Having a strong regulatory system, but no ability to monitor compliance, will ultimately lead to a system that is rife with opportunism. A lesson in the perils of avoiding the regulation of innovation is evident in the problems that resulted in South America following the commercialization of GM soybeans in Argentina. While Argentina was the only South American country to approve GM soybeans in 1996, within 5 years their production had spread to Uruguay, Paraguay, Bolivia, and Brazil. This wildfire of unapproved adoption reached the point that by 2003–2004, it was speculated that 80% of Brazil's soybean production was from GM varieties; yet no GM varieties had been approved for use in that country. Ultimately, this reinforces the point that borders between nations are not walls, but, in fact, they are open spaces, frequently porous.

Some countries and cultures have successfully rejected recent technologies. The Canadian government rejected the use of the biotechnology developed drug rBST, intended to boost milk production in dairy cattle. While some sporadic use may be occurring by dairy farmers in Canada, this is minimized by a series of enforcement measures. Similarly, the Amish farmers in the US have successfully rejected the farming innovation of mechanization. Admittedly though, it is easier to discern when your neighbor is attempting to cheat the system by using a tractor than by using an indistinguishable variety of seed.

Drawing upon the insights and observations of the diverse chapters in Part II, we posit that the following five points should be important factors of consideration regarding any, and all, SEC risk assessments.

First, risk assessment is going to vary according to culture, geography, and national identities. Just as every person is their own unique individual, so too are countries when it comes to managing the affairs of state. Countries have the sovereign right to regulate and govern innovations as they see fit; however, it is important to remind all jurisdictions that all processes of regulation and governance regarding innovation have to be compliant with the existing international obligations of that nation. Countries that insist on adopting SECs as part of their national biosafety strategy will need to be able to rationalize the evidence used as a result of such inclusion.

Second, in some instances, it has been established that methods of assessing an SEC are at best, minimal and inefficient. The justification for including any SEC into a regulatory framework has to be clearly articulated and cannot be a process of establishing a "shopping list" of regulatory requirements. If the rationale for the regulation is not clear, the implementation, the data requirements, and decision-making will all be frustrated by the lack of clear articulation.

Third, the establishment of risk baselines will be a crucial step in the process. Without quantifiable evidence that documents the situation prior to the potential commercialization of a GM crop, it will render the data gathering process worthless as no one will be able to discern a beneficial impact from a negative impact.

Fourth, politics will be an important part of any SEC assessment process, even though it may not be recognized. Numerous chapters highlighted that SEC risk assessments are vulnerable to political involvement or interference. The inclusion of political interference will undoubtedly increase the level of uncertainty for technology developers and could impact investments in innovation in such jurisdictions.

Finally, while standardization of data brings many important benefits, some data collection techniques used in one jurisdiction may not be suitable for others. Science-based regulation has its points of contention, so this should be an expected part of the SEC risk assessment process as well. Without a dispute settlement mechanism in the CPB, the logistics of resolving standardization disagreements may overwhelm the CPB.

References

Isaac G, Phillipson M, Kerr WA (2002) International regulation of trade in the products of biotechnology. Estey Cent Law Econ Int Trade 2:5–158

Morgera E, Tsioumani E (2011) Yesterday, today, and tomorrow: looking afresh at the convention on biological diversity. University of Edinburgh School of Law, working paper series no 2011/21

Mugwagwa JT (2012) Multiple meanings, one objective: the case of biotechnology policy convergence in Africa. Asian Biotechnol Dev Rev 14(3):31–48

Paarlberg R (2008) Starved for science: how biotechnology is being kept out of Africa. Harvard University Press, Harvard, MA

Perron-Welch F (2012) Socioeconomics, biosafety, and sustainable development. Asian Biotechnol Dev Rev 14(3):49–69

Smyth S, McDonald J (2012) Investment, Regulation and Uncertainty: Managing New Plant Breeding Techniques. Paper for 16th ICABR Conference on The Political Economy of the Bioeconomy: Biotechnology and Biofuel, Ravello, Italy

Smyth SJ, Kerr WA, Phillips PWB (2013) Accelerating adoption of GM crops through a trade liability regime. Plant Biotechnol J 11:527–534

Chapter 20
Assessing the SEC Landscape and Moving Forward

Karinne Ludlow, Stuart J. Smyth and José Falck-Zepeda

20.1 Introduction

Present society faces a dilemma of how to feed a global population of 9 billion by 2050 and to meet other biomass needs in the near future. The severity of this dilemma is crystalized in the following quote from the Deputy-Director General of the Food and Agriculture Organization, "Agricultural production needs to increase by 70% worldwide, and by almost 100% in developing countries, in order to meet growing food demand" (Tutwiler 2011). A March 2012 Editorial in *Nature Biotechnology* observes that the rate of population growth exceeds that of commodity yield growth, and advocates that averting a global food crisis will

> …require the full deployment of every plant breeding technology currently available, including the generation of crops via transgenesis. But even more importantly, it will necessitate a reemphasis on innovation,… streamlining and harmonization of regulatory oversight, and an end to the political grandstanding that has characterized the agbiotech debate so far (Nature Biotechnology 2012, p. 197).

Recognizing the need for considerable investments in R&D and in resolving other institutional factors that limit agricultural innovation, therefore, becomes a call to ensure meeting food, feed, and other biomass needs in the long term. With an increasingly complex production, environment that considers climate change, changes in food demand, changes in population's income, and increasingly more environmental production challenges in the long term, the challenge becomes even more critical.

K. Ludlow (✉)
Monash University, Clayton, Australia
e-mail: karinne.ludlow@monash.edu

S. J. Smyth
University of Saskatchewan, Saskatoon, Canada
e-mail: stuart.smyth@usask.ca

J. Falck-Zepeda
International Food Policy Research Institute, Washington, DC, USA
e-mail: J.FALCK-ZEPEDA@cgiar.org

K. Ludlow et al. (eds.), *Socio-Economic Considerations in Biotechnology Regulation*, Natural Resource Management and Policy 37, DOI 10.1007/978-1-4614-9440-9_20,

Sustainable agricultural innovation will therefore be a crucial tool in addressing global food security. However, effective international commodity trade will also be of the utmost importance, as identified by the International Centre for Trade and Sustainable Development (Woolverton et al. 2010, p. 1), "Along with a strategy to increase agricultural productivity and diversity, improved global and regional trade must be part of the solution to provide adequate global nutrition." Trade effectiveness is ultimately dictated by domestic and international regulations. It is not surprising, therefore, that the *Nature Biotechnology* Editorial also emphasizes the streamlining and harmonization of the regulation of agbiotech as another vital component in addressing global food security.

Increasingly, agriculture, innovation, sustainable development, and climate change are intertwined. While increasing agricultural productivity will be an important component of the coming decades, so too will be key issues such as increased biomass production for biofuels, improved animal welfare, and addressing both the political interference in science-based regulation and the trans-Atlantic trade gap in products of biotechnology. Increasing food production cannot be a stand-alone issue. For increased food production to truly be effective and sustainable in a world of 9 billion, it needs to be integrated into these other issues and an interdisciplinary approach used to address these integrated issues to create the most efficient results possible.

Regulatory harmonization may be difficult though in the face of the growing dichotomy in understandings of what makes effective regulation in modern day societies. Increased levels of regulation are often viewed, particularly by individuals, as providing enhanced levels of benefits such as safety or security for individuals and societies. Conversely though, others perceive increasing levels of regulation as raising the cost of doing business in that society, a portion (if not all) of those increased costs being passed on to consumers and that effective regulation therefore requires less, rather than more, regulation.

Based on the global experience, a significant agreement exists that science-based regulatory process focused on accepted biosafety regulatory practices is a robust approach, but one that will need to be modified if it is to accommodate socio-economic considerations (SECs) in some situations. The rationale behind the science-based approach is that regulatory systems base their decisions on a risk assessment, while the market defines the value of any technology released to producers. Experiences from Brazil, and perhaps to a lesser degree Mexico, are valuable examples because they allow science-based risk assessment to proceed and if, after the product is deemed as safe or its risks "manageable," any socio-economic issue is identified a formal study is commissioned. Thus more attention is focused on the implementation process and issues and options related to inclusion of SECs.

The crux of the issue for jurisdictions considering the inclusion of SEC assessment as part of their regulatory processes is whether, given regulatory outcomes are always a trade-off between innovation and risk-acceptance (Kuzma and Tanji 2010), the results obtained from SEC assessment justify the increased costs and difficulties placed on the innovation process. The potential impact of regulatory cost and uncertainty on public sector investment in innovation will be of particular relevance to many developing countries already investing in developing products of their own. The crux of the issue is ultimately one of whether or not the increased cost of regulation actually results in benefits, or simply more expensive products. Simply put, what are the trade-offs between safety, efficacy, and cost?

An accurate assessment of the necessary regulatory trade-offs is dependent upon accessibility of all pertinent information. As regulation moves away from science-based toward SEC-based regulation and the accompanying increasing opportunity for political interference, it will be crucial to ensure that decision-making regarding GMOs is done in the most transparent way possible. Clearly, governments are going to have agendas, but if agricultural biotechnology innovation is to be fostered, then a clear understanding of the reasons why SECs are to be assessed, when and how that assessment is to be made, and the method by which the results will be integrated with those of the traditional science-based assessment process is needed.

Furthermore, any potential course of action will need to consider compliance with a jurisdiction's international obligations. SECs should be used to support a nationally or culturally distinct approach to the regulation of agbiotech, provided they are not disguised barriers to innovation and trade. Although Article 26.1 of the CPB allows countries to include SECs in their decision-making, the scope of such inclusion in the CPB is possibly narrowed—but not limited—to the value of biological diversity to indigenous and local communities. There are several questions about definitions of the terms used in the Article (such as which biodiversity is important: agricultural biodiversity or overall biodiversity? where to measure such impact: in the local community or where adoption occurs?) which are likely to require more negotiations by the CPB parties. Clarification of the theoretical and practical relationship of the CPB and the WTO agreements is also needed. These issues are beyond the scope of this book, but a more thorough understanding of the implications of SEC assessment in agricultural biotechnology regulation, as undertaken in this book, will inform any such clarification.

In summary, inclusion of SECs into biosafety decision-making can have positive and negative impacts. The decision to include SECs into decision-making is a policy/political decision. Policy-makers will need to examine the different issues related to options and alternatives for potential inclusion of SECs in decision-making. If the decision is or has been taken to include SECs, the need arises to define appropriate standards and choices to ensure a feasible and functional biosafety system.

20.2 Objectives of an SEC Assessment

Policy convergence—that is, growth in similarity of policies over time—is generally seen as a positive development from economic, regulatory, technological, and environmental viewpoints (Mugwagwa 2012). For example, cross-national similarity allows countries to share resources, draw lessons from each other, and shorten technology and product approval processes (Mugwagwa 2012). International regulatory consistency is similarly seen by some as a worthwhile aim, particularly because of the resulting enhancement of international trade.

It is to be hoped that convergence would occur around best practice. However, developing best practice, methods, and policy guidelines for SEC assessment implementation and inclusion in decision-making is difficult without a clear understanding of the objectives of the inclusion. Even if such objectives are consistent though, nations can be expected to consider different SECs and orders of priorities as being relevant to their meeting those objectives. Further, although certain fundamental values such as human health and safety will be prioritized by all, even for those values some nations and their citizens may be more willing to take certain risks than others, influencing the objectives they are seeking to address. Achieving regulatory harmonization will therefore be challenging.

For all nations though, scientific and technological innovation is crucial. Innovation crystallizes possibility—what before was only "could be" now becomes real. While agricultural biotechnology is now real, this reality has brought with it its own possibilities and expectations that in turn are relevant to policy objectives. As Borup et al. (2006, p. 286) observe, "expectations link technical and social issues..". Expectations are also of value, and Borup et al. (2006) even suggest that there can be no differentiation between our expectations of things such as biotechnology and what those things in fact are. Whether it is accepted that expectations are part of the value of technology or not, "expectations are culturally managed" (Borup et al. 2006, p. 295) and "[b]iotechnology, particularly plant and animal biotechnology, is perceived differently by different cultures because of different assessments of its prospects and its ethics" (Perron-Welch 2012, p. 49). These different individual cultural understandings and expectations need to be recognized and understood (Mugwagwa 2012).

Borup et al. (2006) argue that expectations are a key element in understanding science and technology change, mainly because they frequently bridge or mediate across different boundaries and otherwise distinct (although overlapping) dimensions and levels. "Expectations are foundational in the coordination of different actor communities and groups (horizontal co-ordination) and also mediate between different scales or levels of organization (micro-, meso-, and macro-vertical co-ordination). They also change over time in response and adaptation to new conditions or emergent problems (temporal coordination)" (p. 286).

Care will be needed to ensure that the inclusion of SECs does not lead to an uncertain and therefore unworkable biosafety regulatory system. An unworkable system is one where the rules and standards for decision-making are not known by all stakeholders. The implication is that the regulatory process needs to define what will be required, how studies will be evaluated, what is the standard or benchmark of acceptability, how will a decision be made and how will the socio-economic study relate individual SECs to each other and to traditional environmental and food safety assessments, among other issues. In fact, one important aspect that a robust system needs to develop is its resilience and ability to minimize the possibility that the inclusion of SECs can be manipulated to support a specific position by an interest or pressure group. Having an unworkable system, one that cannot render a decision in a timely, cost-efficient manner and whose decision is robust, protective and accepted by society, is not a desirable outcome and is a questionable use of scarce societal resources.

Clarification of policy regarding, and the objectives of, SEC assessment will require the exploration of further questions beyond the scope of this book. These include investigation of why addressing the SECs of only one type of agricultural innovation is considered worthwhile. This will assist with thinking around the benchmarks used in any SEC assessment. Agriculture, conventional or otherwise, is a dynamic process continually adopting new innovations of which GMOs are only one group, and these can also be expected to have repercussions on SECs. Any benchmark or baseline used to assess GMOs against will therefore also need to take these into account.

20.3 Determining the Decision-Maker: Process Is Crucial

There is a need to decide who will be the assessor/decision-maker and where the personnel will come from for SEC assessments. In regard to the latter, Mugwagwa (2012) notes that it is difficult for policy implementers to move to become policy developers without the necessary experience. Further, predictability and consistency in decision-making may be difficult, given Mugwagwa's observations that mobility of policy actors, whereby they move from one policy arena to another to pursue new employment opportunities or to fill capacity gaps, means continual fluctuation of understandings of an issue among groups of actors and also blurs distinctions between different categorizations of understandings.

For nations that are struggling with resource deficiencies and that do not have the luxury of a career civil service, or certainly one that has considerably less job mobility than is the case in other nations predictability and consistency in decision-making may be less of a problem. However, Paarlberg (forthcoming) identifies one of the significant challenges for regulators in developing countries, particularly African countries. Policy-makers are often part of the social elite in developing nations, and as such, have been educated in Europe, so their children tend to be European educated as well. This results in a European attitude of sorts toward the regulation of GMO products. Political bias already inherent to a regulatory system is something that will undoubtedly impact the approval process for GMOs within an SEC-based regulatory framework.

The framing of the decision will also affect the identity of the decision-maker. According to Pavone et al. (2011), framing a decision as one on a technical issue has the result of changing what perhaps could be a political question addressed by policy-makers instead into a scientific one, given to an expert committee and diverting responsibility to techno-scientific networks. More importantly, they also note that assessment procedures themselves have assumptions that are value laden such as the significance given to the distribution of benefits and harms, what constitutes a benefit worth taking a risk for and what level of risk is acceptable.

Further, they also note that social and political values are embedded in the very technology because of the processes that led to the development of the technology (public and private investment, involvement of universities, companies and start-ups, promises made, and mobilization of political and social associations) and of the changes it makes to the innovation regime, research agenda priorities, and allocation of public resources, including perception of the problem for which the technology was developed.

One of the major impediments in the regulatory process for GMOs is the fractionated regulatory framework both within and across jurisdictions. Lack of harmonization of regulation significantly impacts the cost of getting a GM crop variety through the system and care is needed to avoid the introduction of SEC assessment adding further fractionation. This problem can be highlighted in a number of places. First, there is a lack of harmony regarding where the particular authority for the regulation of GMOs and/or biosafety is bureaucratically housed within federal-type governments. For example, a 2011 international workshop on socio-economic impacts of GM crops (Lusser et al. 2012) had regulatory officials from 17 EU Member States responsible for the regulation of GM crops. The regulatory affiliation of the 17 representatives provides a sense of the regulatory multiplicity for agbiotech regulation within the EU. The representatives predominantly came from ministries of agriculture, environment, or rural development. This multiministerial regulatory authority compounds the challenges of the regulatory process.

Second, the fractionated regulatory framework in some jurisdictions is seen in the multiple Ministries or committees having oversight authority or regulatory mandates regarding the approval of GMOs. With the approval power within each EU Member State residing with disparate departments when compared with other States, the regulatory focus and expertise shifts from one nation to the next. So, for example, agricultural impacts might be the focus of regulatory scrutiny in one nation, while environmental impacts is key in another and farmers impacts in yet another. The delays and uncertainty caused by the diversified approach to the regulation of GM crops and products in the EU has resulted in an onerous regulatory system for its own region, but more importantly in a global sense, one that as identified earlier by Paarlberg (forthcoming) also restricts the commercialization of GM crop technologies in developing countries, most notably Africa.

The EU's move away from science-based regulation is also triggering the "politicization" of risk, in that the various member countries receive scientific risk assessment data provided by European Food Safety Authority (EFSA), but vest the decision-making authority for GM variety approval with the EC Council of Ministers (political representatives). Decoupling the risk assessment process from the variety approval process within the EU has resulted in grid-lock, damaged harmonization and created a regulatory environment that is incapable of timely decision-making, and is fraught with uncertainly. As commented in the EC (2011) report on socio-economic impacts from GMO production, diversity in the scale and scope of SECs is to be expected. This creates uncertainty for technology developers and as has been documented by Smyth and McDonald (2012), regulatory uncertainty is directly connected with a decline in project investments.

20.4 Knowledge: Public Input and Transparency

There are social variables in the levels of trust attached to decision-makers. Brown and Michael have found that social patterning of expectations across communities often arise from asymmetries in access to information on which expectations are based,[1] with Borup et al. (2006, p. 292) adding that "Elevated levels of expectations and confidence also have the effect of inflaming concerns about risk in different communities based on differing values, knowledge or institutional and organizational forms." Knowledge, expectations, and public input will therefore be influential on any SEC assessment.

One difficult aspect of decision-making as it pertains to SECs is the potential for subjectivity. Subjectivity in a decision-making process detracts from transparency, which means any particular decision cannot usefully serve as a precedent for later applications and creates the potential for inconsistent decisions as different decision-makers place different emphasis on certain SECs in differing orders of magnitude. Without a distinctive method to judge the results of the SEC assessment of a particular GMO, it is difficult to develop certainty and trust. Transparency in the assessment and decision-making processes will also facilitate public input that can be fed back into the decision-making process, enhancing the process's reflection of society's choices.

Much of the current debate regarding the approval of GMOs continues to relate to safety and that in turn influences attitudes to SEC assessment. 2014 represents the 20th year of commercial GM crop production, yet discussions abound about safety. The most poignant observation about the safety of GM crops is perhaps best captured in the EC press release announcing the 2010 assessment of GM crop research funding in Europe, when it is stated that there is "no scientific evidence associating GMOs with higher risks for the environment or for food and feed safety than conventional plants and organisms" (EC 2010). Nevertheless, the question of how safe is safe is a dominant part of the push for greater regulation of GMOs through the application of SEC-based regulations. In Canada and the USA, science-based regulation has done an excellent job of ensuring that the GM products that have entered the market have been safe for humans and the environment. Certainly, there are issues that require monitoring, such as the number of weeds that are showing signs of glyphosate resistance, but the environmental benefits of reduced tillage and herbicide applications that have resulted from the commercialization of GM crops should not be overlooked (National Research Council 2010; Smyth et al. 2011a, b). Clarification of the objectives of SEC assessment will need to distinguish between questions already addressed in science-based regulation and those that are not, as well as those questions that justify the additional regulatory response and expense and those that do not.

---

[1] Mads Borup et al (2006) 'The sociology of expectations in science and technology' 18:3–4, 292 citing Brown and Michael.

20.5 Undertaking an SEC Assessment

Assessment can be undertaken as an integral part of planning, decision-making, and monitoring or as a bolt-on extra (Barrow 2002). Ramatha and Andrew (2012, p. 23–24) note that there are at least four different phases of biotech decision-making where SECs can be taken into account: during development of a domestic biosafety regulatory regime; during risk assessment for particular modified organisms; after risk assessment , e.g., during risk management, when decisions are made as to whether identified risks are acceptable; and appeal, review or renewal of a permit. Perron-Welch (2012, p. 56) identifies the following practical steps that countries must take if SECs are to be taken into account in decision-making:

- Policies mandating integration of SECs into decision-making processes
- Clear definition of SECs and explicit criteria to determine when and where SEC assessments are required
- Identification of the stages at which SEC assessments should take place
- Efficient and cost-effective regulatory processes
- Public participation mechanisms to ensure credible assessments and more widely accepted decisions

It may also be necessary for SEC assessments to be revised during the lifetime and beyond of the particular GMO given that the impacts and utility of technologies change over time (Nuffield Council on Bioethics 2012).

In terms of implementation approaches, it is possible to observe a wide range of potential approaches in dealing with SECs. In a thorough examination of this subject, this book provides examples of countries that do not have a mandate for the inclusion of SECs within the regulatory dossier as is the case in the USA and Canada. Other countries, such as Argentina, mandate the inclusion of SECs but narrow the scope to a very specific impact area. In other countries with mandatory requirements for SEC inclusion, legislation indicates what issues to address in an SEC assessment, but not how the outcome of the assessment will be judged. In particular, these countries' legislation does not provide any indication of methods or decision-making rules for balancing the science-based risk assessment outputs and the SEC assessments.

Those countries that impose a mandate on the inclusion of SEC requirements without sufficient guidance in terms of methods and decision-making rules may introduce unnecessary regulatory delays and uncertainty to biosafety and GMO approval processes. In some situations, leaving the inclusion mandate broad and generic leaves the door open for developers to submit SEC impact assessment studies they may deem sufficient for enabling a biosafety regulatory process. However, unclear procedures for SEC assessments may discourage investments in new technologies by the private sector because it introduces regulatory uncertainty and unpredictability (Smyth and Falck-Zepeda 2013). The need to clearly define the regulatory process is especially critical in developing countries where there may be a greater need for addressing pressing production and productivity needs not easily resolved through conventional means (Falck-Zepeda et al. 2012). In fact, lack of clarity may even be more discriminatory to the public sector in developing countries because they may not be capable of financing additional costs or addressing uncertainty, especially when dealing with international public goods where the rates of private returns are low, although the social returns may be high. In essence, increasing regulatory burdens will make research and development processes in crops, animals, and traits of interest to developing countries harder to invest in, in the long run.

In 2011, the European Commission (EC) undertook a Member State assessment of socio-economic impacts from GMO production (EC 2011). All but two Member States participated in the assessment and two key issues were identified from the report. First, the meaning and scope of socio-economics varied considerably across the Member States and the stakeholders involved in the assessment at the Member State level. Second, the "limited fact-based background" (EC 2011, p. 4) resulted in polarized opinions based on the insufficient evidence. Clearly, not only is baseline data lacking, but so too, are methods of assessing the current impacts.

20.6 What Lies Ahead?

The inclusion of SEC assessments, especially in those systems where there is very little clarity in terms of methods and decision-making rules, can introduce the potential of increasing regulatory lags due to delays, and certainly will increase the cost of agricultural biotechnology regulation (Bayer et al. 2010). In both cases, there are social costs attached that may negatively impact the deployment of agricultural biotechnologies of interest to the developing and developed countries. Irrespective of how countries deal with having more guidance in terms of assessment methods, they will also need to have clarity in terms of objectives and decision-making rules that will guide inclusion of SECs.

SECs are just one aspect of biosafety management. There are other tools which policy and decision-makers could use to make biosafety systems more efficient and protective, including regional approaches to regulation, building up flexible regulatory systems, matching regulatory intensity to the particular organism or trait's objective risk and others. The latter are important issues in terms of biosafety regulatory design and management to which economics and other disciplines can contribute for ensuring the deployment of safe and effective technologies to resource poor farmers and farmers in a more industrialized setting.

An issue that should not be understressed in terms of importance is the scope of SEC assessments. Whether the scope is strictly on economic issues, or whether it is expanded to include broader considerations including ethical, religious, cultural or individual/group expressions of opposition, it will have an impact on the methods and approaches needed to undertake such regulatory decision. Science-based assessments pursuing a narrow scope have a robust history in terms of methods and approaches for conducting such assessments. If the scope is expanded to broader considerations, there are several uncertainties as to the feasibility, robustness, and reliability of such estimations, and baseline measure, especially in an *ex ante* (pre-approval) regulatory process. SEC assessments in these situations may not be even feasible. In terms of methods, the need exists to greatly expand the scope of approaches followed by practitioners to include those that deal with risk and uncertainty, irreversibility, and flexibility. Moving forward, SEC assessment in agricultural biotechnology regulation will require crossing a largely unexplored landscape.

References

Barrow CJ (2002) Evaluating the social impacts of environmental change and the environmental impacts of social change: an introductory review of social impact assessment. Int J Environ Stud 59(2):185–95

Bayer JC, Norton GW, Falck-Zepeda JB (2010) Cost of compliance with biotechnology regulation in the Philippines: implications for developing countries. AgBioForum 13(1):53–62. http://www.agbioforum.org/v13n1/v13n1a04-norton.htm

Borup M et al (2006) The sociology of expectations in science and technology. Tech Anal Strateg Manag 18(3–4):285–298

European Commission (2011) Report from the commission to the European Parliament and the council: on socio-economic implications of GMO cultivation on the basis of Member States contributions, as requested by the conclusions of the Environment Council of December 2008. http://ec.europa.eu/food/plant/gmo/reports_studies/docs/socio_economic_report_gmo_en.pdf. Accessed 20 June 2013

European Commission (2010) Commission publishes compendium of results of EU-funded research on genetically modified crops. Press release. http://europa.eu/rapid/press-release_IP-10–1688_en.htm. Accessed 20 June 2013

Falck-Zepeda JB, Yorobe Jr J, Husin BA et al (2012) Estimates and implications of the costs of compliance with biosafety regulations in developing countries: the case of the Philippines and Indonesia. GM Crops Food: Biotechnol Agric Food Chain 3(1):52–59

Kuzma J, Tanji T (2010) Unpacking synthetic biology: identification of oversight policy problems and options. Regul Gov 4:92–112

Lusser M, Raney T, Pascal T et al (2012) International workshop on socio-economic impacts of genetically modified crops co-organised by JRC-IPTS and FAO. Workshop proceedings. JRC scientific and policy reports

Mugwagwa JT (2012) Multiple meanings, one objective: the case of biotechnology policy convergence in Africa. Asian Biotechnol Dev Rev 14(3):31–48

National Research Council (2010) The impact of genetically engineered crops on farm sustainability in the United States. National Academy Press, Washington DC

Nature Biotechnology (2012) Agnostic about agriculture. Nat Biotechnol 30(3):197

Nuffield Council on Bioethics (2012) Emerging biotechnologies: technology, choice and the public good. Nuffield Council, London

Paarlberg R (Forthcoming) African non-adopters. Chapter 12. In: Smyth SJ, Phillips PWB, Castle D (eds) Handbook on agriculture, biotechnology and development. Edward Elgar, Cheltenham, UK

Pavone V et al (2011) From risk assessment to in-context trajectory evaluation. Environ Sci Europe 23(3). doi:10.1186/2190–4715-23–3

Perron-Welch F (2012) Socioeconomics, biosafety, and sustainable development. Asian Biotechnol Dev Rev 14(3):49–69

Ramatha L, Andrew J (2012) Socio-economic aspects in decision-making in the context of the biosafety protocol: Malaysia's experience and case studies. Asian Biotechnol Dev Rev 14(3):19–30

Smyth SJ, McDonald J (2012) Investment, regulation and uncertainty: managing new plant breeding techniques. Paper presented at 16th International ICABR conference on the political economy of the bioeconomy: Biotechnology and Biofuel, Italy, 24–27 June 2012
Smyth SJ, Falck-Zepeda J (2013) Socio-economic considerations and international trade agreements. Estey Cent J Int Law Trade Policy 14(1):18–38
Smyth SJ, Gusta M, Belcher K et al (2011a) Changes in herbicide use following the adoption of HR Canola in Western Canada. Weed Technol 25(3):492–500
Smyth SJ, Gusta M, Belcher K et al (2011b) Environmental impacts from herbicide tolerant Canola production in Western Canada. Agric Syst 104:403–410
Tutwiler MA (2011) Abstract for keynote address to the GMCC-11 conference. Vancouver, Canada, October 26–8
Woolverton AE, Regmi A, Tutwiler MA (2010) The Political economy of trade and food security. International Centre for Trade and Sustainable Development, Geneva, Switzerland

About the Authors

Leslie Butler is a Marketing and Policy Economist in the Department of Agricultural & Resource Economics at UC-Davis. Butler is a native of New Zealand where he earned his BS degree at Lincoln University. He also holds an MS degree in Agricultural Economics from the University of Reading in England, and a PhD in Agricultural Economics from Michigan State University. Apart from his research in dairy marketing and policy, Butler also has interests in the social and economic impacts of biotechnology and technological change, environmental policy and intellectual property rights issues. He regularly testifies at national and state hearings regarding dairy policy and has published numerous articles on dairy production and economic marketing and policy. Recently Butler has been partaking in research associated with water in California, and particularly in issues associated with the Sacramento-San Joaquin Delta.

José Falck-Zepeda is a senior research fellow in the Environment and Production Technology Division (EPTD) of the International Food Policy Research Institute (IFPRI) and leader of the polity team in IFPRI's Program for Biosafety Systems (PBS). He has a PhD and MSc in Agricultural Economics from Auburn University Alabama, BSc in Animal Science from Texas A&M University, and Agronomist degree from Pan-American Agricultural School (Zamorano University). His relevant publications include the following co-authored articles: 'Estimates and Implications of the Costs of Compliance with Biosafety Regulations in Developing Countries: The Case of the Philippines and Indonesia' (GM Crops and Food: Biotechnology and Agriculture in the Food Chain 2011) and 'Socio-Economic Considerations in Biosafety and Biotechnology Decision Making: The Cartagena Protocol and National Biosafety Frameworks' (Review of Policy Research 2011). Falck-Zepeda is the author of 'Socio-Economic Considerations, Article 26.1 of the Cartagena Protocol on Biosafety: What Are the Issues and What Is at Stake?' (AgBioForum 2009) and is co-editor of the 2011 special issue of AgBioForum 'Farmers and Researchers Discovering Biotech Crops: Experiences Measuring Economic Impacts among New Adopters'. He has made presentations in high level global policy forums and in national level events.

K. Ludlow et al. (eds.), *Socio-Economic Considerations in Biotechnology Regulation,* Natural Resource Management and Policy 37, DOI 10.1007/978-1-4614-9440-9,

Marnus Gouse is currently appointed as Postdoctoral Research Fellow in the Department of Agricultural Economics, Extension and Rural Development at the University of Pretoria, South Africa. He has been studying the farm-level and socio-economic impacts of genetically modified crops in South Africa since 2000, with special focus on smallholder and subsistence farmers. His research over the last 12 years has been supported by the Rockefeller Foundation, ESRC/DFID and the Bill and Melinda Gates Foundation and has been published in the form of peer-reviewed journal articles, book chapters and working papers. Marnus is a part-time consultant to the Bureau for Food and Agricultural Policy (BFAP) and has worked with a number of Southern Africa based NGOs, mainly on informing transgenic crop related biosafety and socio-economic policy and decision making through farm-level research.

Guillaume Gruère was Senior Research Fellow at the International Food Policy Research Institute (IFPRI) in Washington, DC. He has published over 25 peer-reviewed publications on market and trade-related regulations of genetically modified (GM) products, including 2 co-authored articles and chapter contributions to 2 edited books that have received awards from the Agricultural and Applied Economics Association (AAEA). He has written papers on trade of GM products for the International Agricultural Trade Research Consortium (IATRC) and the International Agricultural Trade Policy Council (IPC) and contributed trade sections to policy reports for the World Bank and the African Development Bank. His work in this area was presented to officials from over 15 Asian and African countries as well as in high level policy meetings of the Asia Pacific Economic Cooperation (APEC). Dr Gruère has two MS degrees in agronomic engineering and environmental economics from AgroParisTech (France) and a PhD in agricultural and resource economics from the University of California Davis (USA).

Charles Lawson is Associate Professor at Griffith University, Australia and a member of the Australian Centre for Intellectual Property in Agriculture. Charles researches IP; competition; trade; access and benefit-sharing; and public administration. He has a particular interest in the TRIPS agreement and has recently completed, with Justin Malbon and Mark Davison, a commentary assessing the text and its meanings. He has also published on GMOs and risk assessments in Australia.

Karinne Ludlow's research field concerns the legal responses to, and policy concerning, innovative science particularly biotechnology and nanotechnology. She has a BSc specializing in microbiology and a LLB (Hons) from Monash University, Australia. Ludlow was in private legal practice before joining the Faculty of Law, Monash University. In 2005 she completed a PhD at that University on the legal challenges to the commercialization of GMOs. She has published many refereed papers (including in *Jurimetrics, Trends in Biotechnology* and *European Intellectual Property Review*) and book chapters, an edited book and has been an invited speaker at international and national conferences on the regulation of, and policy concerning, innovative science. Her work has informed regulatory and legal responses to new technologies. She was a senior author (with Bowman D and Hodge G) of an extensive 2007 report for the Australian government on the effect of nanotechnologies on all major Australian regulatory schemes (*A Review of Possible Impacts of Nanotechnology on Australia's Regulatory Framework*). The project put Australia amongst the first governments in the world to have initiated such a comprehensive review of their regulatory frameworks in relation to nanotechnology. Ludlow's work in biotechnology has been referred to by the Australian High Court, the Australian Law Reform Commission and other reports for the Australian Government. Ludlow's other published research incorporates torts, intellectual property and environmental law. She has co-authored two torts law texts.

Vivian Moses was educated first in Cambridge as a biochemist and then at University College London in microbiology. Moses held a couple of junior posts in London before spending 12 years as a postdoctoral research fellow and later as a research director at the University of California in Berkeley. Returning to the UK in 1971, he was appointed to the Chair of Microbiology at Queen Mary College where he remained until formally retiring in 1993. During that period he became heavily involved with microbial enhancement of oil recovery and co-founded Archæus Ltd, a start-up dedicated to the exploitation of that aspect of biotechnology. The 1990s marked the start of a new career for him, directed mainly to researching the spread of information and helping people at large to understand the promise and problems of biotechnology. With visiting professorships at King's College and University College, both in London, he was also the Director of The Centre for Genetic Anthropology at UCL for 10 years. Since the beginning of 2000, he has been the Chairman of the CropGen network (www.cropgen.org), an independent information service which makes the case for crop biotechnology (GM Crops), mainly in the UK. With that capacity he has been working as a coordinator of three EU contracts to explore European consumer attitudes and responses towards GM-products, especially foods actually offered for sale in the respondents' local stores.

With 170 or so research papers, reviews and books to his name, together with Ronald Cape, he co-authored and edited a major textbook, Biotechnology—the science and the business. He is presently the co-editor of GM Crops & Food.

Ari Novy has a PhD in plant biology from Rutgers University in New Jersey, USA. Primarily trained as a population geneticist and evolutionary biologist, he conducts research on evolution in invasive plants, horticultural improvement, beekeeping management, habitat restoration, and endangered species. He also researches several socioeconomic aspects of agricultural biotechnology adoption. Novy is currently the Public Programs Manager at the United States Botanic Garden in Washington, D.C., USA where he spends the majority of his time working to engage the public in every kind of botanical education he can manage.

Peter WB Phillips is Professor of Public Policy in the Johnson-Shoyama Graduate School of Public Policy at the University of Saskatchewan, in Saskatoon, Canada. He earned his PhD in International Political Economy at the London School of Economics and practiced for 13 years as a professional economist and senior policy advisor in Canadian industry and government. At the University of Saskatchewan he has held the Van Vliet Research Chair, created and held an NSERC-SSHRC Chair in Managing Technological Change, was a founding member and director of the virtual College of Biotechnology and was a founding director of the Graduate School of Public Policy. He has had visiting appointments at the University of Edinburgh, LSE, the OECD, the European University Institute and the University of Western Australia, is associate editor of *AgBioForum*, a leading on-line journal, was a member of the NAFTA Chapter 13 expert panel on GM maize in Mexico and was a founding member of the Canadian Biotechnology Advisory Committee. He has been a member of the Board of Directors of the Canadian Agri-food Policy Institute, the Estey Centre for the Study of Trade, Law and the Economy, and Ag West Bio Inc, which operates a biotech venture fund. His current research focuses on governing transformative innovation, including regulation and policy, innovation systems, intellectual property management, trade policy and decision systems. He is a co-lead and principal investigator of a US$ 5.4 million Genome Canada project entitled 'Value Addition through Genomics and GE3LS (VALGEN, 2009–2014)' and has been an applicant and investigator on more than 15 peer-reviewed grants worth more than US$ 150 million. He has been the author of 44 articles and 56 book chapters and author or editor of 12 books, including, *Accessing and Sharing the Benefits of the Genomics Revolution* (with C Onwuekwe), *Innovation in Agri-Food Clusters: Theory and Case Studies* (with J Karwandy, T Webb and C Ryan) and *Governing Transformative Technological Innovation: Who's in Charge?*

Melinda Smale is a Professor in the Department of Agricultural, Food and Resource Economics, Michigan State University. She joined MSU in 2011 after working for a number of years with developing country researchers through the Consultative Group on International Agricultural Research (CGIAR). From 2002, as a Senior Research Fellow at the International Food Policy Research Institute (IFPRI) and Economist at Bioversity International, she led a global research program on the impacts of biotech crops, agricultural biodiversity, local seed markets, and underutilized crops. From 1989 to 2000, while living in Malawi and later in Mexico, she analysed the adoption and impacts of improved wheat and maize seed as an economist for the International Maize and Wheat Improvement Center (CIMMYT). During the 1980s, Melinda worked in Pakistan, Somalia, Mauritania and Niger on shorter-term assignments for CIMMYT, Chemonics International, Volunteers in Technical Assistance (VITA), and USAID. She is an Honorary Fellow with Bioversity International, serving on the Advisory Committee of the Collaborative Crops Research Program of the McKnight Foundation, and on the editorial committees of two international journals. She has won awards for outstanding journal articles from the Agricultural and Applied Economics Association and the Crop Science Society of America, and has published 70 articles in peer-reviewed journals, 5 edited books, and 25 book chapters. She received her PhD in Agricultural Economics from the University of Maryland. She has advanced proficiency in French, basic Spanish, and has studied Chichewa, Swahili, Urdu, Hassaniya and Somali briefly during field work.

Richard D Smart graduated with an MSc Agric degree from the Department of Grassland Science at the University of KwaZulu-Natal in South Africa. His field of research integrated the disciplines of pasture science and agricultural economics by doing an economic comparison of overwintering grazing strategies for Merino ewes and their lambs on selected planted pastures in East Griqualand, South Africa. Thereafter he worked for 2 years in the field of environmental management (concentrating on the rehabilitation of opencast mined areas) for Amcoal Ltd—a coal mining company in South Africa. He then gained 14 years' experience in practical agriculture as a sheep, cattle and crop farmer in South Africa. In October 2011, Richard joined the Chair Group Agricultural and Food Economics, Center of Life and Food Sciences Weihenstephan, Technische Universität München in Germany. His research involves investigating economic issues of biotechnology focussing on the regulation of genetically modified plant crops.

Stuart J. Smyth is a Research Scientist with the Department of Bioresource Policy, Business and Economics at the University of Saskatchewan. He received his PhD in 2005 from the University of Saskatchewan and the focus of his research predominantly addresses the interaction between innovation and regulation within the context of agriculture. Smyth is part of a group of academics that received US$ 5.4 million in funding in 2009 from Genome Canada to examine the genomic, economic, environmental, ethical, legal and social (GE^3LS) issues pertaining to bioproducts and biofuels. He is also the GE^3LS Principal Investigator for two other Genome Canada projects, one on the flax genome and the other on cellulosic biofuels. Smyth is a member of the Board of Directors of the International Consortium on Applied Bioeconomy Research (ICABR) that organizes annual conferences on the bioeconomy. Over the past 5 years, he has published 6 books, 4 book chapters and 26 journal articles.

Debra M Strauss is Associate Professor of Business Law at Fairfield University Charles F Dolan School of Business, where she teaches International Law, the Legal Environment of Business, and Business Law and Ethics. A Phi Beta Kappa graduate of Cornell University and Yale Law School, Debra Strauss served as a law clerk for the Honorable Charles L Brieant, former Chief Judge for the US District Court, Southern District of New York. In addition, she practiced law at Rogers & Wells in New York (now Clifford Chance LLP), specializing in commercial litigation. She returned to Yale Law School as Director of Judicial Clerkships and went on to direct and author, the National Judicial Clerkship Study on behalf of the American Bar Association. Professor Strauss has written numerous articles on the courts, judicial clerkships, and diversity issues, as well as the definitive book in the field—*Behind the Bench: The Guide to Judicial Clerkships*. In her most recent publications, Professor Strauss has been noted for presenting models for civil lawsuits against terrorist groups and the labeling and monitoring of genetically modified organisms (GMOs) in food. As Food and Drug Law Institute Scholar, her articles on the international law and trade, ethical, food safety, legal liability, and intellectual property aspects of GMOs have been published in prominent journals such as the *Food and Drug Law Journal*, *Journal of Food Law & Policy*, *Stanford Journal of International Law*,

American Business Law Journal, Journal of Legal Studies in Business, and *The International Lawyer*. Her work has been cited by legal authorities and the courts as well as by other academics. She is the President-Elect and Program Chair of the North East Academy of Legal Studies in Business and a member of the Biotechnology in Agriculture Group of the Yale University Interdisciplinary Center for Bioethics.

Paul B Thompson, with a PhD in philosophy, holds the WK Kellogg Chair in Agricultural, Food and Community Ethics at Michigan State University, where his faculty appointment is spread across academic units dedicated to philosophy, community sustainability and agricultural, food and resource economics. Thompson has been publishing in peer-reviewed journals on the governance of emerging technologies in the food and agricultural sector since the mid-1980s, including *Plant Physiology, Journal of Animal Science, Reproduction, Fertility and Development, Food Protection Trends, Cahiers d'Economie et Sociologie Rurales* and *The Journal of Agricultural and Environmental Economics*. His book *Food Biotechnology in Ethical Perspective* was first published in 1997 and came out in a second edition from Springer in 2007. He has served on advisory boards for Genome Canada and the US National Research Council, and has spoken on issues in agricultural biotechnology at scientific and policy conferences in over 13 countries. Thompson's research on biotechnology has been supported by grants from the US National Science Foundation and from the US Department of Agriculture. Recent publications of note include: 'Synthetic Biology Needs a Synthetic Bioethics' *Ethics, Policy and the Environment* (2012), 'Agrifood Nanotechnology: Is This Anything?' in *Understanding Nanotechnology: Philosophy, Policy and Publics,* U Fiedeler, C Coenen, SR Davies and A Ferrari, eds (2010) and 'Why Using Genetics to Address Welfare May Not Be a Good Idea' *Poultry Science* (2010). Thompson has held joint appointments in philosophy and agricultural science at Purdue University and Texas A&M University. He is a two-time winner of the American Association of Agricultural Economics Award for Excellence in Communication and in 2012 received the Richard P Haynes Award for Distinguished Lifetime Achievement in Scholarship from the Agriculture, Food and Human Values Society. In addition to Thompson's research on the management of risks from biotechnology, nanotechnology and synthetic biology, he is known as a pioneering figure in the field of agricultural ethics. The books *The Spirit of the Soil: Agriculture and Environmental Ethics* (2010) and *The Agrarian Vision: Sustainability and Environmental Ethics* (2012) have been highly influential in establishing agriculture and food as key topics in bioethics and the philosophy of the environment.

Crina Viju was born and raised in the Southern part of Romania. After she received BSc and MSc in Business Data Processing department at the University "Babes-Bolyai" of Cluj-Napoca, Romania, she moved to the Mediterranean Agronomic Institute of Chania (MAICH), Crete, Greece where she obtained an MSc in the Economic and Management Science department. In September 2002 she started a PhD in Agricultural Economics at the University of Saskatchewan, which she completed in the summer of 2008. For a period of 2 years, from 2007 to 2009, Viju worked as an Assistant Professor at Johnson-Shoyama School of Public Policy, University of

Regina, Canada. In 2009, she moved to Carleton University in Ottawa, where, currently she is Assistant Professor at the Institute for European, Russian and Eurasian Studies. She is also the academic coordinator at the Centre for European Studies (a European Union Centre of Excellence) at Carleton University and the coordinator of the Research Node on 'The Political Economy of European Integration' at Carleton University's Center for European Studies. Viju has a strong background in applied economic theory and contemporary quantitative methods. Her research interests include the economic effects of different trade policies applied within the EU, US and Canada, European economic integration and EU enlargement. Some of her latest publications are "Zero Tolerance and the Rules of Trade: Lessons from the Discovery of Genetically Modified Flax in the European Union" in the *World Economy Journal* (with Yeung M and Kerr WA), "The Trade Implications of the Post-Moratorium European Union Approval System for Genetically Modified Organisms" in *Journal of World Trade* (Yeung M and Kerr WA), "Agriculture in the Canada-EU Economic and Trade Agreement (CETA)" in *International Journal* (with Kerr WA) and "Taking an Option on the Future: Subsidizing Biofuels for Energy Security or Reducing Global Warming" in *Energy Policy* (with Kerr WA).

Justus Wesseler holds the chair in Agricultural and Food Economics at the Center of Life and Food Sciences Weihenstephan, Technische Universität München (TUM) in Germany. He holds a degree in agricultural, environmental and natural resource economics from the University of Göttingen, Germany. Before moving to TUM, he was an Associate Professor at Wageningen University in the Netherlands. He researches the economics of biotechnology, value chain economics, and regulatory economics including issues related to coexistence. His major focus is on the contribution of value chains to improve sustainability, and the impact of new technologies and regulations on value chains. Wesseler is the head of the Research Department of Agricultural Economics at the Center of Life and Food Sciences Weihenstephan, member of the International Consortium of Applied Bioeconomy Research (ICABR), co-editor of *AgBioForum*, and coordinator of the EU funded project: Practical Implementation of Coexistence in Europe (PRICE). In 2013, The Research Foundation of Flanders in Belgium appointed Wesseler for a 3 year term as an external member of the Expert Group on Economics, Business Economics and Management. His research work on coexistence, biotechnology and related regulations has been published in more than 80 contributions to peer-reviewed journals and books.

